Deciding
about Design Quality

Sidestone Press

Deciding about Design Quality

Value judgements and decision making in the selection of
architects by public clients under European tendering regulations

PROEFSCHRIFT

ter verkrijging van de graad van doctor
aan de Technische Universiteit Delft
op gezag van de Rector Magnificus prof. ir. K.C.A.M. Luyben,
voorzitter van het College voor Promoties,
in het openbaar te verdedigen op
dinsdag 24 augustus om 12:30 uur

door Leentje VOLKER

Techniek en Maatschappij ingenieur
geboren te Arnhem.

Printed and bound in Great Britain by Marston Book Services Ltd, Oxfordshire

Dit proefschrift is goedgekeurd door de promotoren:

Prof.ir. H. de Jonge
Prof.dr. K. Lauche

Samenstelling promotiecommissie:

Rector Magnificus, voorzitter
Prof.ir. H. de Jonge, Technische Universiteit Delft, promotor
Prof.dr. K. Lauche, Technische Universiteit Delft, promotor
Prof.mr.dr. M.A.B. Chao-Duivis, Technische Universiteit Delft
Prof.dr. A.R.J. Dainty, Loughborough University
Dr. J.L. Heintz, Technische Universiteit Delft
Prof.dr. H.A. Mieg, Humboldt Universität Berlin
Prof.ir. W. Patijn, Technische Universiteit Delft

© 2010 Leentje Volker

Published by Sidestone Press, Leiden
www.sidestone.com

ISBN 978-90-8890-053-2

Photographs cover: Louvre Pyramid, Paris
Cover design: K. Wentink, Sidestone Press
Lay-out: P.C. van Woerdekom, Sidestone Press

Full page illustrations: Henk Dekker
1) The Rock (2009), Amsterdam - Erick van Egeraat; 2) Addition Jewish Museum (1999), Berlin - Daniel Libeskind; 3) Tate Modern (1947, 1963, 2000), London - Sir Gilles Gilbert Scott, Herzog & de Meuron (conversion to museum); 4) ING Group Headquarters (2002), Amsterdam - Meyer & Van Schooten; 5) City Hall and Central Library (1995), The Hague - Richard Meier; 6) Reichstag building (1894, 1990), Berlin - Paul Wallot, Sir Norman Foster (addition & reconstruction); 7) Stadsschouwburg (1918, 2008), Haarlem - J.A.G. van der Steur, Erick van Egeraat (addition & renovation); 8) Louvre Pyramid (1989), Paris - I.M. Pei; 9) Te Papa Tongarewa (1998), Wellington - Jasmax Architects.

Contents

Preface XI

1 Introduction 1
  1.1 The selection of architects 1
  1.2 Research focus 3
  1.3 Knowledge gaps and scientific challenges 5
  1.4 Research questions 9
  1.5 Research approach 10
  1.6 Audience 11
  1.7 Outline of the thesis 12

2 Assessing quality and value in design 15
  2.1 Introduction 15
  2.2 Quality in the built environment (architectural design approach) 15
    2.2.1 Interpretation of architectural design quality 16
    2.2.2 Assessment of design quality 19
  2.3 Design perception and affect (cognitive approach) 22
    2.3.1 Characteristics of the design perception 23
    2.3.2 Implications for design of the built environment 25
  2.4 Product experience and emotion (interaction approach) 26
    2.4.1 Origin of product experience 27
    2.4.2 Implications for architectural design 28
  2.5 Value systems in design (process approach) 30
    2.5.1 The concept of value 30
    2.5.2 Applications in design and construction 33
  2.6 Integration and implications 34

3 Judgement and decision making 41
  3.1 Introduction 41
  3.2 Three generations of decision theory 42
    3.2.1 Early models of decision theory 42
    3.2.2 First generation of decision theories 44
    3.2.3 Second generation of decision theories 46
  3.3 Rationality versus intuition 50
    3.3.1 Towards a definition of intuition 50
    3.3.2 Factors of influence on the use of intuition 52
  3.4 Individual decision making in organisations 53
    3.4.1 Cognitive processes 53
    3.4.2 Sensemaking 56
    3.4.3 Factors of influence on cognitive processes 58
    3.4.4 Affect, mood and emotion 60

3.5 Decision making in groups   62
   3.5.1 Cognitive and social processes   62
   3.5.2 Effects of group decision making   63
   3.5.3 Expert teams   66
3.6 Conclusion   67

4 The context of architect selections   **71**
4.1 Introduction   71
4.2 The political context   71
   4.2.1 Decision structure of public commissioning bodies   72
Box 1: Public administration in the Netherlands   74
   4.2.2 Participation and stakeholder involvement   76
4.3 The context of design and construction   78
   4.3.1 The concept of design competitions   78
   4.3.2 Partner selection in construction   84
4.4 The legal context   86
   4.4.1 EU Procurement law   86
   4.4.2 The Dutch interpretation of procurement law   91
4.5 The economical context   94
   4.5.1 Dutch market potential   95
   4.5.2 Tenders for architect selection in the Netherlands   96
4.6 Current practice in the Netherlands   100
   4.6.1 Perceptions and expectations   100
   4.6.2 Models and guidelines   104
4.7 Conclusion   107

5 Theoretical framework   **111**
5.1 Introduction   111
5.2 Proposed success factors for a tender design   112
   5.2.1 Reading the decision task   112
   5.2.2 Searching for a match between aims, ambitions, needs, and
      opportunities   117
   5.2.3 Writing the decision process   120
   5.2.4 Aggregating different kinds of value judgements   125
   5.2.5 Justifying against different rationalities   130
5.3 Research design   133

6 Three empirical tender cases – cross case analysis   **139**
6.1 Introduction and research questions   139
6.2 Research methodology   139
6.3 Framework for data analysis   141
6.4 Case descriptions   142
   6.4.1 A pragmatic process: A School with Sports facility   142
   6.4.2 A democratic process: A City Hall with Library   145
   6.4.3 A political process: A Provincial Government Office   149

6.5 Findings about actors — 151
    6.5.1 Steering committee — 152
    6.5.2 Project team — 153
    6.5.3 Jury panel — 154
6.6 Findings about project characteristics — 155
    6.6.1 Project characteristics — 155
    6.6.2 Client governance — 156
    6.6.3 Social context of stakeholders — 158
    6.6.4 Project management — 160
6.7 Findings about the tender design — 161
    6.7.1 Tender brief — 162
    6.7.2 Process procedure — 167
    6.7.3 Stakeholder involvement — 176
    6.7.4 Process of decision making — 182
6.8 Reflection — 192

7 Process design - a competition case and validation — 195
7.1 Introduction and research questions — 195
7.2 Research methodology — 195
7.3 Case characteristics — 198
    7.3.1 Description of the context — 198
    7.3.2 Description of actors — 200
    7.3.3 Case description — 201
7.4 Competition design — 206
    7.4.1 Competition brief — 206
    7.4.2 Process procedure — 211
    7.4.3 Stakeholder involvement — 220
    7.4.4 Decision procedure — 224
7.5 Discussion — 238
    7.5.1 Case reflection — 238
    7.5.2 Findings in the context of procurement law — 240
7.6 Validation of findings — 241

8 Conclusion, recommendations and reflection — 247
8.1 Introduction — 247
8.2 Conclusion — 247
    8.2.1 Five sensemaking processes contributing to an interplay of rationalities — 247
    8.2.2 Reading the decision task — 250
    8.2.3 Searching for a match between aims, ambitions, needs and opportunities — 253
    8.2.4 Writing the decision process — 255
    8.2.5 Aggregating different kinds of value judgements — 260
    8.2.6 Justifying against different rationalities — 263

8.3 Recommendations 265
    8.3.1 Recommendations for the design of a tender 266
    8.3.2 Suggestion for change in Dutch tender practice 268
8.4 Reflection on the research approach 270
    8.4.1 Research tradition 270
    8.4.2 Research methods 270
    8.4.3 Case selection 271
    8.4.4 Generalisation of the results 272
    8.4.5 Scientific relevance 273
8.5 Future research 275

References 281

Publications related to this research 305

Summary 307

Samenvatting 313

Acknowledgements 321

Curriculum Vitae 323

Preface

"If you please--draw me a sheep!" [....] So then I made a drawing. [...] My friend smiled gently and indulgently. "You see yourself," he said, "that this is not a sheep. This is a ram. It has horns." So then I did my drawing over once more. But it was rejected too, just like the others. [...] By this time my patience was exhausted [...]. So I tossed off this drawing. And I threw out an explanation with it. "This is only his box. The sheep you asked for is inside. "I was very surprised to see a light break over the face of my young judge: "That is exactly the way I wanted it! [...] And that is how I made the acquaintance of the little prince.

From: 'The Little Prince' by Antoine de Saint Exupéry, translated from the French by Katherine Woods

The architect selection process of clients is an intriguing decision process which shows a lot of resemblance with decisions in our daily life. We make numerous decisions every day. Some decisions are based on a range of explicit and physical alternatives; others are based on implicit ideas about the options we have in mind. In many of these situations interaction takes place between people and between people and their alternatives. Somehow we have to find a way to communicate about our preferences and the direction of our decisions. My research interests focus on the interaction between people and the options that are available when they are making decisions that relate to the built environment. This interaction has a lot of do with expectations, perceptions, emotions, and the use of our senses and intuition. Many of these issues are intangible by nature.

During my research I frequently experienced an intriguing tension between tangibility and intangibility of the characteristics that shape our built environment. I performed my PhD research at the Faculty of Architecture of Delft University of Technology in the Netherlands. In general technology is considered as a tangible aspect of our environment. Many engineers assume that, in the end, every tangible can and will be measured and explained. Some of them even think that everything that cannot be measured should not be included in research. In psychology subjectivity and unconsciuss thoughts are a given and scholars try to deal with it as well as possible. Designers often believe that every situation is unique and should therefore be handled as such. I think that we need all these different perspectives to increase the level of scientific knowledge about the built environment. To discover the 'truth' in such a multidisciplinairy environment we have to try to stand on the shoulders of our diverse ancestors, but also identify and cherish our differences. Because a gap still exists between ratio and truth it might be the best option just to listen to our intuition in finding the way how.

In this research I tried to find out how public clients decide about 'their' architects. I used insights from social sciences, in particular social and environmental psychological, as well as architectural design to successfully answer the research questions. Nevertheless I also realise that the more you know, the more you become aware of the things you do not yet know. Especially in the connections between the different fields of science I think there is a lot to discover still. Science is never finished and that is why I think it is so fascinating. I hope you will get inspired by reading this thesis and we can continue our scientific journey together.

Leentje Volker
August 2010

Chapter 1

INTRODUCTION

Architect soap opera in Delft
It starts to gradually show features of a soap opera. The municipality organises a competition for a new Railway Station/Municipal Office, and in no time all competing architects are angry.
Volkskrant - Machteld van Hulten on 11 January 2007 (translated from Dutch)

Slip-up: architect cannot build library of Utrecht after all
The municipality of Utrecht made a miscalculation, and appointed the wrong architect.
Volkskrant - Jochem Lybaart on 23 August 2008 (translated from Dutch)

Public in Rotterdam wants Search, but OMA will build town office
Not the public darling Search, but Rem Koolhaas' office OMA will provide the design for the new Rotterdam Municipal Office. The professional jury considered the four green cones proposed by SEARCH in collaboration with Christian Müller to be 'too eccentric'.
Cobouw – Edo Beerda on 22 October 2009 (translated from Dutch)

Jo Coenen: `We were an hour and forty-five minutes late'
The entry of Jo Coenen & Co Architects for the new town hall of Almelo has been rejected because it was submitted almost two hours too late, says Jo Coenen in a response his office being excluded from further participation in the selection procedure. The architect blames himself, but a bitter aftertaste remains.
Architectenweb.nl – 20 January 2010 (translated from Dutch)

1.1 The selection of architects

The above quotes are prototypical examples of newspaper headlines that have dominated the image of tender procedures of Dutch public clients selecting an architect in the past few years. The search for an architect can be characterised as an interactive selection process in which a client tries to find an architect who can visualise and implement the clients' needs and ambitions best. It is a challenging process of surprises and unforeseen circumstances in which legal and social obligations have to be considered. The diverse roots of the selection process appear to cause conflicts in current Dutch tender practice. In this research I explore the origin of these problems as currently experienced by public commissioning clients in architect selection in order to propose implications for future practice.

The current practice of architect selection by public clients has its roots in three distinct systems: 1) tendering for services and works, 2) the selective search to identify a suitable architect or design team, and 3) the architectural competition (Strong, 1996). Changes in the traditional regulations for architect selections influenced by European procurement law have encountered resistance from the architectural community (e.g. Atelier Kempe Thill, 2008; Postel, 2001). Until a few years ago Dutch architects did not, as contractors do, go to court when decisions during the tendering phase were not as desired. This practice has changed and we find more and more case law on tender procedures in which architects are the suing party (van Wijngaarden & Chao-Duivis, 2010a, 2010b). A dispute over aggregated rating points and a difference of a tenth of a percent led in several cases to a repetition of the entire selection process (Houben, 2007; Lybaart, 2008). In one case several Dutch firms withdrew their proposals from a tender because of excessive assignments in the award phase and involved the media to start a counter demonstration (Heijbrock, 2008). Committee decisions that used to be an enrichment and source of conflict of the architectural profession as well as a matter of professional judgement among fellow architects have become subject to judicial scrutiny of procedural correctness (Mieg, 2006). The community of architects, which takes pride in professional rigor and aesthetic judgement, feels that it has become prey to the community of lawyers, consultants and unprofessional risk averse clients (Architectuur Lokaal, 2009a; Kroese, Meijer, & Visscher, 2008).

Architects who participate in tenders, and political parties as well, increasingly demand additional justifications from public clients for their decisions, as for example in the case of the new Municipal Office in Rotterdam (van Geels & Kriens, 2009). It is felt that the design quality of Dutch buildings and of the Dutch cultural environment will eventually suffer because of a lack of diversity, creativity and innovation (Atelier Kempe Thill, 2008; van der Pol, Brouwer, Jansen, Mensink, & Geertse, 2009). Decision makers representing a public client have to justify their decisions to their stakeholders, the architectural profession and society at large. They seem more and more reluctant to expose their professional judgements in such a litigious atmosphere. The current regulations encourage the use of rating schemes with criteria and weighting factors, instead of more qualitative assessments as were traditional in design competitions, because a rational(-istic) decision is often seen as more acceptable and easier to defend than a decision based on intuition or professional judgement (Sinclair & Ashkanasy, 2005). The official procedures thus work to deprive the client of the explanations of professional quality judgements they were intended to bring to light. The pressure to comply with transparent and objective procedures seems to dominate over the need to select maximum design quality (Maandag, 2007). This leads to the idea that the process is dominated by political decision making rather than a desire to make the right judgement about the quality of a design (van Hardevelt & Schönau, 2009).

Architectural competitions are part of the design tradition. Since the ancient Greeks organised a design competition in 448 BC for the design of a war memorial on the Acropolis, the quality and the value of architectural designs have been a matter of public debate (de Haan & Haagsma, 1988; Strong, 1976). In a pursuit for excellence a design competition offers a choice between different design concepts to those representing the public or the client (Spreiregen, 1979). In architecture it is generally accepted that the quality of design proposals is judged by panels (juries or committees) of specially commissioned experts working alongside representatives of the client. However, the choices made by architectural juries are often controversial and could lead to public debates or scrutiny (de Haan & Haagsma, 1988; Strong, 1996), whereas the choices made by laypersons are often considered poor and ill informed by architects. A design competition does not only fulfil the pragmatic aim of selecting an architect, but fulfils also more political and societal goals such as discovering new talent, creating a dialogue on design, marketing a project, or coordinating different interests (Spreiregen, 1979; Svensson, 2008). For public clients this has led to pressure to make jury decisions more transparent and to include the public more directly in decision making (Sagalyn, 2005). Theories about decision making confirm that experts are generally better decision makers than laypersons and that they prefer to use intuition over rational analysis (Hogarth, 2005; Rosen, Salas, Lyons, & Fiore, 2008). However, expertise is often limited to a particular domain. Even with the same level of expertise, committee members often vary in their judgements of concepts, and even greater differences appear between expert and lay judgement of architectural quality (Collins, 1971; Gifford, Hine, Muller-Clemm, & Shaw, 2002; Nasar & Kang, 1989).

In an attempt to address issues of fair competition, the European Union has imposed strict rules for the tendering of public contracts. The selection of an architect is part of the regulations for contracting architectural design services (European Parliament & Council of the European Union, 2004). The EU rules are intended to enhance equal treatment, integrity, objectivity and transparency of the selection process. Procurement law offers several procedures to organise a tender process. Even though a design competition (referred to as 'contest' in legal terms) is a possible option under the EU Directives, most public commissioning bodies choose the restricted tender procedure to select their architects. In this clients break with the tradition of design competitions. At the same time they include elements of the traditional design competitions in their tender procedure, such as the submission of a design proposal and an open debate about design quality. The anonymous assessment of proposals and expert jury panels are often replaced or augmented by other procedures, while it is these very elements that secured fairness in design competitions. Legal and management consultants are often hired to support the commissioning client in organizing the tenders. Yet the growing field of case law on EU and national procurement law does not seem to provide enough support to guide clients on compliance with the regulations. In Dutch practice these developments have led to "a confrontation between the need for certainty and the desire for creativity" (van der Pol, et al., 2009, p. 10). The effect has been to encourage defensive and risk avoiding strategies of public clients in the design and realisation of tenders in architecture. This is particularly disconcerting since architect selections could offer valuable opportunities that could benefit governance and cultural heritage.

1.2 Research focus

This research focuses on the process of architect selection from the perspective of public clients in the context of EU procurement law. The project started in 2005 when the Dutch architectural community began to feel uncomfortable about the manner in which public clients selected their architects. EU tenders were still seen as design competitions in which a jury panel assessed the submissions and determined a winner. However, the legal context created additional awareness of the consequences of such a decision. Public clients realised that they were obliged to set up a tender procedure before they could select an architect (van der Pol, et al., 2009). They also started to realise that such a project needs to be embedded in their organisational structure and requires specific knowledge about architecture and procurement. Responsibilities have to be assigned to different kinds of actors. A building project takes place in a dynamic political context with legal as well as social obligations towards users, citizens, the professional field of architecture and society in general. The aims of a building project depend on the ambitions, means and accommodation needs of the client. By selecting an architect the material direction of the project is chosen and the project acquires its physical shape by design. A tender for architectural design services offers opportunities to a public client that range from involving citizens in order to increase political support to positioning a city. Public administrators also have personal and strategic aims. A tender is therefore an important part of the initial stage of a building project.

Social science can contribute to the understanding of psychological processes of decision making in the context of architectural design. Such an analysis can help to identify situational characteristics in the dilemmas that public clients face in the selection of an architect. In this research I empirically investigated the origin of these dilemmas. The research is based on two research traditions: judgement of design quality, and cognitive and social processes of decision making. In both traditions the concepts of perception (e.g. Bell, Greene, Fisher, & Baum, 1996; Tversky & Kahneman, 1981), intuition (e.g. Dane & Pratt, 2007; Sadler-Smith & Sparrow, 2008), judgement (e.g. Gifford, et al., 2002; Hogarth, 1988), expertise (e.g. Hekkert & van Wieringen, 1998; Hutton & Klein, 1999), and emotion (e.g. Desmet, 2002; Simon, 1987) play an important role. My definition of design quality is grounded in the comparison of the character and elements of quality in design from the disciplines of architecture, environmental psychology, product experience, and value management. Design quality is consequently determined in the interaction between the individual and the object of study. The results of this interaction are part of the value judgements that are used during decision making. Decision making is a way to align expectations and needs to reach goals (Hodgkinson & Starbuck, 2008a). Individual decisions are made by processes that are range from rational to intuitive and conscious to unconscious (e.g. Chaiken & Trope, 1999; Hogarth, 2005; Sadler-Smith & Sparrow, 2008). The concepts of bounded rationality (Simon, 1997), heuristics (Gigerenzer, Todd, & ABC research group, 1999; Tversky & Kahneman, 1974), and naturalistic decision making (e.g. Gore, Banks, Millward, & Kyriakidou, 2006; Lipshitz, Klein, & Carroll, 2006) could explain the decisions that are made about design quality. They show that not every decision is rational by nature (Sinclair & Ashkanasy, 2005). At the organisational level the concept of sensemaking (Balogun, Pye, & Hodgkinson, 2008; Weick, 1995) and image theory (Beach, 1990) attribute to the understanding of decision processes and the importance of the justification of a decision (Vidaillet, 2008). It is within this strategic, dynamic and elusive organisational context that public clients have to select their architect.

There are two conflicting models to frame the process of architect selection in the context of European tendering regulations. On the one hand, procurement law appears to be based on a rational managerial decision model that assumes that decision criteria for comparison of alternatives are known beforehand (Beach & Connolly, 2005; Harrison, 1999). On the other hand, Kreiner (2006) suggests that decision making during a design competition is actually a process of sense-making instead of choice. Jury members see it as an obligation to select a winning entry by consensus and usually try as hard as possible to reach this (Kazemian & Rönn, 2009; Svensson, 2008). Criteria on which the final decision is based are typically developed or specified during the process because such flexibility is needed to create room for negotiation (Kreiner, 2006). This is in contrast with the legal obligations to apply the same criteria and their relative weights as those published in the call for proposals in order to be as transparent as possible. This kind of transparency would enable competing architectural teams to be aware of the potential importance of the preferences and biases of individual jury members, but essentially winning a competition appears to be a matter of 'pushing the right button' or 'finding the hidden key' (Kreiner, 2007a).

According to Rönn (2008, p. 10) most difficulties in decision making during the selection of an architect result from "a number of legitimate interests that need to be weighed against each other, such as p. .l members who have different roles, interests and judging qualifications, the cc ·tion programme that describes the goals, brief, conditions, and requirement   resent different solutions for the assignment and compet.   set the general rules." In my opinion the selection proces   ld indeed be considered as a process of sensemaking in which   ..im to select the most suitable designer to deliver architectural d   quality for a future building of a client. Rational decision processes that are implied by procurement law often do not reflect the way in which selection decisions are made in practice but do provide the context of the decision process. Architectural judgement and the interpretation of legal requirements clearly need to be shaped accordingly. Both architects and public clients lack knowledge of the 'right' way to implement the EU legislation regarding the selection of service providers in architectural design.

1.3 Knowledge gaps and scientific challenges

Several knowledge gaps can be identified concerning the topic of architect selection that also frame the scientific challenges in undertaking this research. In the long history of design competitions hardly any attempts have been made to observe, analyse or evaluate the process of selecting architects (van Wezemael, 2008). Most publications on design competitions show the rich diversity of submissions accompanied by a statement of the jury about the relevance of competitions for the architectural profession (e.g. de Haan & Haagsma, 1988; Glusberg, 1992). Others describe the aims, procedures, potentials and pitfalls in a historical perspective (Lipstadt, 2005; Spreiregen, 1979; Stichting Bouwresearch, 1980; Sudjic, 2005) or the relevance of competitions for the architectural profession (e.g. Collyer, 2004; Larson, 1994). In almost every publication the competition is experienced a unique phenomenon. Studies either focus on one competition, the winning entries from a range of comparable competitions or on input from a relatively small amount of experts. The experiences and interests of the client as future user of the building are usually neglected (Nasar, 1999; Nasar & Kang, 1989). Manzoni (2010) found only 58 serious publications to include in her review study about design competitions, which shows the limited research tradition in the field. Manzoni distinguishes four major research themes over the past 35 years: the role of competitions and their (dis)advantages; their history; their systems and their adoptions across countries; and the management and organisation of competitions. She found an increasing interest in competitions from the management and organisation studies in the last five year (e.g. Ewenstein & Whyte, 2007; Jones & Livne-Tarandach, 2008).

In the early 1970's Collins discussed the similarities and differences in legal and architectural judgements without noticeable scientific consequences (Collins, 1971). Possibly the only publication that explicitly addresses EU procurement law in architectural competitions was written by Judith Strong (1996), a former Competitions Director for the Royal Institute of British Architects. Recently a few scholars have studied the judgement process of jury panels in the current context of design competitions (Kazemian & Rönn, 2009; Kreiner, 2006, 2008; Silverberger, 2010; Spreiregen, 2008; Svensson, 2008), sometimes directly related to the assessment of design quality and expertise (Rönn, 2010; Svensson, 2010) or the strategies of architectural teams that join competitions (Kreiner, 2007a, 2007b; Manzoni, Morris, & Smyth, 2009). Only Kreiner (2006) explicitly addresses the sensemaking character of the decision process of the client while selecting a winner of a competition.

None of these publications address the impact of procurement on the procedure and decision processes in a competition. Competition and tender publications and models have been written from the perspective of the architect (e.g. van Campen & Hendrikse, 1997) or the legal advisor (e.g. Jansen, 2009), but not from the perspective of the client or the management of a complex project in a governance context. The available procedures are often taken as given and no interest in shown in the considerations of the client or organiser made during preparation and realization of the competitions. The long tradition of design competitions could have opened up numerous possibilities if the competitions had been well documented, structurally evaluated and scientifically compared. Connections to fields of decision making, quality perception or strategic management could easily have been made and could have supported the development of current practice.

The initial ease with which the traditions of design competition, procurement and partner search were merged implies that no real obstacles were experienced. However, the increasing number of problems in current practice suggests that underlying conflicts do exist. On the one hand, these problems appear to be caused by the removal of essential elements that used to guarantee the proper functioning of the selection processes, such as an expert jury panel. On the other hand, the merge seems to stress the inadequacies of the long standing tradition of design competitions. Although the context of architect selection has changed, discussions in practice do not properly address these changes. Since the introduction of the procurement directives in 2004 (Directive/18/EC of the European Parliament & Council of the European Union) the selection of an architect is suddenly out in the open. Architects participating in the selection process have the opportunity to take their dissatisfaction to court. The increasing numbers of law cases shows a cultural change in the attitude of architects in relation to their clients. This means that the architectural profession is gradually looked upon from a legal perspective. Terms such as fair competition, objectivity, and transparency now start to have different meanings than they had in the era of design competitions. Case law deals with legal reasoning rather than architectural reasoning. It is based on extreme cases where practice led to a serious argument that could not be solved in a dialogue or by mediation. The dialogue that used to be perceived as essential for the development of design quality is now conducted in presence of a judge, and the focus has shifted from the product to procedural issues. There is a general lack of insight into the preceding phases that led to an escalation of opinions.

Clients and consultants are remarkably absent from the discussions that tend to be dominated by architects (e.g. Atelier Kempe Thill, 2008; Postel, 2001). Most architects portray themselves as victims of the tender system rather than responsible actors, even those who also serve as members of jury panels themselves (van der Pol, et al., 2009). With the exception of Atelier Kempe Till (2008), neither architects nor clients have approached this problem from a research perspective. In general discussions about tenders do not address cause and effect. They tend only to describe the problems instead of critically analyzing them (see for example Architectuur Lokaal, 2009a; van der Pol, et al., 2009). For example, current literature does not discuss the differences between a design contest in which the EU regulations are applied, and a design competition without this strict legal framework. As far as I am aware of the evaluation process of jurors, neither the differences between an expert jury panel and a mixed jury panel, nor the benefits and disadvantages of the available procedures have ever been seriously addressed scientifically. Without a proper analysis of the origin of problems, a solution cannot easily be found.

The lack of a proper problem analysis is rooted in both the culture and the multi-disciplinary character of the research field. In spite of its international character and the numerous journals and books on architecture, the field of architectural design still has a rather limited empirical research tradition. Architects tend to publish books about their own body of work in order to promote their practices. The few books about the architectural practice that are widely acknowledged are written from a managerial rather than a scientific perspective (see for example Allinson, 1997; Blau, 1984; Coxe, et al., 1987; Gutman, 1988). On the one hand, the focus on the profession shows the pragmatic character of an applied discipline aiming at improving design and design processes. On the other hand, theoretical and philosophical discourse would provide input for continuous debate about values and relevance of architecture on societal level. The lack of critical debate and evaluations of precedents is a potential threat for the development and standards of the profession (Collins, 1971). Methodological attempts have been made to link design and empirical research (e.g. de Jong & van der Voordt, 2002; Groat &

Wang, 2002) but by and large the field is still scattered and unstructured. The importance of scientific research does not seem to be fully acknowledged yet in the field of architecture. The field of architecture is part of continuing debate about the position in between arts and (applied) sciences.

In my opinion architecture is essentially a multi-disciplinary field of history, philosophy, engineering, psychology, economics, management, politics, sociology and more. However, insights from these related fundamental fields of science are not widely adopted in architecture. An immense research potential arises from the interaction between applied and fundamental sciences in architectural design (Groat & Wang, 2002). This research therefore aims to implement knowledge from the fundamental sciences of judgement and decision making in the context of design. It also tries to link the various design fields in their search for quality and excellence in design.

There are also limitations within the fundamental sciences that prevented the development of common base for research in the field of architecture. Within science there are many different disciplinary perspectives that each focus on their own area of interest. In this sense science can be considered as the blind men touching the elephant failing to develop an integrated understanding of the phenomenon as each man is locked into his own particular view of the elephant. I believe that we can agree about ideas on how to perceive and decide about the 'elephant' of architecture. Apart from the people-environment studies, few scholars from outside the discipline of design have adopted architects or architectural design as research objects. A few exceptions can be found in the areas of marketing and communication (e.g. Day & Barksdale, 1992; Jones & Livne-Tarandach, 2008), cognitive science (e.g. Hamel, 1990), and in the ethnographic study of architecture (e.g. Yaneva, 2008).

Research in people-environment studies has often focused on one or two aspects or individualistic approaches to how people view the built and natural environment (Buijs, 2009). This is related to the common tradition in the use of mainly quantitative research methods. The fragmentation of knowledge makes implications difficult to understand for people from other disciplines and different levels of expertise. It leaves out the historical and social context, and therefore the dynamics, diversity and complexity of the built environment on a holistic level. The field of organisational behaviour usually aims at decision quality for the effectiveness and efficiency of the organisation instead of the benefits for society. In decision making and design implicit knowledge and intuitive actions appear to be essential. Not many scholars observe people in their natural environment and reflect on their (often inconsistent) behaviour. The field of public administration studies decision makers in their political arena but does not focus on the temporarily external effects that a tender project causes by the direct interaction between the parties on the demand side and those on the supply side of the market. Sociologists study groups in their natural environment but are not particularly interested in individual traits or the economically based mechanisms that people apply in their behaviour. Economical studies are based on the interplay between demand and supply on a market level in order to predict future behaviour and do not show much interest in the people behind the numbers, fact and figures they study. I therefore believe that both a connection between the fundamental fields of science and the field of architecture as well as between the scholars in (architectural) design is needed.

The context of this research did not support the lifting of research boundaries either. The selection of architects takes place in a very closed setting, just as panel discussions in other areas such as grant allocation or the recruitment of professors (Lamont, 2009; van den Brink, 2009). "The major part of the organiser's or jury's knowledge remains secret: it is neither communicated in the program brief nor in protocols" (van Wezemael, 2008, p. 9). Expertise is usually applied in high risk or complex situations with multiple interests and no clear distinction between right and wrong. Such decisions have a high chance of being condemned. The few scholars that were able to enter the scene were either actively involved themselves or were granted only limited access. Information from both supply and demand side is very sensitive by nature and its elicitation requires trust between these parties and the researcher. Architectural firms are not very open about their strategies for participation in competitions either. This has had important implications for the approach of this research. Yet I still believe that the roots of conflicts in the behaviour of public clients can best be studied in their natural context.

1.4 Research questions

The aim of this research project is to empirically study the process of value judgement and decision making of public commissioning bodies during architect selection in the context of European procurement law. The main objective is to describe, understand and explain the design and implementation of procedures by means of which the quality of design proposals is judged in order to award a contract to the architect who will deliver design services for a particular building project. My main aim is to attempt to resolve the current conflicts that arise from the perception of the legal obligations of the actors of a tender project and their natural behaviour in making decisions about the quality of architectural design.

The research questions in this study are:

1. How do public commissioning bodies decide on the selection of an architect in the context of EU procurement law?

2. Which situational characteristics influence the process of decision making of public commissioning bodies in this context?

3. What are the implications for the design of procedures for the selection of architects?

The first question relates to the activities that take place during the selection process and the decisions that are made to select an architect. The selection process is divided into a preparation, selection, and award phase. The research addresses which actors take which decisions in which phase and how these decisions are related. It also analyses the kinds of decisions that are made in assessing design quality, the kinds of decision processes that can be recognised, and how the decisions are motivated and legitimised.

To answer the second question situational characteristics were identified that influence the process of architect selection. The research identifies the external and internal contextual factors and the interrelating or conflicting character of these factors in relation to European procurement law. It also shows the differences between laypersons and experts in decisions about architectural design quality and what other characteristics play a role.

By providing a clear understanding of the decision making behaviour of public commissioning bodies, a comprehensive picture of the problems was derived that could support changes in current practice. These insights could increase the awareness among actors and hopefully replace myths by facts. Based on the results of this research success factors are proposed for improving the selection procedure of an architect and preventing further conflict between public clients, architects, their advisors and other stakeholders in the public domain.

1.5 Research approach

The lack of a clear problem analysis, theoretical and empirical insights, and openness for research made me choose for a qualitative research approach. This approach was based on the roadmap proposed by Eisenhardt and consists of conducting real life case studies simultaneously with a reflection on constructs and theories found in literature (Eisenhardt, 1989; Eisenhardt & Graebner, 2007). In order to increase the external variance of the data and possible generalisation of the findings, I selected three cases in which the restricted tender procedure was applied by a Dutch public client. The restricted tender procedure has been the most commonly used tender procedure in the Netherlands (Geertse, Talman, & Jansen, 2009). Selected restricted tender cases also presented the best opportunities to gain access to data. The three tender cases were complemented with a fourth case about an ideas competition for a new Faculty Building for Architecture that was destroyed by a fire in May 2008. The research methodology is described in more detail in Chapter 5, together with the possible success factors for the design of a tender project that were extracted from the literature.

In the first phase of the research (mid 2006 – beginning 2008) three case studies were performed on tender cases in the Netherlands:

1. A large elementary School with a Sports facility.
2. A City Hall with Library for a middle-sized city.
3. A new Provincial Government Office with office facilities.

These three cases all concerned a restricted European tendering procedure of a Dutch public client and were selected based on the clients' willingness to participate in the research. In two cases the client committee eventually selected one architect among five or six tenderers and awarded a traditional Design Bid Build contract. The third case was cancelled after the selection phase in which seven tender candidates competed for a Design-and-Build contract. A variety of data (observations, interviews, documentation) was collected for each case to allow for triangulation between self-report, observed behaviour and official documents. The research method and results of these cases are described in Chapter 6.

Subsequent to data collection of the third case, the opportunity arose to be involved in an international open ideas competition for a Faculty Building. In this ideas competition 471 international participants competed to win € 60.000 of prize money. Most of the data were collected by participatory observation, interviews and documents analysis in the period of June 2008 – May 2009. The opportunity to actively participate in a case similar to a selection procedure made it possible to gain more insight in the strategic and pragmatic decisions that need to be taken in designing a selection procedure. It enabled me to compare the strict European tendering procedures to a design competition as a start of the official tender. Contrary to the restricted tender procedures the design proposals were judged anonymously in the competition by a professional jury. The commissioning client body of the ideas competition was more experienced than the commissioning clients of the three real life tender cases. The research methods and results of the Faculty Building case are portrayed in Chapter 7.

The theoretical framework and the results of the four cases were successfully validated in a workshop with professional clients, legal professionals, and architects. Chapter 7 also includes the results from this validation workshop.

1.6 Audience

This research contributes to knowledge in the areas of architectural design, the psychology of making judgements, and organisational decision making. It is therefore of interest to public commissioning clients, management consultants, architects, policy makers and legal advisors in practice, but also to scholars in the field of design management, product experience, environmental psychology, or decision making. The main audience of this thesis is public commissioning bodies that have to organise a tender, their advisors, and governmental authorities that develop and implement regulations and policies and scientific scholars in this multidisciplinary area. Because the research shows insights into the client perspective that have never been studied before and are usually not open to the public, I believe that the results of this research also offer an interesting story for those interested in architectural design and competitions.

This research connects the academic disciplines of psychology, organisation science, law and architecture by focusing on actual behaviour of clients in the context of design. The scientific relevance lies in a theoretical contribution to the understanding of the processes of value judgement and decision making in the context of architectural design. The architect selection process was empirically studied through the eyes of the actors (on both the individual and group level) that represent a public commissioning body during a tender procedure. The psychological perspective shows the potential conflict between cognitive and social decision processes and rationalities from the fields of law and design. The research compares different perspectives on design quality and explains underlying cognitive and social processes of decision making. In this sense the findings are not specific to the selection of architects but can be used to create insights in other kinds of decisions, in either the public or private sector, about issues with an elusive and subjective nature. It contributes to the debate on rationality versus intuition and shows the effects of affect, sensemaking, politics and perception on different rationalities. The decision processes as studied show great similarities with grant allocations, student evaluations, and tenders or other kinds of purchases in other sectors. International publications suggest that the problems with architect selection in the Netherlands are similar to those in other countries, but that national policies and culture could change perceptions on the seriousness and nature of the problems. The research therefore could also be of interest for international practitioners and researchers. Directions for further research lie in the field of public administration, strategic management and experimental psychology, or in a comparison to different sectors, private contracting authorities and other procurement procedures and methods.

1.7 Outline of the thesis

The thesis consists of several parts: Chapter 2, 3 and 4 provide the theoretical background and the context from which a theoretical framework is drawn in Chapter 5. This frame is used to design the research approach and reflect on the empirical data from the four case studies as described in Chapter 6 and 7. Chapter 7 also includes the results of a validation workshop. The results of this research led to a conclusion, recommendations and reflection in Chapter 8. The structure of the thesis is displayed in Figure 1.1. Those who do not wish to read the details of the theory and empirical studies are advised to start with Chapters 5 and 8, as they provide a summary of the theoretical and empirical findings of this research.

The theoretical foundation of this thesis is incorporated in two chapters. In Chapter 2 four perspectives on the origin and aspects of design quality are analysed. At the end of this chapter these perspectives are integrated to characterise the process of selecting an architect and measuring design quality in this context. A definition is given of value judgement that originates from the interaction between the individual and the object of design. Chapter 3 introduces three generations of decision making theories and describes the cognitive and social decision processes as described in these theories. Emotion, affect, expertise and intuition are distinguished as factors that affect processes of decision making. The theoretical insights are related to the level of expertise needed to make decisions about architectural design quality and the selection of an architect.

The context of architect selections is described in Chapter 4. It addresses the governance structure of a public client and the international tradition of design competitions in architecture and construction. Then the most relevant issues of procurement law are summarised and the Dutch market potential and current situation is analysed to show the economical dimensions of the problem. The fourth chapter ends with the current perceptions, expectations and supportive models and structures for the selection of an architect in current practice. Chapter 5 describes the theoretical research framework of the thesis and bridges the theory and context with the empirical work.

The empirical data are reported in Chapters 6 and 7. Chapter 6 focuses on the first two research questions about the processes and situational characteristics of influence on decision making. It consists of a description and the results

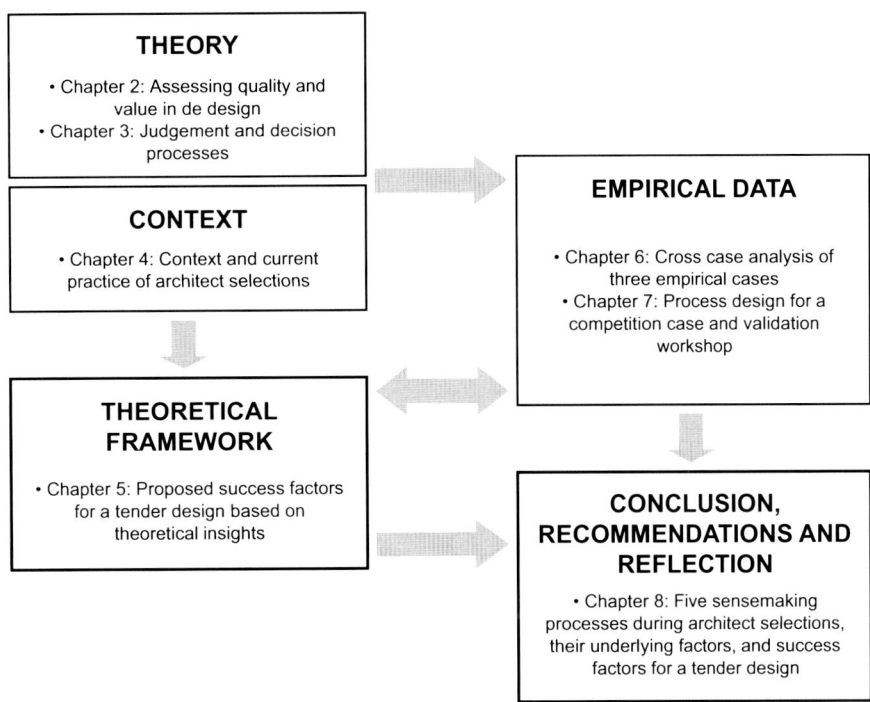

Figure 1.1 Thesis outline

of the cross-case analysis of three tender cases of Dutch public clients. The cases were first analysed individually and then compared based on a framework of actors (steering committee, project team, and jury panel), project characteristics (governance structure, stakeholders en project management), and four elements of tender design (tender brief, process procedure, stakeholder involvement, and decision process). Chapter 7 aims to answer the third research question about the implications of the findings for the design of an architect selection procedure. The theoretical and empirical findings of the previous chapters are used to analyse the data collected during the organisation of an open international ideas competition. The results show which decisions were taken in order to design the competition brief, procedure of the process, involvement of stakeholders, and final decision process and which difficulties and dilemmas were experienced during this selection process of the winner. Based on the results of all four cases a validation workshop with experts from the architectural community was organised. The results of this workshop are also included in Chapter 7.

In Chapter 8 conclusions are drawn about value judgement and decision making in order to select an architect. The conclusions are based on the theoretical insights as well as the results of the cases and the validation workshop and show five sensemaking processes and eleven situational characteristics that could cause conflicts between the psychological and legal rationalities of architect selections. These conflicts are related to the most important principles in procurement law. The chapter also contains a reflection on the research approach, scientific relevance and validity of the findings. Based on the insights from the theoretical framework, the results of the empirical cases and the validation workshop, fifteen success factors for practice are proposed. The chapter ends with a suggestion for a change of the current practice and several directions for further research.

Chapter 2

Assessing quality and value in design

2.1 Introduction

For evaluating the quality and value of architectural design, different perspectives can be adopted depending on the disciplinary focus and purpose of the discussion. The definition of architectural quality has been the issue of a long-standing theoretical debate. This chapter adds new perspectives from related disciplines to rejuvenate the debate on architectural design quality. It considers several perspectives on perception, assessment, evaluation and judgement of design quality that in my perspective contribute to the understanding of assessing design quality. In the final section of this chapter these perspectives are integrated.

The work on perception and evaluation is based on the psychological concept that what we perceive is the result of an interaction between the physical environment and the person. External and internal factors could cause differences in the outcome of this interaction. Theories about how design quality is evaluated have also been developed in the area of product design. Because a building can be seen as a product, this literature is also taken into account. Literature on value management and value engineering is used to link the debate about architectural design to the development and realization of design quality. A research perspective is taken, which means that most literature refers to empirical findings from a design product or building process perspective. Philosophical and paradigmatic debates about 'true' value of design and design processes are deliberately left out, as they do not contribute to the focus of this thesis.

The chapter addresses the debate about design quality from the fields of architectural design, environmental psychology, product experience, and value management. In each section the conceptualization and assessment of design quality are discussed and visualised in a figure. In most sections the differences between individuals and groups of people, and options for participation are discussed. By integrating the above perspectives, a working definition and framework about the judgement of architectural design quality is developed in the final section.

2.2 Quality in the built environment (architectural design approach)

The first perspective considered in this chapter is the architectural design approach. This approach builds on an ancient tradition of the assessment of the built environment. Design quality distinguished tangible and intangible characteristics that are part of the famous three dimensions of a building as mentioned by the Roman architect Vitruvius (Vitruvius & Morgan, 1960). There is not much systematic empirical research in the architectural design tradition about design quality, but the topic of design quality assessments often features in design competitions. In this section the most common definitions and approaches on design quality are discussed.

2.2.1 Interpretation of architectural design quality

Architectural quality embraces all the aspects by which a building is judged. The oldest known operationalisation of architectural quality (about 25 BC) is that of the Roman architect Vitruvius, which distinguishes three aspects: 'Utilitas, Firmitas, and Venustas' (most commonly translated as 'commodity, firmness and delight' or sometimes as 'utility, durability and beauty') (Gann, Salter, & Whyte, 2003; Vitruvius & Morgan, 1960). This trilogy has been a source of inspiration for architectural theory since then and continues to be so for several contemporary researchers. Macmillan (2006) for example distinguishes between exchange, use, image, social, environmental and cultural value for the built environment while Gann and Whyte distinguish 'functionality', 'build quality' and 'impact' (see Figure 2.1) (Gann, et al., 2003; Whyte, Gann, & Salter, 2004).

In the context of design competitions, Kazemian and Rönn (2009) and Heynen (2001) propose the most general basic criteria in quality assessments of architectural projects in the context of competitions that include: the context and its surrounding, the coherence and its totality, the functionality, the technical solutions, and the development potential of a design. Every scholar appears to create his or her own version of the list of characteristics of a design, while at the same time these lists cover basically the same things. Prasad (2004b) argues, with reference to the development of the Design Quality Indicator (DQI), that design quality can only be achieved when the three quality fields of functionality, build quality and impact all work together as overlapping areas of concern. True excellence is reached when quality is achieved in all three fields synergistically.

FUNCTIONALITY

Functionality is concerned with the way in which the building is designed to be useful and is split into Use, Access and Space. Issues assessed are for example: Does the product support the required functions? Does it do what it is supposed to do?

BUILD QUALITY

Build quality relates to the engineering performance of a building fabric and is split into Performance, Engineering and Construction. Issues are for example: Will the structure be stable? Are the lighting levels, thermal climate and acoustics appropriate for its use?

IMPACT

Impact refers to the building's ability to create a sense of place, and have a positive effect on the local community and environment. It is split into Character and innovation, Form and materials, Internal environment and Urban and social integration. Issues are: How beautiful is the building? Does it excite the people who use it, making them want to be in or around it?

Figure 2.1 The design qualities within the framework of the DQI (retrieved June 22nd 2007 from www.dqi.org.uk)

According to Kazemian and Rönn (2009) a judgement about design quality will include a comprehensive totality, surprising excellence, an expression of appreciation, and timeless values. Authors who not only consider the building as such but also include the development and management of a building often expand the trilogy with contextual factors, which include finances, time and resources including the future perspective of a building (Gann & Whyte, 2003; Gerritse, 2008; van der Voordt & van Wegen, 2005; van Rossum & de Wildt, 1996). According to Collins (1971) the problem of establishing architectural ideals today is not so much due to the difficulty of weighing the three aspects but more so about a realistic understanding for non-architects of the difference between price and value. So by extending the traditional trilogy of Vitruvius with contextual factors the discussion about design quality in architecture changes from 'the true value of design' to 'a balance between costs and quality'.

Discussions in architecture relate to physical buildings as well as to designs as representations of future buildings. Holistically design quality can also be seen as the achievement of an integrated totality that is more than the sum of the parts (e.g. Bártolo, 2002; Dijkstra, 2001). However, during communication and discussion design quality often seems to be decomposed. In everyday usage 'features', 'properties', 'traits', 'characteristics', 'attributes' and some other terms could be substituted for 'qualities' or 'values' of design (Dijkstra, Rijksgebouwendienst, & Ministerie van VROM, 1985; Gerritse, 2008; Macmillan, 2006; Pultar, 1996; Thomson, Austin, Devine-Wright, & Mills, 2003; van der Voordt & van Wegen, 2005). In general, qualities of products may be classified under two general categories that in practice often interrelate and overlap:

1. technical, physical, hard, functional, objective or tangible qualities, in this research referred to as 'tangible characteristics';

2. perceptual, soft, subjective, judgemental or intangible values, in this research referred to as 'intangible characteristics'.

Tangible characteristics can most easily be measured and quantified by an assessment system which is generally acknowledged. An example of this is the assessment of size in meters or the measurement of sound in decibels. In the absence of other information business cases in construction are usually based on quantities and costs (Gerritse, 2008). The tangible characteristics fit into the principle of 'manage and measure' as identified by Gann and Whyte (2003), which is based on a belief that designers can make rational responses to social, economic and environmental needs. Research in this tradition has been focused on achieving better design by measurement, management and integration of the process. The Dutch Normalisation Institute developed several standards (NEN-norms) about the technical, physical and functional performance specifications of a building. In case of a future building, demand specifications are usually included in a 'brief' (UK terminology) or 'programme' (US terminology). Often these norms or specifications refer to tangible qualities or local Buildings Decree. In the construction industry, quality is - in line with the quality and standardization movements such as ISO 9000 - associated with competency and proficiency levels as a route to customer satisfaction to control the conformance of the product against predetermined goals (Bártolo, 2002). Until recently quality control in construction was, however, not really concerned with the appropriateness of the product characteristics themselves and not directly related to the customer's perception of service (Allinson, 1997; Thomson, et al., 2003). The shift from a sellers market to a buyers market has made contractors and developers focus increasingly on their customers' perceptions of quality.

'Intangible characteristics' refer to a personal response to built form, people's perception of space, texture, colour and light, the meanings and associations attached by people to places or the way by which people assign aesthetic qualities to their surroundings (Bártolo, 2002; Vitruvius & Morgan, 1960). According to Gerritse (2008) and Macmillan (2006) intangibles are vital to architectural design but often suppressed in discussion about the realisation of a building. They are essentially a question of perception, and consequently a question of characteristics and preferences which should be qualified. Although less easy to quantify, fulfilling the intangible requirements is essential "to make the object unique, recognizable and give it a meaning that exceeds the right to exist based on the function of the building" (Dijkstra, 2001, p. 17), to provide a 'sense of place' (Canter, 1977) and to develop 'cultural as well as future value' (van Rossum & de Wildt, 1996). These characteristics would therefore fit best into the 'judgement-based' approach, which is adaptive, focusing on the experts' abilities to evaluate the design product (Gann & Whyte, 2003). In the context of intangibles, the term 'values' is used more often than the term 'characteristics'. In a brief the intangible characteristics of a building design usually comes to the fore in the architectural ambition of the client and the expected appearance or character of a building. There are no standards or performance specifications for intangible characteristics, but that does not mean that clients do not have expectations them.

Figure 2.2 depicts how design quality is commonly discussed in architecture by using the classification of the DQI for the Vitruvian trilogy. Although Thomson et al. (2003) and Cold (1993) argue that the product value of a building appears to be a result of the personal and contextual perspective of the user and the physi-

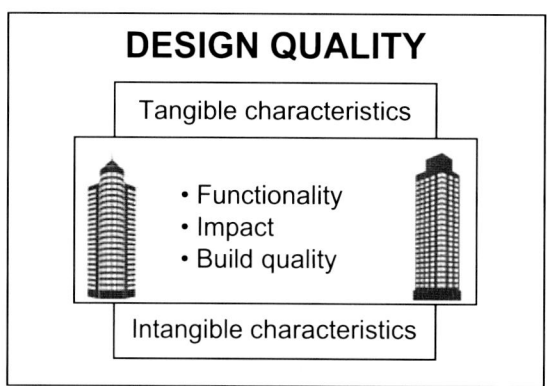

Figure 2.2 Design quality as tangible and intangible product characteristics from an architectural perspective

cal qualities of a product, discussions about architectural design quality tend to focus on the qualities of the product itself and about architecture on a societal level. Functionality, impact and build quality imply a judgement about a mixture of tangible and intangible product qualities from the perspective of a person experiencing objects in the built environment.

There seems to be no real difference between discussions about realised objects or designs as projections of future buildings. The person who experiences building qualities features only implicitly in this theoretical debate. Criteria that decision makers bring to the table appear to be somewhat different than the criteria by which architectural quality is assessed in the professional field. While functionality, build quality and sustainability would all seem to be equally important aspects of architectural quality as aesthetics, the discussions on aesthetics, impact on the surrounding environment and added architectural value are most prominent. In practice discussions with clients and developers tend to strive for a balance of different kind of qualities within a frame of time, money and other contextual factors that determine the integrative quality of architectural design (Gann, et al., 2003).

2.2.2 Assessment of design quality

The measurement of design quality has been discussed for some time without resolution (Prasad, 2004b). However, these discussions have not lead to substantial empirical research about assessing design quality. Most of the assessments in architecture occur in the context of design competitions, tender situations or part of quality management within organisations without much reference to science. Competition jury reports often reflect the considerations of the jury members in relation to the design brief and general state of architecture (e.g. de Haan & Haagsma, 1988; Svensson, 2008). Journals in the field of architectural design tend to describe and laud design projects or philosophise about phenomena in the built environment (Collins, 1971). Although these activities offer great potential for a decent link with design and political debate about design quality, systematic assessment appears to be missing (Nasar, 1999; van Wezemael, 2008).

One of the main reasons for the lack of systematic building assessments could be that architecture is judged from different perspectives. The question is therefore whose judgement and whose values should prevail in assessing design quality. According to Benedikt (2007) there are at least four venues in which architecture is currently evaluated: architects among each other, the public, clients, and members of allied professions. Each venue uses its own values to evaluate architecture which leads to a motley collection. Marans and Spreckelmeyer (1982) and Macmillan (2004) distinguish several user groups, such as office workers, public visitors and the community at large, to take into account during evaluation. Macmillan (2005) takes finances, design and construction, occupants' organisation, public realm and visitors as a starting point for the sustainable development of buildings. The most common conflict of values in architectural culture seems to be the one between individual artistic expression and a service to users and clients; but "however common this dichotomy may be in practice, it is artificial in principle: serving a client's programme could lead to an artistic building which functions superbly" (Saunders, 2007, p. 133). Existing building performance measurement systems often focus on specific aspects of buildings, such as use and satisfaction (e.g. Maarleveld, Volker, & van der Voordt, 2009; Vischer & Preiser, 2004), air quality in relation to productivity (e.g. Clements-Croome, 2005) or environmental assessments (e.g. BRE Global, 2009; Van den Dobbelsteen, 2004). A Post Occupancy Evaluation (POE) is generally focused on a systematic evaluation of occupant satisfaction with respect to several physical attributes (e.g. lighting, temperature, parking) and peoples feelings about privacy, social interaction, and occupation of spaces on a particular moment (Macmillan, 2004; Preiser, Rabinowitz, & White, 1988; Vischer & Preiser, 2004). This kind of evaluation is intended to provide an empirical basis for quality improvement. Yet few POEs address the aesthetics or overall architectural quality of the building in a substantial matter, which might be the reason for the substantial gap between research, theory and practice in architecture. Green and Moss (1998) and Kelly, Morledge and Wilkinson (2002) suggest that if organisations are to achieve the full potential of the learning cycle it is essential that post-occupancy evaluation findings are fully integrated into the brief and design of new building projects. However, due to lack of resources after occupation as well as a lack of awareness about the value of such knowledge for future real estate developments POEs remain scarce (Bordass, 2003). The integration of POE results into the value management process would be a first step in filling this gap and increasing the validity of POE research (Green & Moss, 1998).

In architecture experts are often assigned to assess design quality, either on paper or in an existing building. A common method to aggregate expert judgement on creative tasks is to ask judges to assess the results on several dimensions (Amabile, 1996). The validity of the assessment is determined by the inter-rater reliability of independent judgements about creativity. Hekkert and van Wieringen (1993, 1996) used this technique in a research project on the appraisal of art. Both studies found that judges who have experience in the domain but are not considered as experts, assess quality as reliably as experts do. The level of expertise does seem to cause differences in the values attached to certain dimensions of quality. Limitation of these studies is the failure to consider the possible influence of the dynamic context of real life settings. Van Rossum and de Wildt (1996) found that experts used four levels of architectural quality: 1) insufficient or not fulfilling the standardised quality, 2) average architecture with basic quality level, 3) superior recent architecture, and 4) distinctive and honourable architecture that attracts attention, worth a trip. The results of this study suggest that experts first make a selection based on the fulfilment of basic qualities, and then compare the remaining options on the level from basic to excellent design quality. The conclusions imply that in architecture the categories of over-performance to the basic qualities of a building are more important than categories of under-performance.

While the public and commissioners appear not to be actively involved in the general debate about design quality, the involvement of users and visitors during the design of a building varies significantly from project to project. In general experts from the architectural community are seen as the most appropriate agents to develop and guide the development of a building project (Blau, 1984).

However, it remains unclear if this is simply a habit or a result of consistently high levels of client satisfaction. According to Heynen (2001) the assessment of quality is a matter of communication and use of language. She proposes that the relations between the context, project definition, designed space, and design potential should be judged as part of the design assessments of clients. Nasar (1999) shows by evaluating several design competitions from the perspective of environmental psychology that jurors do not always have the same preference as the public. He proposed a POE-like method to systematically involve visitors and users in the evaluation of designs during competitions to be able to include the future preferences of users. Marans & Sprenckelmeyer (1982) indicated that while most members of the public would agree that the building in question was worthy of its particular architectural awards, a significant minority would not concur with the views of the professionals judges, nor would the building's occupants, whose ratings of architectural quality are often less than favourable. The length of stay and the role (owner, designer, user, employee, visitor, public) in which a juror is in contact with the building could be an influencing factor for the appreciation of a building (Marans & Spreckelmeyer, 1982).

Participatory design (Sanoff, 2006) or co-design (King, 1983) could help to reduce the differences between the users and other stakeholders groups (see also Chapter 4.2.2). The Construction Industry Council in the United Kingdom has developed the Design Quality Indicator (DQI) as a tool to measure and discuss design quality with the numerous stakeholders in a building project (see Construction Industry Council, 2009). The DQI was developed based on a 'rational-adaptive approach', which accepts that quality is a difficult and uncertain aspect to measure but that the development of tools to think about the impact of the design could be beneficial and has compulsory status for specific kinds of projects in the UK (Gann, et al., 2003; Gann & Whyte, 2003; Prasad, 2004a; Whyte & Gann, 2003). The framework of the Design Quality Indicator as described by Gann and his associates takes the three basic qualities of functionality, impact, and build quality, as well as a resource envelope with finance, time, natural and human resources, into account. Results can be visualised by using a spider diagram with scores on the design aspects per stakeholder on a scale from 1 to 5 (see Figure 2.3).

Experience shows that the DQI stimulates discussions about quality among stakeholders (Cardellino, Leiringer, & Clements-Croome, 2009; Dewulf & Meel, 2004; Eley, 2004; Slaughter, 2004), although it cannot be used as a performance measurement system (Markus, 2003) and does not offer actors a concrete solution to the problem of formulating requirements. The framework and measurement system of the DQI, or related evaluation models and systems, display little room for the additional holistic values praised in the literature. According to Prasad (2004b) there is no point in discussing quality unless the very basics are fulfilled and Saunders (2007) states that an architecture of 'maximum aliveness' is likely to satisfy several (if not most) evaluative criteria at once, or to satisfy one or two criteria to an extraordinary degree. It seems that the totality is also a quality on its own, but that this holistic impression is lost in the measurement and analysis of the parts. According to Cardellino, Leiringer and Clements-Croome (2009,

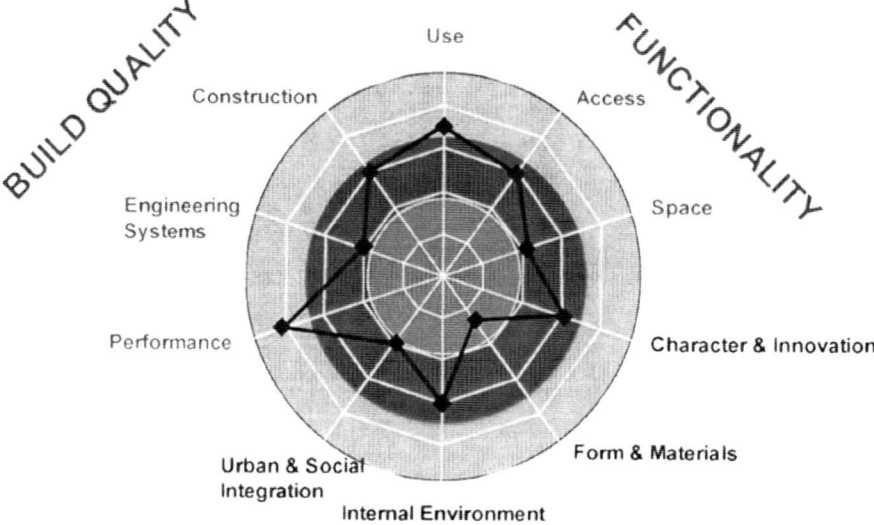

Figure 2.3 Spider diagram at output of a DQI assessment (Gann et al., 2003); The closer to the middle the score, the closer the quality would reach excellence

p. 260) "the compulsory use of the DQI has successfully cemented the commitment to design quality [....]. However, the architecturally biased approach seemingly underestimates the value of intangible aspects of design and chances are that the tool becomes a 'tick the box' exercise". Devine-Wright, Thomson and Austin (2003, p. 51) also state that the DQI "will only give an index of just that, a collection of individual responses". It therefore does not facilitate and development of the values in a project team. Because intangible qualities are not easily measured, decision making and communication about design quality are complicated. There is a need to establish a common vocabulary for the various types of values for different stakeholders in order to consider intangible and tangible costs and benefits as equal in decision making about design (Gerritse, 2008; Macmillan, 2006). The value approach of Mills, Austin, Devine-Wright and Thomson (2009) might offer that kind of communication frame, but implementation of this frame will require a change in thinking about value and quality in construction.

2.3 Design perception and affect (cognitive approach)

The second perspective I would like to address in this chapter is the cognitive approach of design perception and affect. Environmental psychologists began to study the built environment in the second half of the twentieth century. Their work is based on the assumption that perception is a result of the interaction between an individual and objects in the built or natural environment. This perception can be measured on a scale with several dimensions. Within this approach several attempts have been made to explain the potential differences between novices and experts. This section deals with the most important findings about the perception of design quality and some implications for the design of the built environment.

2.3.1 Characteristics of the design perception

Most research about perception of the built environment has been carried out on the aesthetic preferences for buildings exteriors and natural landscapes (Bell, et al., 1996; Gifford, 2002; Karmanov, 2009). The general aim of the field of environmental or architectural psychology is to reduce stress levels and increase the wellbeing of people by exploring which kind of environmental cues are preferred. Discussions in this field are less holistic and less practice-related than in architecture and generally focus on theoretical concepts about the perception of intangible characteristics of the built environment. So far only a few of the concepts developed in the field of environmental psychology have found their way into architectural practice (Philip, 1996). Recent publications in journals such as 'Environment and Behaviour' and the 'Journal of Environmental Psychology' show that most current studies focus on specific issues related to the built environment such as sustainability, natural disasters and virtual reality, rather than on the perception of physical buildings only (Giuliani & Scopelliti, 2009).

According to Nasar (1994) an individual's experience of a building depends on an interaction between its features and the individual's previous knowledge structures of experience related to the particular class of building. Through interaction with the environment and development of the organisational structures of knowledge, individuals from different places, cultures and subcultures would develop different meanings and preferences across categories. For a full understanding of architectural appraisal, it is important to learn how appraisals vary between different groups (Gifford, Hine, Muller-Clemm, Reynolds, & Shaw, 2000). Two kinds of aesthetic variables can be distinguished: those that relate to the structure of forms - formal aesthetics - and those that relate to the content of forms - symbolic aesthetics (Lang, 1987; Nasar, 1994). Formal and symbolic aesthetics can interact. Attributes of formal aesthetics include dimensions such as novelty and complexity. Symbolic aesthetics are experienced through mediating content variables which reflect the meanings associated with buildings (Nasar, 1994). Judgements with denotative meanings place objects into content categories without evaluating them, for example a type (church, prison) or a style (modern, traditional). Judgements referring to connotative meanings relate to the quality and character of buildings and its users, for instance friendly or unfriendly. Content variables have both denotative and connotative meaning; someone can for example recognise a style (denotative meaning) and like or dislike the style (connotative meaning).

Studies in the field of environmental psychology often include the use of a semantic differential to measure individual preferences of different groups of people in a laboratory setting in combination with personality and physical tests. The influence of contextual or process factors is often left out or randomised. Pictures, cards, videos, virtual reality or sorting tasks are used to simulate the real environment. The semantic differential techniques measure connotative meanings that people attribute to different concepts (Osgood, Suci, & Tannenbaum,

1957). This means that the reaction that people show after the confrontation with a physical environmental is measured by the three independent dimensions of 'evaluation, activity and potency' or variations thereof (Gifford, et al., 2000; Nasar, 1994). These dimensions correspond to the early insights of Vitruvius that architecture consists of order and arrangement, proportion and symmetry, décor and distribution (Brenders, 2008; Kelly, 2007).

Much of the later work in the field of environmental psychology has focused on validating and expanding the three original dimensions of aesthetic preferences. Russell, Ward and Pratt (1981) found two independent, bipolar factors of affective quality of the built environment – pleasure and arousal –, which have been validated by many other scholars in different settings (Gifford, et al., 2000; Karmanov, 2009). Based on these factors Russell et al. (1981) developed verbal scales of pleasant-unpleasant, arousing-sleepy, exciting-gloomy, and distressing-relaxing dimensions to be used as dependent variables in studies aimed at understanding which objectively specifiable properties of environments influence affective reactions. The theoretical framework for environmental preference as developed by Kaplan (1987) uses the predictors complexity, mystery, coherence and legibility. Interesting about this approach is the concern with the outcome of the information about the environment (understanding or exploration) and the availability of information (immediately available or inferred), which seems to relate to the different aims of people experiencing the built environment and whether the physical objects would fulfil the requirements and expectations. Latest research of Gifford et al. (2002) show six key cognitive properties in assessing the aesthetic quality of large contemporary building: complexity, friendliness, originality, clarity, meaningfulness and ruggedness. Jacobsen and Hofel (2002) found in an experiment with novel graphic patterns that especially symmetry and complexity correlate highly with aesthetic judgements of beauty. The variables empirically found to predict preference can be analysed in terms of their evolutionary significance (Kaplan, 1987). Some of the predictors appear to require fairly extensive information processing. This supports the hypothesis that a rapid, unconscious type of cognition may precede certain affective judgements and confirms a strong tie between cognition and affect.

In summary, the assessment systems as developed in environmental psychology focus on the possible outcome of the interaction between an individual and an object in the built environment. Figure 2.4 depicts the underlying assumption of this approach. This perspective considers an individual as a passive actor in judgement about design quality without taking into account much of the social and organisational context of the decision maker.

The underlying assumption of this interaction is that the aesthetic preference for an environment or building (judgement) can be measured on a scale with several dimensions, such as originality and complexity. Although researchers are aware that these processes can be influenced by individual, situational, social, and cultural factors, the exact effects and the link to the practice of designing have not yet been identified. To assess buildings in terms of their symbolic or formal aesthetics people use the denotative and connotative meanings of buildings. The pleasant and arousing quality of an environment seems to imply a strong connec-

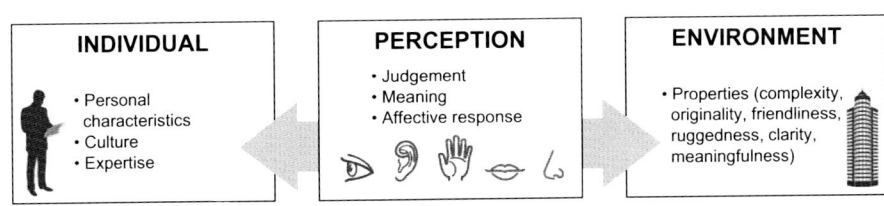

Figure 2.4 Perception as a result of the interaction between an individual and the built environment.

tion to affect. Intangible characteristics are used to address meaning and classify buildings, which could lead to an affective response in certain cases. The different characteristics of individuals explain differences in perception between groups of people.

2.3.2 Implications for design of the built environment

Although the field of environmental psychology has examined important underlying principles about the interaction between the individual and the built environment, limited practical value and use of difficult language seem to be the main reasons for architects to claim an "unfulfilling marriage between architecture and psychology" (Philip, 1996, p. 277). However, as for example the research of Macmillan (2006) shows, architects, occupant organisations and real estate developers could benefit from the insights in the field of environmental psychology and increase the attractiveness of the built environment by implementing existing knowledge more systematically in design. Several studies in environmental psychology focused on the differences of preferences between groups, especially between architects and non-architects. Most scholars conclude that these differences are related to the affective component of meaning and can be explained from a different weight assigned to design features (Gifford, et al., 2000). According to Karmanov (2009) these differences are probably less distinctive than is usually thought. The results of Groat (1982) and Hubbard (1996) suggest that architects and non-architects employ different sets of criteria and stimuli for evaluating buildings, but these differences do not lead to awareness of differences in styles of buildings. This means that differences in building styles are not always recognised by non-architects and have therefore limited power to predict preferences. Within the research sample of Brown and Gifford (2001) and that of Fawcett, Ellingham and Platt (2008), several buildings were given high aesthetic ratings by both groups. The models of preference used by Devlin and Nasar (1989) revealed that both architects and non-architects favoured novelty and coherence (or clarity) in building features, but the non-architects favoured simplicity and 'popular' attributes, while the architects favoured complexity and 'highly architectural' attributes. Nasar and Purcell (1990) and Fawcett et al. (2008) also found that for both groups preference related to increased complexity and decreased novelty, but the architects placed greater weight on complexity. These results suggest that architects favour aspects that are discrepant from the existing architectural character of a building, while the non-architects favour complexity within the recognizable knowledge structure that contributes to the basic attributes of a building. The research of Gifford, Hine and colleagues (2000) indicates that in contrast to the pleasure and global rating results, architects and non-architects agreed significantly about the arousal-eliciting characteristics of the 42 buildings. A cultural comparison between Australian and American groups showed both similarities and differences (Nasar & Purcell, 1990).

The results of the study of Gifford et al. (2000) indicate that non-architects produce more heterogeneous ratings (as a group) than architects in terms of emotional response. Non-architects' ratings of architecture are known to be subject to various influences, which make their ratings more diffuse. Architects' aesthetic standards on the other hand have been focussed through their selection and training. Architects are socialised by their professional education in ways that widen the aesthetic gap between themselves and the public (Purcell, 1986; Wilson, 1996). Valadez (1984) attributes a great part of these differences to the research instruments, which often require verbal responses about stimuli and less within-group variance due to a relatively narrow range of traits compared to the general public. The results of Valadez's research indicate that architects only differ from other professional groups in their quantitative judgements of the habitats studied, but not from all groups in their qualitative judgements. Novice architects seem to base their judgement more on affect (Hubbard, 1996) while professionals can retrieve more information from a building (Wendte, 2004). It seems hard for experienced architects to drop their own criteria for conceptual properties in favour of those of non-architects when they predict public evolutions (Brown & Gifford, 2001). The best-predicting architects seemed slightly less experienced but were better able to understand the matter in which the general public thinks about aesthetics of buildings. These architects related their evaluations to buildings' conceptual properties in a manner similar to that of laypersons.

Although architects are not obliged to design buildings to suit the public rather than themselves, Brown and Gifford (2001) suggest that architects could use their understanding of lay thinking to create buildings that delight both themselves and their public. To facilitate this process of integration of preferences, Gifford et al. (2002) suggest a broader form of architectural education or socialization that stresses both the creative extension of the great aesthetic trends and a better understanding of public taste. A broader education will not reduce the richness and diversity as long as different schools train architects to appreciate and design different styles of architecture (Wilson, 1996). Non-architects could be trained also: "Rather than restricting designers to styles that the users already understand, environmental education for non-architects could result in greater appreciation of other styles. Once people understand the styles of architecture that make up their cities, there would be a variety of different tastes to be catered for by variety of different architects" (Wilson, 1996, p. 42).

2.4 Product experience and emotion (interaction approach)

The field of product experience and emotion originates from industrial design and is based on several traditions in the field of psychology. It shows that products can elicit emotions when a user interacts with a product and therefore adds an important perspective on the aspect of design quality in the relation to architect selections. The implications for the field of architectural design as described in this section relate to the levels of measurement and the establishment of a link between the perception of the product and an affective response in the judgement of its value.

2.4.1 Origin of product experience

In the field of product design the findings from environmental and cognitive psychology are applied to support the development of successful products and capture a consumer's reaction to a product. The field of product experience originates mainly from the field of cognitive psychology, emotion and perception (Schifferstein & Hekkert, 2008). In their book on product experience, Schifferstein & Hekkert (2008) provide an overview of literature from the human perspective, the interaction perspective and the product perspective.

From the interaction perspective three components or levels of product experience are distinguished (Desmet, 2002; Desmet, Porcelijn, & Dijk, 2007; Hekkert, 2006): aesthetic pleasure, attribution of meaning, and emotional response. These components are strongly influenced by the verbal scales of Russell et al. (1981) for pleasure and arousal as described in the previous section. At the aesthetic level a product's capacity to delight one or more sensory modalities is considered. The degree to which a person manages to detect structure, order, or coherence and assess a product's novelty or familiarity typically determines the affect it generates. Just like with the work of Kaplan (1987) these effects can be explained by examining the evolutionary basis of the human perceptual system. The experience of meaning concerns the cognitive processes of recognition, interpretation, association and assignment, which attach a meaning to a product. These processes are subject to individual, cultural and physical differences (Hekkert, 2006). The emotional response relates to the result of a cognitive, though often automatic and unconscious process caused by the interaction of the human with the product. An emotion involves a relation between the person experiencing an emotion and a particular object; e.g. one is in love with something. Desmet (2002) states that an emotion is elicited by an evaluation (appraisal) of an event or situation as potentially beneficial or harmful in relation to the person's product concerns. It is the interpretation of an event (or product) rather than the event itself that then causes an emotion. Because appraisals mediate between the products, concerns and emotions, different individuals who appraise the same product in different ways will experience different emotions.

Figure 2.5 shows the interaction of the user with the product that causes product experience. This perspective considers the user as a rather passive actor in the interaction with the product. It focuses mainly on characteristics of the outcome of a product experiences instead of the appropriateness of the product characteristics. It does take characteristics of the user and the product into account to explain differences between groups of people.

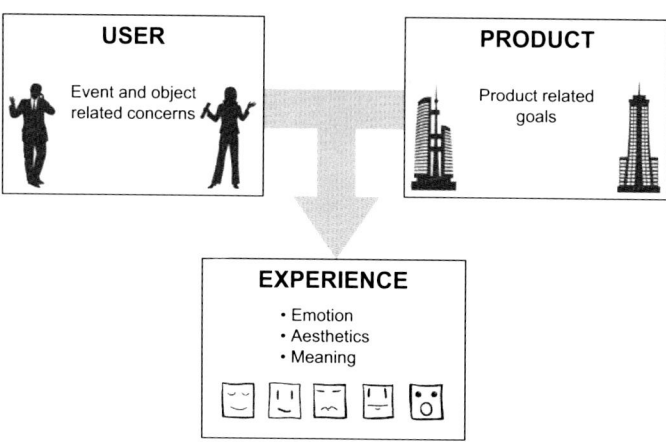

Figure 2.5 The origin and components of product experience as interaction between a user and a product

2.4.2 Implications for architectural design

In order to understand responses to human-product interaction, one must understand the user's concerns given the context in which a user interacts with the product. Khalid and Helander (2006) conclude that a product should be designed to support customer needs, including the customer's personal or personality attributes. This can be done by providing flow or ease of use, and inducing feelings or emotions in interacting with the product. They also state that customers tend to make decisions based on their feelings, perceptions, values, and reflections that usually come from gut feelings rather than logical or rational thinking. As such, designers and manufacturers should consider making emotional design a bottom line in product design. The Product Emotion Measurement Instrument measures the different kind of emotions that users experience and can be applied during product development (Desmet, 2002). In this manner designers can try to 'design for emotions'. The Kano model of product development deals with the customer requirements in relation to the characteristics of the product (Kano, 1984; Walden, et al., 1993). In this model customer satisfaction is not based on functionality alone – it includes product performance that transcends the stated performance (see Figure 2.6). Kano used the ideas of the Motivation-Hygiene theory of Herzberg, Mauser and Snyderman (1959) to distinguish basic qualities (hygiene factors that prevent dissatisfaction) and surprising qualities (characteristics that exceed customer expectations) and can form the basis of love for the product.

Desmet, Porcelijn and van Dijk (2007) distinguish different kind of product characteristics that relate to levels of product quality: 1) basic features that the product type is expected to deliver, 2) performance features that differentiate between competing products, and 3) excitement features that the consumer did not expect to see in the product and is excited to find. Those who want to design products that evoke feelings of 'wow' should find and apply one or more 'excite-

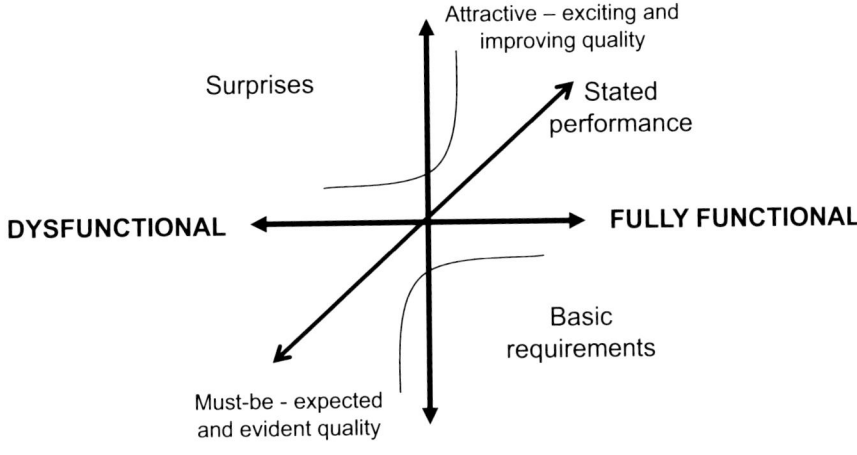

SATISFACTION –
Love the product

Attractive – exciting and
improving quality

Surprises

Stated
performance

DYSFUNCTIONAL

FULLY FUNCTIONAL

Basic
requirements

Must-be - expected
and evident quality

DISSATISFACTION –
Dislike the product

Figure 2.6 Representation of product qualities according to the Kano principles (based on Walden, 2003)

ment features'. Based on patterns found with their framework of product experience, the 'wow'-experience could be elicited by products that are appraised as unexpected, unfamiliar, promising, and fit for possession (Desmet, et al., 2007). These kinds of categories strongly relate to the ones identified by Kano, in the architectural debate and the developers of the DQI.

The position of usability in product experience seems to be disputed. Usability can be defined as the extent to which a user can employ a product in order to achieve a particular goal (Norman, 2002). Often the dimensions used to operationalise usability are effectiveness (the degree to which the particular goal can be satisfied), efficiency (the amount of time it takes to satisfy the goal), and ease of use (the amount of effort it takes to satisfy the goal). Correlations have been found between usability and aesthetics and meaning (Schifferstein & Hekkert, 2008). According to Boztepe (2007b) user value is, just as aesthetic pleasure, an attribution of meaning and emotional response, created as a result of interaction between what the product provides and what the user brings in terms of goals, needs, limitations etc (product concerns). Just as in architecture, in product design the level of experienced satisfaction is often used as a measure for usability (because products that are usable will more likely elicit positive emotions). But according to Desmet and Hekkert (2007) usability should not be considered as a level of product experience but as a source of product experience (a product characteristic) because it does not cause a change in the core affect attributed to the product human-interaction. This difference in opinion might relate to the perspective of the research studies (short term/sale perspective or long term/use perspective) or the perception of the role of the user during product development (active or passive).

2.5 Value systems in design (process approach)

In discussing design quality the term value is just as important as quality and perception. In this section the concept of value is discussed from the perspective of the construction industry. This perspective focuses on the realisation process of a building and shows the multiple stakeholders' perspectives. I therefore consider this as the fourth perspective that is relevant for assessing design quality.

2.5.1 The concept of value

In construction the use of the term 'value' seems to be preferred over the use of the term 'quality'. Distinctions can be made between value, value judgement and value system (Kelly et al., 2004; Pultar, 1996). Pultar (1996) defines value as "any stated descriptor which forms the subject matter of a belief in the correctness or preferability of a choice". According to Thomson et al. (2003) and Le Dantec & Y-Luen Do (2009) values are the principles, standards and qualities by which we live and guide our actions. The two types of values as defined by Rokeach (1973) – terminal values (end goals) and instrumental values (means to reach goals) – seem to explain the differences between the 'product concerns' of Desmet (2002) and the 'process values' of Wandahl (2005). Value has theoretical, economic, aesthetic, social, political and religious dimensions (Allport, Vernon, & Lindsey, 1960) and could be of influence on perception (Bell, et al., 1996). Values can develop at different stakeholder levels: individual, organisation, or society (Lepak et al., 2007). Mills, Austin, Thomson and Devine-Wright (2009) successfully applied the slightly adjusted Schwartz Values Survey (e.g. Schwartz & Boehnke, 2004) in order to develop an approach that supports (project) stakeholders in understanding organisational and individual values and priorities. Their approach differentiates between ten values categories, such as Universalism, Others oriented, Conformity, Power, Security and Stimulation, and provides insight into possible alignments between cost consultants, engineers, value management consultants and building maintenance and operations professionals.

People make value judgements when they asses an object with regard to their beliefs and expectations. Desmet & Hekkert (2007) refer to 'product related goals' in this matter. According to Collins (1971) value judgements should always take into account their environmental, psychological, procedural and historical context of precedents of the people involved in the judgement process. When individuals collaborate to realise a common goal projects are formed. A value system can emerge if values are expressed and shared between them (Kelly, Male, & Graham, 2004; Wandahl, 2005). A value system may be defined as a collection of value judgements held by a person or a group regarding various values involved in a phenomenon (Pultar, 1996) and is identified by the relative importance assigned to values such as freedom, pleasure, equity (Robbins & Judge, 2008). Value similarity produces a social system or culture that facilitates the interactions necessary for individuals to achieve their common, event oriented goals (Desmet & Hekkert, 2007; Robbins & Judge, 2008; Wandahl, 2005). Within a value system, value judgements do not exist independently of each other; there are bound to be interactions and conflicts between the various value judgements. In this context Wandahl (2004) argues that parties have difficulties creating common goals and a tangible background for decision making because of pre-existing differences between the parties. This does not mean that it is not possible to share values in construction. The study of Mills et al. (2009) showed for example differences between organisations in engineering, design, consultancy and maintenance, but also found values that were aligned between the organisations, such as responsibility, honesty and loyalty.

As described in Chapter 2.2 a product has tangible and intangible characteristics, which in themselves do not determine how people form their preferences. In line with the perspective of product experience and environmental psychology, Wandahl (2005) defines product value as the value which an individual places upon an object or outcome. In this manner, the act of valuing is dissociated from the value itself. Also, it becomes possible to allow for the fact that different people may hold different beliefs or preferences based on the same value. Boztepe (2007a) presents four major categories of user value based on an ethnographic study on the assignment of use and value to kitchen appliances: utility value, social significance value, emotional value, and spiritual value. Macmillan (2006) distinguishes several comparable main types of value of a mixture of tangibles and intangibles and as much stakeholder groups between whom value is exchanged in construction projects. These value types are related to the different building types described by Loe (2000): use, exchange, image and business value buildings. Bucciarelli (1994) concludes that design is a social process and the objective reality of a technological artefact is in fact a social construct. The design object is therefore per definition alive and laden with uncertainties and ambiguity.

According to Gerritse (2008) product value is the extent to which product characteristics fulfil the requirements. When the outcome of a value assessment is attributed to a product as one of its qualities, product value becomes a measurable product attribute. Dreschler et al. (2005) conclude that in construction all definitions of value (except for the ethical one), compare some level of performance, functionality, utility, benefit or quality (perception) with the associated level of price or cost. Also Prins (2009), who defines architectural value as a complex triadic system between planet, spiritualism and profit with mutually dependent elements, and Kelly et al. (2004) denote the relationship between function, cost and worth. In value engineering one speaks of exchange value, esteem value and use value to denote the relationship between costs and benefits (Best & de Valence, 1999). Based on the economic principle of welfarism, Dreschler, Beheshti and de Ridder (2005) consider value as the amount to which all persons involved are influenced in their well-being as a consequence of the construction project. This means that value is defined as an absolute quantity to be used in the process of decision making and all values need to be weighed and added up to determine the total monetary value or total score (Dreschler, 2008; Jansen, et al., 2007). The point system, monetization system and ratios are three methods to combine price and value (Jansen, et al., 2007).

According to Thomson et al. (2003) the distinction between quality and value lies in the objective or subjective character of its judgement; value assessments can be subjective when framed against an individual's values, while quality assessments can be considered objective when the relationship between benefit and expense is compared on a level of fulfilment of requirements. They conclude that value is the relationship between positive and negative consequences for an individual (output and input, or benefits and sacrifices). The allocation of a monetary sum to express perceived value is a common means of setting the price of a product. "When a consumer asks him or herself 'is it worth it?', they are making a value judgement in light of their own, often tacit, values and comparing this with the market value assessment (typically expressed as a price)" (Thomson, et al., 2003, p. 337). In this sense value judgement are seen as rational considerations of individual decision makers among the different value types that can be addressed to (future) buildings.

Figure 2.7 shows the concepts of values, value judgement and the value system in relation to the product value as argued by the value approach in construction management. In this approach people are seen as stakeholders (groups) that act as active participants in the development process of a building project.

Stakeholders can be defined as groups or individuals who have stake in or expectation of a projects performance (Newcombe, 2003). During ongoing dialogues of stakeholders with the project team both subjective and objective assessments of value are constructed which shape the final outcome of the project. The Design Quality Indicator is a helpful tool in this process of interchanging perceptions of stakeholder groups. However, the question remains if stakeholders are actually capable of rational considerations about benefits and sacrifices, especially in case of the long-term consequences of certain alternatives. Research and theories underlying the assumptions in the process approach seems absent, and little use is made of existing theories and models in other fields of science.

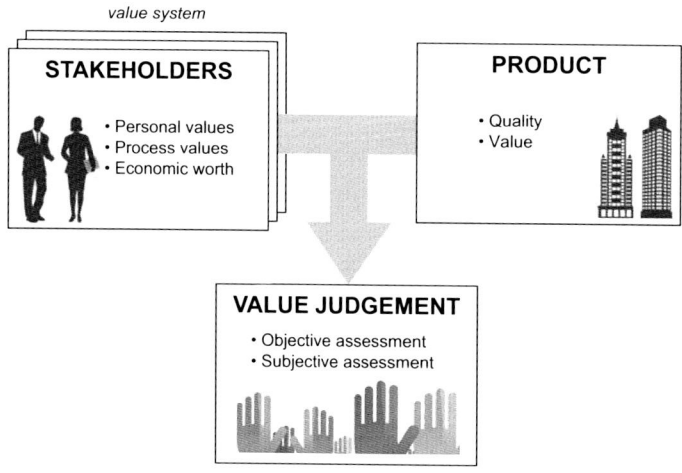

Figure 2.7 Value judgement as a results of the clients values and the product value

2.5.2 Applications in design and construction

Through an ongoing dialogue of stakeholders with the project team both objective and subjective assessments of value are constructed. The DQI provides a value framework to develop a common language to engage stakeholders and could be an appropriate tool to structure and summarise these assessments (see section 2.2.2). Another example of an system in which the ongoing dialogue between stakeholders about values plays an important role is the concept of value thinking (Thyssen, Emmitt, Bonke, & Kirk-Christoffersen, 2010; Wandahl, 2005). During the construction phase of a building the design is subject to debate because ideas have to be translated into physical products. Value thinking has two phases: 1) Value Design by Value Management, finalised in a contract for construction, and 2) Value Delivery by Value Engineering and finalised in a construction product. Value Management always involves a relative and balanced consideration of tangible and intangible costs and benefits to value design and a willingness to make trade-offs in order to make gains (Best & de Valence, 1999). A distinction can be made between internal (process) values of the participants of the delivery team and external (product) values of the client, which the delivery team should be focusing on achieving (Emmitt et al., 2005; Kelly, 2007). Wandahl (2004) found in three experiments that the different parties in construction (users, clients, architect, engineers, contractors) have only a slightly different perception of value in a construction project. Within the framework of Value Management both the internal and external value can be separated into process value and product value. Process value is about giving the client the best experience during the design and construction of the project and comprises soft values (communication, conflict solving etc), hard values (time, budget) and process values (learning experience). Product values are based on the Vitruvian values, augmented by contemporary concerns for harmony with the surroundings, environmental issues and buildability (Emmitt, Sander, & Christoffersen, 2005). Product and process values can interact, and especially when the product becomes visible, it could mean changes in the values or rather the interpretation of the values. Howes and Gifford (2009) therefore distinguish assigned values (situational values) and held values (deep-seated, enduring values) that are part of a dynamic value hierarchy. In conflictive situations they found that people who have less firmly held values were more influenced by situational differences. According to Kelly (2007) the value system of a client is not a hierarchy but an order of preference for variables at the same level. The results of a workshop in which the Kano system was applied seem to suggest that a valid variable is subject of a continuum against which the client can indicate a point of satisfaction (Walden, et al., 1993). The experiences of the Thyssen et al. (2010, p. 29) indicate that a skilled process facilitator is required to develop a language in which client values are understandable for building professionals because values cannot be made explicit "once and for all by writing it down in a fixed value system for subsequent design evaluation".

Value delivery and value engineering might have a negative connotation because it is sometimes interpreted as cutting costs instead of raising values, which could lead to a focus on the lowest costs instead of the highest quality (Loe, 2000). In relation to value engineering Loe (2000, p. 52) concludes that "despite this impressive level of debate and activity, reconciling the value of good design in architecture remains an elusive concept. [....] we are adept at exchange value but still have to weld to this technique the means of measuring the benefits that well designed buildings bring to the social, political, urban, and image values." In practice costs do change the values of the stakeholders, especially when confronted with the first prototypes of the product (Wandahl, 2005). This could be due to the fact that they then realise the actual costs of their ambitions. It is through the process of design that values are exposed and negotiated in the search for potential solutions (Le Dantec & Y-Luen Do, 2009).

Boztepe (2007b) points at the distinction between pre-purchase and post-purchase value in marketing research. Pre-purchase refers to expectations regarding the value a product is going to deliver that are formed prior to purchase of the product. Post-purchase value involves value realised through the use of a product. During the life cycle of a building the value of the design changes from expected value of the client, to promised value of the designers and then delivered value of the contractors. The interaction with prototypes means that expectations and experiences get mixed. The difference between a client and a customer is related to these values concept. "A customer is someone who purchases a commodity or service from a merchant who supplies an existing specified product for a price, and acts in his/her own interest. If during the moment of purchase the performance standards are met, the contract and the customer is satisfied. A client, by contrast, is someone who engages the professional advice to act in the client's interest in a condition of dependency and direct contact." (Tombesi, 2006, p. 276). A client assumes influence on the development of the product because he/she buys a not yet existing product (Doree, 1996). Collins (1971) ascribes the layperson's inability to appreciate the nature of the architect's dilemma to the fact that a layperson normally only purchases finished articles that they have inspected before any decision is made as to the price they are willing to pay. "Show him an old Office which is being offered for auction, and he will bid for it with full confidence that he knows a good bargain when he sees one. But show him the sketches and working drafts of a building he has commissioned, and he will find it inconceivable either that so few drawings are worth 5 per cent of the estimated cost of the building, or that the architect cannot make immediate alterations so as to halve the contractor's prices without sacrificing any of the amenities" (Collins, 1971, p. 196).

2.6 Integration and implications

The perspectives discussed in this chapter all deal with the perception and interpretation of product qualities. Perception occurs when an individual is confronted with the tangible and intangible characteristics of a product. In this research I therefore define design quality as an overall value judgement by an individual person based on the interaction between this person and an object in the built environment. I assume that value judgement includes an assessment about the level of quality as well as an affective response. After a judgement a product is associated with a certain value, which could relate to all kinds of product values, such as use value, environmental or economic exchange value. In this sense the product receives its value after its meaning is interpreted by an individual. Design quality relates directly to different dimensions of product value because these dimensions include a judgement about certain characteristics. Most of the characteristics of a building are considered as evident and only lead to conscious dissatisfaction in case basic requirements are not met. Cases of extreme positive or negative experiences could lead to a surprise or 'wow'-experience.

Individuals act in the context of their personal value system, which is applied during the judgement. This value system includes the goals and expectations of the individual. If assessments are made in groups, a value system is developed among the members of the group, which could influence the individual value system. The individual value system could also be influenced by personal, social, and external variables of the context. The different perspectives that are addressed in this chapter show that differences in value judgements can be assigned to a difference in the weight of values, the amount and use of information during the judgement, or a difference in training of the expression or verbalisation of product experience. I presume that the stage in which a product is in – a representation of a future product or a physical object – is part of the information that is available during value judgement. Most of the perspectives considered in the previous sections do not link value judgement to decision making, but implicitly suggest that a decision can be seen as a consensus among value judgements of the group members about a product in the built environment. I therefore conclude that decision making is an iterative process of different kinds of value judgements, resulting in different kinds of product values. These values are not easy to sum up and justify as one 'truth' because they are based on perceptions of the group members. Eventually there could be a consensus between group members about design quality. Assessing design quality is consequently part of a process of sensemaking among the decision makers. Figure 2.8 sketches the context of value judgement about design quality.

As a result of the interaction between the individual and the product, a value judgement always includes an experience that is accompanied with an affective response. Based on the theory addressed in this chapter I propose that this affective response consists of three components: aesthetic experience, experience of meaning, and emotional experience. The degree to which a person manages to detect structure, order, or complexity determines the degree of preference and delight for a product on an aesthetic level. The approach of environmental psychology focuses on this aesthetic experience, which seems to have an evolutionary basis. In architecture one often focuses on the connotative aspects of a building that provide meaning, and considers the social and societal impact of a product in the built environment. The approach of product experience focuses on the emotional appraisal of an event or situation as potentially beneficial or harmful to a person. From the literature it remains unclear in what sense and how exactly this affective response influences the assessment process of design quality and therefore the process of decision making of clients.

Architects could take advantage of the knowledge in the field of environmental psychology, product experience, and findings from post occupancy evaluations to try to better meet the needs of the client with their designs (Philip, 1996). For

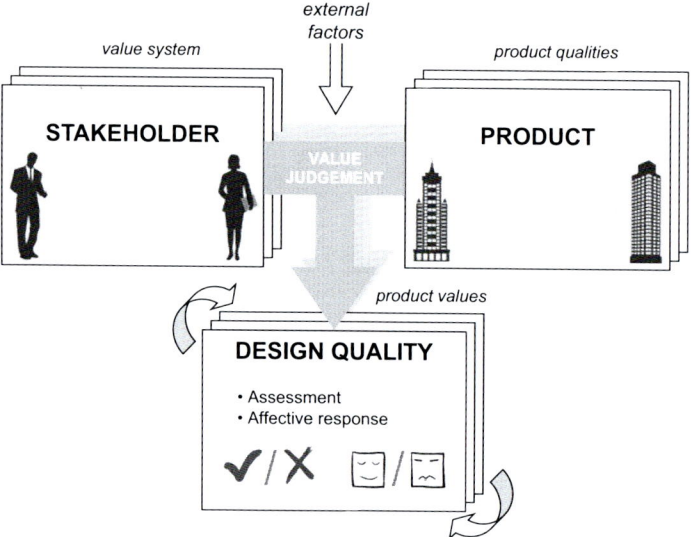

Figure 2.8 The concept of design quality defined as value judgement as a result of the interaction between a product and an individual in the context of a value system of a group

example, Nasar (1994) suggests adjusting the level of order, familiarity, complexity and discrepancies to the aim of the competition. Design review seeking excitement should encourage high complexity, atypicality and low order; design review seeking calmness should encourage high order and naturalness. 'Wow'-experiences seem important to increase sale of a product. A 'wow'-experience is probably based on a perception of highly appreciated and integrated physical attributes. It could also be based on only one striking feature for the right value system. In the context of architect selection design for emotion could be an interesting strategy in order to improve the chance on winning a tender. This however requires additional research.

A value judgement includes an assessment about the level of quality or value of a product. In my opinion most perspectives discussed in this chapter deal with four general levels in assessing quality: under-performance, basic performance, added value, and excellence. Some authors state that the distinction between quality and value lies in the objective or subjective character of its judgement and the frame that is used to make the comparison. Others attribute the subjectivity and objectivity to the character of the characteristics of the product. In this research I assume that the physical character or tangibility of a product quality make it possible to develop a commonly accepted, assessment system. Results of assessments based on the use of quantitative assessment systems are usually more consistent and therefore perceived as objective. For product characteristics that are less tangible an assessment system is often built among a system of (expert) assessors whose value judgements are commonly accepted. The validity of these kinds of assessments can be measured by the agreement among the assessors, which of course does not have to lead to the same judgement for every assessor.

The outcomes of an assessment of an individual (expert) assessor is often perceived as subjective by nature and therefore of lesser value than the objective judgements. In case of intangible characteristics the greatest objectivity is often created through inter-subjective assessment by a group of experts which can be considered. In today's society, perceived objectivity and quantification seem to be valued higher than validity and reliability. It appears that people assume an objective judgement within a standardised assessment system to have more predictive power than a subjective judgement. A quantitative assessment produces a numerical deviation between judgements, which can then be statistically examined to test its significance. However, small deviations within a measurement system are no guarantee that the value judgement is valid, reliable and generalisable. Therefore we cannot definitively choose between objectivity or subjective assessments: good assessments require good assessment systems adapted to the task of judgement (Hogarth, 2005). This means that the decision task should determine the assessment system. If subjectivity is part of the decision task, the system should take this into account.

Based on the theory in this chapter I conclude that assessment systems for design quality need to take into account that affective responses occur in the interaction between a product and an individual. These responses come to the fore only after the confrontation with the proposals or presentations of the entrants of a tender. This means that such a response cannot be predicted and the exact aspects by which the assessment is made cannot be known beforehand. Research has produced insights on the dimensions of the value of a product, such as social, user and economic value, which can be traced back to the essentials of Vitruvius and appear to be a constant factor in design assessments. Therefore they could be used to develop an appropriate assessment system. In my opinion an assessment of design quality should always include an individual consideration of costs and benefits. A remaining challenge is to design a system that includes holistic as well as decomposed quality and takes into account the changing value hierarchy of assessors as individuals as well as on a group level.

The literature from psychology suggests that the ease of the assessment of the physical attributes (characteristics) of a product relates to the potential deviation of the outcome of the assessments between different individuals. The assessments within groups of different levels of expertise appear not to differ as much as between groups, but even these differences appear to be restricted to both ends of the scale. Especially in extreme cases of modern or traditional design, experts and novices seem to come to different value judgements. Experts often represent the architectural or a design related professional community. But because of the high impact of the built environmental I propose that a value judgement should also represent other stakeholders, such as users or clients/customers. Then the questions arises who would be able to represent the stakeholders best and which competences are needed to be a good juror. This could require additional education or support for some jurors. Even people with a design education might need training in order to be able to judge design quality outside the context of their own interests. Further research is required on the different levels of expertise of stakeholders and their abilities to assess design quality in relation to the costs and benefits of user participation.

The position of usability and functionality of a product in relation to the affective response is still a topic of debate among the scholars addressed in this chapter. This could originate from whether usability or functionality are seen as product characteristics or to imply assessments about the use of the product. It could also depend on the difference in weights of assessors or developers of the product. The difference between pre-purchase value (product expectations) and post-purchase value (realised by use) appears to be interwoven in the different approaches to design quality. In the built environment value management theoretically links pre-purchase value to stakeholders with post-purchase value to users during the dialogue. In product design researchers mostly focus on pre-purchase value, whereas environmental psychologists concentrate on evaluations during and after use. The difference between a client and a customer is related to these values concepts. The architectural and value approach both address the active involvement of a client and other stakeholders during the process of development, whereas from the perspective of perception and product experience the users or customers seem to have a rather passive role. Depending on the phase of development, the term product is used for a future physical object (a design) or an existing physical object (a building or consumer product). In architecture a product is 'bought' based on expected or intended design quality. In that sense the client determines the quality standards, although these inevitably change during the course of the project. It is very likely that the manifestation of the product is also reflected in the affective response. The kinds of considerations made during assessments reflect the active and passive role of the client cum customer; the balance between costs and qualities strongly relates to an active role in product development, while the experiential response implies a passive role of the client/customer.

For future research it would be interesting to know more about the relation between the dimensions of complexity, pleasure and arousal in relation to the building characteristics which cause surprise or contribute to the 'wow'-effect to different stakeholder groups in the built environment. Hogarth (1988) introduced the difference between compensatory and non-compensatory models of value judgement to address the differences between a total sum of aggregated scores and compensating aspects. This concept seems to be applicable to the issues in architectural design and needs further exploration in the field of design assessment. The lack of empirical data in the fields of architecture and construction management might be explained by the fact that products can be more easily piloted than buildings. Mock-ups make product development more realistic and dynamic, but similar problems did not prevent other fields of science from developing a research tradition. The hands-on and pragmatic culture with limited attention for systematic knowledge development might be another reason for the lack of empirical work. In relation to process management the fields of architecture and construction management could benefit greatly from studies in the field of political science, governance, public decision making and user participation in product design. These fields might also shed light into the differences between a contracting authority in a legal sense and a client in real life. All perspectives discussed in this chapter do not address the issues of liability about the consequences of a value judgement and the different identities represented by an individual. Insights from decision making theories could contribute to knowledge about the level of experience that is required to make appropriate assessments and the development of sound assessment systems in order to make decisions and pursue action in design. In Chapter 3 several decision making theories are discussed in relation to the process of architect selection and value judgements about design quality.

Chapter 3

JUDGEMENT AND DECISION MAKING

3.1 Introduction

The field of decision making offers a broad range of perspectives on individual decision making as well as decision making in groups or organisations. The field has a long research tradition with a diversity of research methods including experiments, surveys, observations, interviews, and multiple research disciplines such as psychology, administrative behaviour, economics, sociology and anthropology. While the first generation of theories treated decision making as rational process akin to making choices in a game, more recent theories study decision making in real-life contexts and include intuition, affect and emotion. At the same time, the scope also broadened from investigating individual, cognitive processes of decision making to addressing social and organisational influences. Attention to the interaction between individual, group and the organisational influences is needed to increase the understanding of decision making. Weick (1969) introduced the concept of sensemaking, which explained that people enact their environment when they make sense of it and act upon their interpretation. Each generation of theories builds on the work of the previous generation. This means that theories do not compete on every aspect and often are complementary. The preference for a certain kind of decision theory seems to depend on the field of science. In economics and engineering, the early models of decision theory and first generation theories seem to be preferred, while most of the work in social and management sciences is based on first and second generation theories.

This chapter provides an overview of the psychological aspects that seem most relevant for judgement and decision making in the context of selecting an architect. It elaborates on the definition of design quality as a value judgement from Chapter 2. The political, cultural and legal context in which public clients enact will be discussed in Chapter 4. Together the first four chapters provide the basis for the theoretical framework in Chapter 5. The chapter starts with an overview of three generations of decision theory (Beach & Connolly, 2005; Hodgkinson & Starbuck, 2008a). The relevance of each generation of decision theory to selecting an architect is discussed. The underlying the differences between the generations of decision theory reflect potential sources of conflict between rationality and intuition. This is explored in the second section. Then the actual processes of making decisions are described. The decision processes and factors of influence are first discussed on the individual and then on the team level. The chapter ends with a short reflection and conclusion in relation to the process of architect selection.

3.2 Three generations of decision theory

This section addresses the three generations of theories in the field of decision making. The early models of decision theory focused on how individual people should make decisions. These models assume that decision makers are capable of making rational decisions because they are fully informed and able to estimate the risks of maximum utility of a decision. When Simon found that the rationality of decision makers is actually bounded (Simon, 1997, 1st edition published in 1947), a new generation of behavioural decision theories started to develop. This generation focuses on the use of heuristics as decision rules. Scholars in the area of naturalistic decision making study decision makers in real life in order to observe how decisions are made when information is absent, what the influence of the organisational context is, what role emotions, intuition and affect play in decision making etc. This became the second generation of decision theories. In this section all three generations are shortly described and applied on the situation of architect selections.

3.2.1 Early models of decision theory

Early models of decision theory can be considered as generation zero in decision making. Most of these models were prescriptive in nature, i.e. they describe how people (individuals or organisations) should behave in order to fulfil the requirements of rational decision making. Prescriptive theories assume that an ideal decision maker is fully informed, able to compute with maximum accuracy, and acts completely rationally in order to strive for maximization of the subjective expected utility in a decision (Beach & Connolly, 2005). Choices are seen as bets with a probabilistic range of outcomes of various utilities. This line of reasoning started with the utility theory of Bernoulli (1738) and Bayes's (1763) theorem of probability theory. The expected utility theory of von Neumann and Morgenstern (1947) can be considered as one of the first prescriptive decision theories and the major decision paradigm since World War II.

The prisoner's dilemma (von Neumann & Morgenstern, 1947) and the basic decision dilemma (Behn & Vaupel, 1982) are examples of games against opponents based on the gambling analogy in prescriptive decision theory. The utility function assumes that preference reflects both the value of the outcome of a decision to a decision maker and his or her feelings about risks. There are various versions of utility theory but they always connect, interrelate and sum up values and risks (von Winterfeldt & Edwards, 1986). Three kinds of real world probabilities exist that make it possible to utilise the options in decision making (Beach & Connolly, 2005; Gigerenzer, 1991): necessary probability with theoretical chances (e.g. a dice has six sides), frequentistic probability that predicts long-run probabilities (e.g. the results of the previous throws), and subjective or personal probability that estimates probabilities of unique events (e.g. personal thoughts about the odds). Research has shown that subjective probability depends on the kind of assessment, the stated probability, and relative frequency in the wording of the questions, while for the other kinds of probabilities accuracy and coherence are most important.

Decision makers seem to use different strategies for different judgemental tasks in different judgemental environments. The practical application of prescriptive theory is called decision analysis and is aimed at developing tools and methods to help people making better decisions, such as decision trees and decision matrices. Systematic software tools are available as decision support systems in business as well as in design and engineering. Wierzbicki (1997) found that multi-criteria decision aids based on early decision models do not support the actual choice because this choice is often made intuitively. Therefore he suggests that instruments should concentrate more on supporting earlier stages of decision processes that precede choice.

Utility theory and probability theory are theories about assigning numbers to events and not about what is valuable to decision makers. The benefit of the research in this tradition is the attention given to repetitive and small decisions common in daily life and the learning effects that occur in these situations (Betsch, 2005). Empirical research has generally shown that the actual preferences and decision behaviour are not in line with utility and probability theory (Beach & Connolly, 2005). The main reasons for this are probably that most real life decisions cannot be seen as gambling problems. Although people sometimes perceive risks and estimate their chances, they do not utilise their options as rationally as is assumed. In daily life decisions are far more complex than proposed in laboratory experiments. Usually information about alternatives is absent and contextual factors influence the situation. Therefore it is hard for decision makers to gain an overview the information available and the consequences of actions. Recent developments of decision support systems in design use different stakeholder interests (van Loon, Heurkens, & Bronkhorst, 2008) or fuzzy logic (Bittermann, 2009) to imitate human reasoning in the context of groups to support decision making. But whatever the qualities of such systems, it still does not change that fact that the actual decision cannot be taken by the system but has to be made by the decision makers themselves.

In general tendering theory assumes that the optimum bid has two components: the estimated costs of executing the project and a strategy for maximizing profit which constitutes a constant mark-up (Runeson & Skitmore, 1999). However, as shown by Runeson and Skitmore (1999), tendering theory does not perform well in terms of accuracy of forecasts and informative content. It shows inconsistencies in the logic of assumed profit maximization behaviour and is not based on a sound theoretical framework. This seems, for example, due to the fact that builders actually do not sell products but skills and services, that these skills are sold on different markets, that learning takes place, that humans are boundedly rational, that prices and mark-ups are affected by changes in demand, that contractors sometimes act strategically or pragmatically, and that the accuracy of cost estimates is generally very low. Despite these inadequacies decision trees and decision matrices can support the basic choices of a tender procedure, the award mechanism or other principle choices of the tender procedure.

Different associations (e.g. Architectuur Lokaal, 2009b; Chao-Duivis, Koning, Spekkink, & Sauerwein, 2007; CROW & Balance and Results, 2009) have developed competition and tender models for the Dutch context, which reflect the opportunities provided by procurement law and the underlying considerations in selecting an alternative for a tender project. The doctoral thesis of Marco Dreschler (2009) proposes for example a decision tree with options for the award mechanism for the economically most advantageous tender. In current tender projects decision matrices are often used to justify a selection or award decision by showing the scores of the proposals on the criteria and their relative weights. In this sense the matrix is used as a Multi-Criteria Analysis (MCA).

3.2.2 First generation of decision theories

The first generation of decision theories was largely influenced by Simon's work on 'bounded rationality' (Simon, 1997, first edition 1947). The concept of bounded rationality is based on the principle that because the capacity of the human mind is far too small to formulate and solve all problems and solution alternatives in a fully rational manner, people choose a path in decision making that 'satisfies' their needs rather than search for an optimum as assumed in utility theory. This notion became the start of the development of the theories that took the actual process of decision making and the organisational context of the decision maker into account (March & Simon, 1958). These first generation decision theories consist on the one hand of concepts based on judgement and decision making, such as prospect theory, and on the other hand on behavioural theories based on the use of general rules of thumb in decision making called heuristics.

Research into so-called reference dependency has shown that the utility concept was actually more complicated than originally thought because people tend to value different categories of money differently. Kahneman and Tversky (1979) incorporated this approach in their prospect theory. Prospect theory assumes that the value of an uncertain 'prospect' is determined by a weighted average of the decision maker's valuations of the various consequences of the prospect, where the weights reflect his or her assessment of the likelihood of each consequence. In prospect theory the value follows an 'S' pattern with steep parts in which loss or gain are perceived as more extreme than would be expected in a linear function. The function does not count total wealth, but calculates the sum of gains and losses from an anchor point and decision weights instead of formal or subjective probabilities (Beach & Connolly, 2005). The results of prospect theory are remarkably robust, and account for a wide range of choices in laboratory settings. They are commonly used in economic analysis. The main criticisms of prospect theory concern the assumption of a passive attitude of decision makers and other psychological effects such as luck, getting even and excitement in real life events.

Starting with Edwards (1961), behavioural decision theorists have focused on finding ways in which human choice deviates from maximization of subjective expected utility by identifying a variety of rules of thumb called 'heuristics' (Kahneman, Slovic, & Tversky, 1982). These general rules of thumb reduce the time and effort required to make reasonable judgements and decisions, especially in routine decisions (Cyert & March, 1963). The three main heuristics proposed by Kahneman and colleagues are the 'representative heuristic', the 'availability heuristic' and the 'anchoring and adjusting heuristic'. The representative heuristic means that decisions tend to reflect characteristics of underlying processes and events that are similar but not the same. The heuristic based on availability describes that during decision making easy examples are used that 'happen to be available' because of their frequency or probability. When the anchoring & adjustment heuristic applies the decision makers uses an initial guess for the first estimation and makes adjustments to this anchor value.

It is still not clear what exactly the value of the concept of heuristics and biases is. It clearly inspired a lot of researchers to analyse biases, the mistakes people make in probability estimation. The representativeness heuristic could for example cause a wrongful belief in the law of the small numbers (the belief that random samples of a population will resemble each other and the population more closely than statistical sampling theory would do), a neglect of the base rate information (the tendency to ignore the relative frequency with which an event occurs), and a neglect of diagnosticity of information (the tendency to make 'non-regressive' predictions when not applicable) (Plous, 1993). These heuristics could strongly influence the probability estimation within decisions. Research on heuristics seems to prove that without help or attention some apparently simple probability problems can be tricky. Heuristics also help to build a causal model for understanding decision situations. Also in law the role of heuristics during judges' deliberations, the difference between legitimization and decision making and de-biasing techniques during cases are discussed (Arkes, 1989; Nieuwenhuis, 1976; Visser, 2009). Gigerenzer and colleagues (2001) suggest that an alternative category of heuristics, the 'fast and frugal' heuristics, match the informational structure and demands of decision makers in a more ecological and better way. They developed the adaptive toolkit which makes people make 'smart' decision in terms of time, knowledge and cognitive computation.

Management and organisational scientists borrowed a plethora of terms from the basic cognitive sciences (mental models, scripts, cognitive maps, schemata) to improve their understanding of decision makers. Still the concepts as found in the organisational behaviour literature lack theoretical substance and are too general to explain decision situations (Beach & Connolly, 2005). The limitations of this type of research lay in its considered context dependency, its reliance on laboratory experiments with students, its lack of cross cultural orientation and the fact that researchers are vulnerable to the same biases they do research on (Plous, 1993). According to Hodgkinson and Starbuck (2008b) behavioural decision theory has made no significant contributions that take meaningful account of social interaction or organisational complexity. A need for more behavioural research in practice exists because "problems identified in practice seldom correspond to only one or a few scientific disciplines and one or only a few organisational specializations" (Kieser & Wellstein, 2008, p. 509). Similarly Beach and Connolly (2005) plead for more research that focuses on implications for practice instead of analyzing biases to explain deviations of heuristics.

Following a legal perspective, tender situations can be understood in terms of first generation decision theories of rational decision making. Procurement regulations require that decision criteria for comparison of the alternatives are known beforehand. The proposals submitted by the architects form the decision alternatives. These proposals include information that could be explicitly compared, such as the price, sustainability score or total area in square meters, in order to make a choice among the alternatives. Therefore procurement decision making seems to be a matter of estimation of the probability for future quality. The adaptive toolkit of Gigerenzer and colleagues (2001; 1999) with fast and frugal heuristics seems suitable to apply in these kinds of situations. Therefore biases easily occur. According to Simon (1997) a decision can be considered rational as long as the perspective in which the decision is taken is clarified, which is the case for legal requirements that assume a rational comparison of alternatives on the basis of weighted criteria. However, tenders occur in organisations acting in a political environment with several groups of stakeholders who want to be included. This means that decision makers do not act as individuals, and processes are often incremental in order to create decision support. In this sense decisions cannot also not be compared with value judgement of independent experts, which is for example the case during grant submissions in art (Hekkert & van Wieringen, 1998). Decision criteria are developed during the whole tender process (Kreiner, 2006) and are part of the sensemaking process (Weick, 1995). Additionally, designs also include less tangible aspects such as aesthetics, integration in the surroundings, style, and future value. "Issues of design quality judgement are often tested in a field of tension stretching from techniques and functions within the proposal to aesthetic experiences" (Kazemian & Rönn, 2009). This means that these aspects are subjective by nature and cannot be solely based on an independent assessment system.

3.2.3 Second generation of decision theories

The second generation of behavioural decision theory, also called naturalistic research and theory, offers alternatives to the gambling principle by describing more realistically how decisions are actually made (Beach & Connolly, 2005). These theories have, until now, mainly focused on professional decision makers and the cognitive processes they engage in while making decisions in pre-choice processes (Zsambok, 1997). Intuition, affect, mood and emotions have become more prominent than in previous generations of decision theory. To a large degree second generation research and theory is also based on the extensive work of Herbert Simon. Simon found that decision making in organisations is strongly influenced by the structure and norms of the organisation, and that decision makers do not use the full array of options that an outsider might consider available (Simon, 1997). While studies based on first generation theories were often done in laboratory settings with inexperienced decision makers about non-contextualised situations, research in the second generation tried to simulate context-related factors such as time pressure, incomplete or unreliable information, and ill-defined goals (Hutton & Klein, 1999) and explore organisational processes that are far from rational (Lipshitz, et al., 2006). Second generation theories build on the experimental work of the previous generations, but also use observational and interview methods to gain insights. At present several pieces and fragmental theories are beginning to coalesce into more general theories (Beach & Connolly, 2005).

Theories of the second generation consider recognition theories, narrative theories, incremental theories and moral/ethical theories (Beach & Connolly, 2005). Recognition theory is based on the concept that knowledge about previous situations guide behaviour in new situations, for example in the case of fire fighters deciding about strategies to put out the fire. Scenario theory, story theory, and argument theory are the three most noteworthy narrative theories. In these theories mental causal models and knowledge about the past and the situation are used to understand the past or predict the future. Decisions are derived from these models. Examples are the stories similar to a jury verdict or military decision. Incremental theories have been developed in the light of policy making. The analogy of 'tree-felling' and 'hedge-clipping' could explain the difference between incremental and regular decision strategies (Simon, 1997). Contrary to a tree, in a hedge the elements are interdependent, changes can be made incrementally, corrections are possible and various approaches in clipping are possible. Therefore incremental decision making is very flexible and fluid. The moral and ethical theories deal with the fact that behaviour is very strongly influenced by an individual's beliefs about what is moral and ethical and therefore proscribed or prescribed. The moral/ethical theories deal with a different kind of utility that cannot be so easily combined and measured on a scale as assumed in prescriptive theory. Beach and Connolly (2005) propose the 'image theory' as a promising second generation theory. Image theory describes decision making as a social act in which the decision maker takes into account the opinions and preferences of other people (Beach, 1990). This means that organisations do not make decisions themselves but decisions are made by individual members of an organisation by the use of three images: a set of values and beliefs, specific goals, and operational plans.

Rosen, Salas, Lyons and Fiore (2008) view naturalistic decision making (NDM) as the third of three general paradigms in decision making, next to the formal-empiricist approach and the rationalist approach of the first two generations. NDM is concerned with the way people use their experiences to make decisions in field settings, which emphasises the level of expertise and the context of the decision maker and the actual process of decision making (Zsambok, 1997). NDM rejects the notion of decision making as choosing among alternative courses of action, and hypothesises sequential option generation and evaluation based on pattern matching, situation awareness and story construction (Lipshitz, et al., 2006). The context studied in NDM typically has ill-structured problems, dynamics, time pressure, multiple players, complex tasks, and other issues dealt with in practice. By analysis of descriptive models of how expert decision makers function, NDM research develops realistic actions and strategies for improving decision making (Rosen, et al., 2008).

The best known theory in NDM is that of recognition-primed decision of Klein and colleagues (1993, 1998). This theory has four main components: recognition of the situation, understanding of the situation, serial evaluation of the potential of various sets of actions for solving the problem, and mental simulation of the possible results of using an action in order to evaluate its potential. In research with fire-fighters, military commanders, police officers and design engineers Klein (1998) found that the more experience a decision maker has in the specific area, the greater the role of recognition. The benefit of this theory is that it is very easy to understand because one can easily recognise the situations that are described. The shortcomings of this theory seem to be that it never progressed beyond the level of general description, and that a link with psychological theories on memories and full recognition processes is not taken into account (Beach & Connolly, 2005; Betsch, 2005). There is still debate about the degree to which NDM truly represents a paradigm shift (Gore, et al., 2006). Because of the scattered character of the work in this field and the lack of an organisational framework, this kind of research seems less systematic than traditional research. Balogun, Pye and Hodgkinson (2008) call for more research in NDM that focuses on making sense of deciding, and Beach & Connolly (2005) plead for a better theoretical embedding in existing cognitive, social and organisational theory. Mosier and Frasier (2009) raise interest in the role of affect in NDM.

Recently several researchers realised that even when rationality is important in decision making, feeling, affect and emotion cannot be ignored (Plous, 1993). In previous research emotions and intuition were usually seen as distraction or irrationality, which is not the case anymore (Robbins & Judge, 2008). Current research on emotions show that mood, regret, disappointment, attachment, overconfidence and risk perception influence decision making in several ways (Beach & Connolly, 2005). The concept of intuitive judgement is traditionally associated with the heuristics and biases proposed by Kahneman, Tversky and others (see for example Kahneman, et al., 1982) but has never been developed into a field of research. Advances in cognitive neuroscience and managerial and organisational cognition have rejuvenated the attraction of the construct of intuition (Hodgkinson & Starbuck, 2008a). Dual process theories (Chaiken & Trope, 1999; Dane & Pratt, 2007; Kahneman, 2003; Sinclair & Ashkanasy, 2005) assume that two modes of processing - automatic unconscious processing and conscious analytical processing - are necessary for many tasks. Elsbach and Kramer (2003) found evidence of the dual categorization processes among experts judging the creative potential of others. During this process 'catchers' used cues to match each 'pitcher' with one of seven well-known prototypes of screenwriters. The authors also found that catchers are influenced in their judgements by cognitive and affective cues in relation to collaborative potential.

In his book 'Blink', Gladwell (2005) provides a convincing story of how people unconsciously make the right judgements and decisions based on the work of several scholars such as Gigerenzer et al. (1999), Damasio (1994), Dijksterhuis et al. (2006) and Klein (1998). Hogarth and Schoemaker (2005) conclude in a review of Gladwell's work that he only focuses on situations in which judgemental performance is high and that suggestions for education and control are absent – similar critiques have been voiced about the findings on heuristics and biases. Therefore they call for more research in this area on specific conditions. Gigerenzer's work (2007) about gut feelings pursues the concept of Gladwell in explaining the power of intuition while applying an adaptive toolbox with search, stop and decision rules of thumb for decision making. Sadler-Smith and Sparrow (2008) argue, however, that heuristics clearly differ from intuition because they are, in contrast to intuition, neither spontaneous, nor domain specific and are based on the application of predetermined rules. It can be concluded that theory building around intuition and emotions in natural context seems to be at the beginning of an interesting era.

Selection processes of architects show great similarities to the situations described in the second generation of decision theories. In architectural tradition experts evaluate the proposals that have been submitted. Based on the few publications that describe the jury processes (Kazemian & Rönn, 2009; Kreiner, 2006; Spreiregen, 2008), it can be concluded that pattern recognition takes place. In competitions judgement tasks are complex, moral, ethical and aesthetic by nature and only limited information is available. Next to that there is time pressure and social pressure. Therefore decision makers need to be aware of their situation and the possible consequences of their actions. Experience and teamwork enhance these kinds of skills that can be understood in the tradition of sensemaking. Architectural design is a professional skill based on education and experience gained through practice. The research on designers and comparable professionals has shown that experienced practitioners interpret and manage complex and demanding situations faster and more accurately by using tacit memory schemes (Mieg, 2006). Their domain relevant experience enables them to make intuitive decisions based on their tacit knowledge and unconscious memory systems. Members of the same profession share this code, and will accept peer review from within their discipline.

The selection of an architect should be seen in the context of a client organisation that aims at realizing ambitions in the field of architecture. A politician can be held accountable for a decision and needs to take national and international policies, stakeholder interests, and laws into account. This responsibility is often accompanied by mixed feelings of anxiety, enthusiasm and fear. According to Kreiner (2006) the process of architect selection is actually a process of sensemaking rather than managerial decision making. Information processing and screening appears more important than the actual choice, which confirms that image theory and the concept of sensemaking could be applied in this context. In a traditional design competition jury panels judge the quality of the designer anonymously by the quality of the design. The designs could act in that situation as boundary objects: artefact that serve as an intermediary in communication between two or more persons or groups who are collaborating in work (Boland & Collopy, 2004). A tender aims at the selection of a service provider in design, not about the purchase of a design product. For client decision makers it is very hard to make sense out of the multifaceted message that an architect sends (Jones & Livne-Tarandach, 2008). They have to make a distinction between the competences of the architect, the capability of company and the quality of the design proposal. In current practice jury panels often consist of politicians and other stakeholder representatives. Although politicians are experienced decision makers, tender decisions about architectural quality require domain specific skills in the area of the built environment. Not much research has been conducted in the second generation theories that take into account stakeholder participation and the role of external consultants within the limitations of the law.

3.3 Rationality versus intuition

One of the main issues underlying the differences between the decision theories is the level of rationality of decision makers. This section explores the concepts of intuition in relation to rationality in decision making and the factors that could influence the use of rationality.

3.3.1 Towards a definition of intuition

According to Simon (1997, p. 84) "rationality is concerned with the selection of preferred behaviour alternatives in terms of some system of values whereby the consequences of behaviour can be evaluated". This means a decision can be objectively, subjectively, consciously, deliberately, organisationally, or personally rational. Etzioni (1988, p. 136) defines rationality as "the concept of a man who acts wisely, and who chooses efficiently the means that advance his or her goals. It entails deliberations; it is not automatic or non-conscious and can vary by degree. It is based on openness to evidence (an empirical orientation) and on sound reasoning (logic)". According to Etzioni (1988) all decisions can be seen as more or less rational. But because of thoughtless rational rules of thumb that are deeply infused with values and other social factors and because decision makers are unable to judge the logical-empirical merit of the rules, decisions are rarely highly rational. In the debate about the degree of rationality of decision makers, March (1997) concludes that rational theories commonly assume that every decision maker knows all alternatives for action, is able to do a probability estimation of all consequences of every alternative action, has a consistent preference ordering for alternative preference courses of action, and uses decision rules that can select a single action to take.

In the context of organisations Dean and Sharfman (1993) argue that a rational action is feasible if decision makers are in agreement about goals and cause and effect relationships and if they are aware of the environmental and other constraints. This means that in other situations, which is often the case in everyday life, decision makers cannot rely exclusively upon rational methods. Most of these findings are based on laboratory or simulation studies. Understanding conscious choice requires the knowledge of the perspective of the actor at the time of choice: an action that appears irrational after the fact might have appeared perfectly rational when the actor chose it (Hodgkinson & Starbuck, 2008a). In real life people appear to actively try to influence events in order to make their choice the right choice (Balogun, et al., 2008). Making reasons explicit could lead to inferior decisions and less satisfaction about the decision (Wilson & Schooler, 1991). Affect, mood and emotion are other reasons that people could behave different from existing theories and models (Beach & Connolly, 2005).

Dijksterhuis (2007) distinguishes three kinds of decision processes: fast decision making without thinking, unconscious decision making, and conscious decision making. Rational analysis and intuitive judgement seem to be complementary components of effective decision making, which make it possible for managers to apply a range of management skills whenever they become appropriate (Sadler-Smith & Sparrow, 2008; Simon, 1987). Dual-process theories (Chaiken & Trope,

1999; Epstein, Lipson, Holstein, & Huh, 1992; Epstein, Pacini, Denes-Raj, & Heir, 1996; Kahneman, 2003) attach a great significance to affect and share the common view that two separate processes of rational analysis and experiential intuition are involved in reasoning. The intuitive system is often associated with unconscious processing of tendencies and preferences while the rational system enables people to assess information deliberately, to develop ideas, and to engage in analyses. The perspective of Dane & Pratt (2007, p. 36) leads to a definition of intuition as a "non-conscious process involving holistic associations that are produced rapidly which results in affectively charged judgements". This would mean that affect and emotion are an integral component of intuitive judgements.

In their attempt to define intuition Sadler-Smith and Sparrow (2008) identified six perspectives: 1) heuristics & biases (e.g. Gigerenzer, et al., 1999; Kahneman, et al., 1982); 2) intuition-as-ability (e.g.Klein, 2004; Simon, 1997); 3) information processing (e.g. Bowers, Regehr, Balthazard, & Parker, 1990; Mintzberg, Ahlstrand, & Lampel, 1998); 4) dual processing (e.g. Epstein, 1994; Stanovich & West, 2000); 5) cognitive-affective/neuro-scientific (e.g. Damasio, 1994; Dane & Pratt, 2007; Sinclair & Ashkanasy, 2005), and 6) intuitionist & alternative epistemology (e.g. Amabile, 1996; Moore, 1903). According to Sadler-Smith and Sparrow (2008) there seems to be a consensus among these perspectives that intuition is an experiential phenomenon based upon implicitly stored constellation of knowledge, skills, perceptions and emotions. Intuition and intuitive processes draw upon a complex interplay of cognitive and affective processes operating below the level of conscious awareness and complex domain-specific schemas of which one is unaware (Dane & Pratt, 2007; Klein, 1998, 2004). Most researchers view intuition as quite fast with limited effort (Dane & Pratt, 2007; Simon, 1947/1997) consisting of involuntary or automatic processes (Sadler-Smith & Sparrow, 2008). In conscious awareness affect ('hunch' or 'gut feeling') and a degree of certitude are the manifestations of an intuition (Sadler-Smith & Sparrow, 2008).

Research on experienced chess players shows that expert's intuition is based on experience, which allows people to recognise the pattern in a situation and draw on previously learned information associated with that pattern to arrive at a decision choice quickly (Chase & Simon, 1973; de Groot, 1946). An expert can therefore decide rapidly based on what appears to be very limited information. These findings inspired Klein (1993) to develop the recognition primed decision theory and made Kahneman et al. (1982) explore the use of heuristics during decision making. Gigerenzer et al. (1999) elaborate on the insights about heuristics and bounded rationality for the development of their toolbox of fast (computationally simple) and frugal (sparing information requirements) rules for decision making. According to Epstein et al. (1996) intuition is the outcome of pre-conscious interpretations of previously experienced events which could determine for a great part the use of recognition heuristics. Gigerenzer and Selten (2001) suggest that strategies to decide on which heuristic to use (meta-heuristics) are guided by unconscious, experience based intuition. However, according to Sadler-Smith and Sparrow (2008), (meta) heuristics clearly differ from intuition because intuition is not based on predetermined rules to be deliberately applied in response to a specific scenario but includes an involuntary response to perceived stimuli. The likelihood of this response might be influenced by for example the acquisition of domain specific knowledge but cannot be forced. The context-dependent use of intuition is one means of distinguishing between expert and novice performance (Sadler-Smith & Sparrow, 2008).

3.3.2 Factors of influence on the use of intuition

Context and problem structure play a crucial role in determining the appropriateness and efficacy of intuitive judgements (Burke & Miller, 1999; Klein, 1998, 2004). According to Dane & Pratt (2007) there are two sets of factors that influence intuition effectiveness: domain knowledge factors (development of schemas by explicit and implicit learning) and task characteristics (intellective versus judgemental tasks, environmental uncertainty). Well structured problems might be compared to tasks with objective criteria for success within a particular conceptual system, while ill-structured problems seem similar to judgemental tasks for which there is no objective criterion or demonstrable solution (Dane & Pratt, 2007). In tightly structured, intellective tasks in data rich, objectively quantifiable, and computationally complex domains, statistical models perform better than human judges (Sadler-Smith & Sparrow, 2008). However, based on several studies it can be concluded that intuition is favoured over analytical approaches in loose decision structures with moral, political, ethical, aesthetic or behavioural judgemental tasks, ill-structured strategic problems with little precedent and information to draw on, and in situations with time pressure, dynamic conditions and experienced participants (Dijksterhuis, et al., 2006; Hogarth, 2002; Kahtri & Ng, 2000; Klein, 2004; Robbins & Judge, 2008; Shapiro & Spence, 1997). Possible reasons for this phenomenon include the ability of intuition to sense changes, to detect failures, to make sense, and to include the company's culture and values that are difficult to describe. These individual, organisational, and cultural factors that seem to influence the use of intuition need additional research before these insights can actually be applied (Hammond, 1987; Hogarth, 2002; Sinclair & Ashkanasy, 2005). Researchers found that managers of small and medium sized companies prefer qualitative data-gathering methods that require judgemental tasks, such as decision trees, force field diagrams and portfolio analysis, over the use of benchmarks and other quantitative forecasting technique in strategic planning (Wright & Geroy, 1991). Still rational theories are commonly accepted and used in economics to predict individual and firm behaviour in the long run. But as Hogarth (2005, p. 68) states "whether tacit or deliberate processes are more valid than the other is not the critical issue. Rather, this is to make valid responses in which both systems are implicated".

More research is needed to explore the differences in task conditions and contexts. Because of the significant developments in cognition, neurology and neuroscience, a fruitful dialogue between intuition researchers from several fields is to be expected (Sadler-Smith & Sparrow, 2008). Those studies should not only be based on self-report and experimental research, but should include new kinds of methods to measure decision performance. Promising areas of interest for empirical research proposed by Dane & Pratt (2007) are the practical implications of the interplay between intuition and analysis, the link between creativity and intuition, ethical decision making in organisations, and the transfer of domain related knowledge schemes in order to keep up with the dynamic character of individuals in organisations. The selection of an architect could be considered as a situation in which interplay consists between intuition and analysis, and which could be studied in real life.

3.4 Individual decision making in organisations

A lot of theories in decision making focus on processes that occur on the level of the individual. These processes are cognitive by nature and can be influenced by the structure of the information that is used, the use of heuristics, expertise or other personal characteristics. Decision makers can however also be influenced by other, non-cognitive factors such as affect, mood, emotions or the organisational context. This section addresses the individual decision processes and the factors that influence these processes, such as sensemaking.

3.4.1 Cognitive processes

The three generations of decision theories are all based on psychological insights about cognitive and social processes of the human mind. A judgement takes place within a system composed of the person, the task environment and the actions that result from the judgement and can subsequently affect both the person and the task environment (Hogarth, 1988). Perception is important in decision making because people's behaviour is based on their perceptions of reality, not on reality itself (Robbins & Judge, 2008). Decision making occurs as a reaction to a problem of which people have to be aware and which they perceive as needing a solution. People often gather data from multiple sources and then screen, process, and interpret the information in order to make a decision. Perception is influenced by factors in the perceiver (such as attitudes, experience and expectations), factors in the target (such as novelty, background and similarity), and factors in the situation (time, work and social setting) (Robbins & Judge, 2008). Not everything can be perceived explicitly. Creativity allows people to indentify options that are viable but also the alternatives that are not visible in the first place or by others. According to Amabile (1997) creativity is composed of intrinsic task motivation (enhanced by a stimulating culture, sufficient resources, freedom, good supervision and supporting work group members), expertise (enhanced by abilities, knowledge, proficiencies and similar expertise), and creativity thinking skills (enhanced by personality characteristics, ability to use analogies and talent to see things differently). Robbins & Judge (2008) distinguish similar factors of influence, namely ability (capacity to perform the various tasks in a job), attitudes (evaluative statements concerning objects, people or events), and learning (any relatively permanent change in behaviour that occurs as a result of experience) as the foundation of individual behaviour in organisations.

Framing is an important process in decision making because it concerns the process of embedding perceived events in a context to provide for meaning (Kahneman & Tversky, 1984). A frame is a mental construct or scheme consisting of elements and the relationships among them that are associated with a situation that is of interest to a decision maker (Beach & Connolly, 2005; Hogarth, 1988). The frame of a situation is called an 'image', which implies both visual and narrative representation. A frame "guides the decision maker's interpretation of what is going on. It derives from the decision maker's knowledge about events that led up to the situation in question and his or her private theories about how people behave and what makes things happen. It therefore tells the decision maker what to expect." (Beach & Connolly, 2005, p. 23). The wording of the situation and the characteristics of the judgemental task influences the framing, and therefore also the process and outcome of the decision (Plous, 1993).

Etzioni (1988) distinguishes three sources of influence on decision making that all must be taken into account to understand human decision making: utilitarian (utility as studied by economics and normative theory), social (social influences as studied by anthropologists and sociologists), and deontological (as studied by ethics). Table 3.1 shows an overview of the decision strategies based on the different perspectives which align with the theories from the different generations. The table exposes the field of tension that decision makers experience during decision making.

When a situation is framed, a person can decide about it by recognition, inference or choice. Recognition can be applied when the situation is so similar to one encountered before that behaviour can be duplicated by the use of 'policies', 'habits' or 'scripts' based on prior experiences. Inference can be applied when an educated guess can be made about the right decision, such as proposed by Brunswik (1947) in his 'lens model' with cues. His lens model uses psychological insights to argue that people experience objects and events by constructing mental models based on cues and making inferences. These cues help people to identity targets and to make comparisons among targets. Beach and Connolly (2005) use the example of the selection of a good salesperson to explain the lens model: first one has to find out what makes a good salesperson (which cues), then one has to use this knowledge to develop a policy about potential job applicants (policy of how to handle the cues), and finally one has to apply the policy to select a new salesperson (assess the cues). However, in most situations not all information is

Perspective	What is the decision maker trying to do?	How do decision makers choose the means to advance the goals?	Who are the key actors?
Utilitarian	Maximise pleasure or self interest (utility).	Selection of action with greatest net utility by weight of costs and benefits.	Free standing individuals make decisions on his or her own.
Social	Conformation to social norms and cultural demands in order to avoid punishment.	Selection of course of action that conform to the expectations of reference group or community.	Decision maker conforms to rules of group or community.
Deontological	Evaluation of moral and ethical considerations (pleasure and morality) in light of utilitarian and social considerations.	Use of emotion and value judgements to select or reject courses of action that are compatible or violate with or prescribed by moral or ethical codes.	Decision maker is guided by own moral and ethical principles derived from groups and communities.

Table 3.1 Overview of possible decision strategies from different perspectives (based on Etzioni, 1988)

available to assess the cues. Next to that the lens model assumes that cues are additive, independent and linear (Beach & Connolly, 2005).

If neither recognition nor inference provides adequate guidance, a decision maker must choose the most promising option after exploration of the alternatives. Researchers have approached choice in two ways: 1) assume that an underlying orderliness exists in the confusion and complexity and trade realism for simplicity (prescriptive decision theory), and 2) assume that confusion and complexity are integral part of the process and trade simplicity for realism (first generation decision theories) (Beach & Connolly, 2005). Both these approaches include gambling principles which means that the probability that a certain situation or option will be preferred has to be estimated. According to Lipshitz, Klein, Orasanu and Salas (2001, p. 334-335) "matching (recognition) differs from concurrent choice in three respects: 1) options are evaluated sequentially at a time - even when presented with several options, decision makers quickly screen most of them by comparing them against a standard, rather than with one another, and then focus on one, or at most two options which are compared; 2) options are selected or rejected based on their compatibility with the situations or the decisions values; 3) the process of matching may be analytic but more often it relies on pattern matching and informal reasoning".

Harrison (1999) provides an overview of the steps of the rational decision-making model for making choices: problem definition, identification of the decision criteria, allocation of the weights to the criteria, development of alternatives, and evaluation of the alternatives. There are strategies for choice that confront conflicts inherent in the choice situation and strategies that avoid these conflicts (Hogarth, 1988; Mieg, 2001). Conflict-confronting strategies are compensatory – they allow people to trade off a low value on one dimension against a high value on another. Conflict-avoiding strategies are non-compensatory which means that they do not allow trade-offs and require a certain minimum or maximum on every dimension.

Because of their bounded rationality people create simplified models that extract the essential features from problems which does not mean that they either act irrationally or optimise (Gigerenzer & Selten, 2001). This offers an alternative to current norms of optimization, probabilities and utilities which takes into account the cognitive limitations of people and the structure of the environment. Once a problem is identified, people search for criteria and alternatives in order to evaluate the alternatives. The search stops when an alternative is found that meets an acceptable level of performance. During the search for matching alternatives to the criteria, people tend to use heuristics. Heuristics are 'ecologically rational' because they are adapted to particular environments. This means that they tend to be domain, culture and time specific in order to save time, but they can be as accurate as complex statistical models while demanding less information and computational power (Gigerenzer & Selten, 2001; Gigerenzer, et al., 1999). The mechanism of how people decide about these rules is not yet well understood.

According to Beach and Connolly (2005, p. 23) "decision making is essentially social behaviour, even when there is nobody else present, because one anticipates how others will react and factors this into the decision. [...] Organisations per se do not make decisions, but individuals in organisations do. And when they do they must take others into account." This context influences the process and outcome of the decision (Balogun et al., 2008). When people make decisions, they suppose that this decision is turned into action (Brunsson, 1989), but there are numerous connections between the process, decision, and action that should act in isolation, sequentially and in coherence to directly link the intention to the outcome. Therefore decision outcomes should rather be seen as the consequences of actions and interaction of multiple issues than of specific identifiable decisions (Vidaillet, 2008). When decisions and decision makers are identified, it is possible to assign responsibility for a course of action to decision makers (Brunsson, 2007). Because decision makers are aware of this, decision makers appear to gather information that they might not necessarily use to show that they are good decision makers or to justify their decision in an unexpected event (Feldman & March, 1981) or they might choose the option that will be easiest to justify (Tetlock, 1992). According to Tetlock (1983) a decision maker can be considered as a politician who is accountable to their 'constituents' and who is constantly concerned with questions regarding the justification of the decision and reaction of others. Etzioni (1988, p. 4) presents the 'I & We paradigm', which "highlights the assumption that individuals act within a social context, that this context is not reducible to individual acts, and that the social context is not necessarily or wholly composed. Instead, the social context is to a significant extent perceived as a legitimate and integral part of one's existence, a We, as whole which the individuals are constituent elements."

3.4.2 Sensemaking

Sensemaking is the process of making something sensible and involves the ongoing retrospective development of plausible images that rationalise what people are doing (Weick, 1995; Weick, Sutcliffe, & Obstfeld, 2005). It has its genesis in disruptive ambiguity and its mixture of retrospect and prospect, is embedded in interdependence and based on a dialogue among people who act on behalf of larger social units. "Sensemaking pays attention to how people 'deal with' (whether unconsciously or otherwise) constraints imposed by the information processing limitations and their organisational context, delving into the socio-political nature of organisations to show that the answer to better decision making does not necessarily lie with the provision of greater quantities of 'more accurate', 'objective' and timely data, but rather requires an understanding of the social processes of negotiation involved in decision making" (Balogun, et al., 2008, p. 235). Central questions are how an event comes to happen and what does an event mean. Different interpretations of decisions could cause ambiguity and confusion about the actions required to implement a decisions.

According to Weick (1995) and Vidaillet (2008), building stories with arguments among decision makers, opportunistic interpretation, spending time on actions, and linking issues all increase sensemaking about decisions and the attribution of outcomes to specific individuals and actions. Christianson, Farkas, Sutcliffe and Weick (2009) showed for instance that organisations can learn through rare events that occur outside the everyday experience of an organisation because rare events trigger leaning by auditing the existing response repertoires, by disrupting or strengthening organizing routines, and by redirecting organisational identity. Balogun, Pye and Hodgkinson (2008) include a sociologic lens on decision making and define sensemaking as a social process of construction and reconstruction of meaning that enables individuals through interacting with others to collectively create, maintain and interpret the world. The intertwined concepts of 'framing' (shaping the meaning of a subject and sharing it with others), 'sensegiving' (attempts to influence sensemaking and construction of meaning toward a preferred redefinition of social reality), 'sensereading' (perception of circumstances and aligning of interpretations), and 'sensewrighting' (inheriting, shaping and reflecting the understanding of the world) are all related to the resource, process and meaning of power effects in organisational decision making (Balogun, et al., 2008). Weick, Sutcliffe and Obstfeld (2005, p. 409) emphasise that sensemaking is about the interplay of action and interpretation rather than the influence of evaluation on choice – "it is a process that is ongoing, instrumental, subtle, swift, social and easily taken for granted".

In the tradition of sensemaking, image theory suggests that decision makers use their images (a individual store of knowledge) to set standards that guide decisions about goals and plans (Beach & Connolly, 2005). In image theory a frame consists of the principles, goals and plans (the constituents) of the images and is used to set the standards that influence the decision (for example by recognition). Potential goals and plans that are incompatible with the standards are quickly screened out and the best of the survivors is then chosen. In the organisational version of image theory, the constituents of the images are called organisational culture (shared values and beliefs), organisational vision (goals agenda and time line), and organisational strategy (blueprint for goal achievement). The framing assures that when people make decisions for and about organisations relevant images are taken into account. Research on the individual version of image theory shows that screening of suitable options seems more important than the actual choice, that compatibility of images is linked to the number of violations to the images (a violation threshold) and that the more members of an organisation agree about the principles, the more they agree about the appropriateness of a plan (Beach & Connolly, 2005). In organisations, compatibility with the company's culture influences the support for the decision while discrepancy between ideal image and actual image could influence job satisfaction. Results of application of the image theory seem promising for issues such as job turnover, organisational justice and clinical treatment selection, issues that show a lot of resemblance with the selection of an architect. According to Balogun et al. (2008) little research has been done that integrates the managerial and organisational cognition. This kind of research probably requires an ethnographic method but it is essential that "the study of deciding not just considers the information processing focus on personal preferences, biases and heuristics, but also on decision maker's identities and their social skills and capabilities" (Balogun, et al., 2008, p. 243).

3.4.3 Factors of influence on cognitive processes

Heuristics may be useful but they can also lead to severe and systematic errors (Tversky & Kahneman, 1974). The concept of heuristics inspired a lot of scholars to do research about the most common systematic biases and errors in applying heuristics in decision making: the overconfidence bias, anchoring bias, confirmation basis, self-fulfilling prophecies, availability bias, representative bias, escalation of commitment, randomness error, and hindsight bias (Hogarth, 1988; Plous, 1993; Robbins & Judge, 2008). Hogarth (1988) identified inconsistency, learning, memory, computational capacity, failure to appreciate randomness and several other cognitive aspects as origins of judgemental biases related to the acquisition, processing, output and feedback of information. Other factors of influence of deviations in decision making by heuristics are personality, gender and cultural differences (Robbins & Judge, 2008). According to Hutton & Klein (1999) more recent research shows that the shortcomings of decision makers have been greatly exaggerated, which means that there is in principle nothing wrong with using heuristics. Books on effective decision making usually explain the biases and propose tools and methods to prevent them (e.g. Russo & Schoemaker, 2002). It is not yet clear whether research on biases and heuristics can be used to guide designers of decision aids in the reduction of human error. Wright and Goodwin (2008) provide an overview of current insight about the effectiveness of methods based on the use of heuristics (such as SMART, Delphi, decision analysis and scenario planning) that are designed to provide structure and support to individual and group decision making.

March & Simon (1958) argue that the basic function of a decision is to solve a problem and thereby reduce uncertainty, but later insights showed that every decision may produce new uncertainties. Uncertainty is intimately linked with error: the greater the uncertainty, the greater the probability of making an error. Lipshitz et al. (2001, p. 337) propose to define uncertainty as "a sense of doubt that blocks or delays actions". They identified three principle forms of uncertainty: inadequate understanding (a sense of having insufficient situation awareness), lack of information (a sense of having incomplete, ambiguous or unreliable information), and conflicting alternatives (a sense that available alternatives are insufficiently differentiated). The RAWFS heuristic combines five principle strategies of coping with uncertainty: Reducing uncertainty, e.g. by collecting additional information; Assumption based reasoning by making assumptions that go beyond directly available data to fill gaps; Weighing pros and cons of at least two alternatives; Forestalling by developing an appropriate response or response capabilities to anticipate undesirable contingencies; and Suppressing uncertainty, e.g. by ignoring it or relying on unwarranted rationalization (Lipshitz, et al., 2001, p. 338). Applying these strategies in a chronological order could be interpreted as a formula for successfully coping with uncertainty.

Experts seem better able to see the significance of information and pay attention to important cues that tell them what to do in critical situations (Brenner, Tanner, & Chesla, 1996). Daake, Dawley and Anthony (2004) found that highly educated professionals making strategic decisions in a hospital relied more on informal data and tacit, experience-based information than on facts and formal data. Hogarth (2002) summarises the most important findings on expertise in the use of decision strategies: 1) expertise is limited to specific domains and acquired through exposure to and activity within them, 2) outstanding performance in any domain takes years of dedication and feedback from teachers, 3) there is limited relation between expertise and predictive ability, 4) experts and novices process information differently by the use of patterns and tacit knowledge. Experts see and represent a problem at a deeper level, have a better ability to see typicality, antecedents and consequences, and spend more time trying to understand the problem (Hutton & Klein, 1999). Dreyfus and Dreyfus (1986) propose that skilled performance passes through five levels of proficiency in decision making: novice, advanced beginner, competent, proficient and expert. Brenner et al. (1996) found that when a novice moves to becoming an expert (s)he first shifts from relying on abstract principles to using concrete experience, (s)he then starts to perceive the situation more as a complete whole with relevant parts instead of a compilation and then, instead of being an observer (s)he can start acting as an participant in the situation. Experts just do the right thing to achieve a goal because they have a larger 'response repertoire' and therefore experience fewer occasions of overload. Parsons (1987) suggests a model that includes several stages of expertise for experiencing art. In the first stage the experience is intuitively pleasurable. In the next stage the experience will also be interesting because of the content. In the third stage one also feels that the experience is very interesting and in the next stage the experience is put into a historical context. Jurors in the fifth and final stage are able to judge autonomously and feel the need to discuss and harmonise with the professional field. According to Mieg (2001) it is not only the psychological skills that make people an expert. From a more sociological perspective someone is considered an expert because someone else attributes expertise to a person. Expertise is therefore always relative and their performance should be carefully checked.

Soane and Nicholson (2008) distinguish several levels of personal differences that could influence the goals that direct decision making, the content of decision making, and the direction of attention to available information. Personal differences occur on the level of biographical variables (age, gender), internal processes (affect, emotion, attribution, motivation, personality, risk orientation, self-monitoring, self regulation, automaticity, self-efficacy), attributes (attitudes and values), and abilities (competences, uncertainty, interactionism, person-job fit). However, the influence of these individual differences is not random because people have systematic and consistent preferences and traits. In general people choose to be in occupations and situations that suit their tendencies. Therefore an overall homeostatic process of self-identity regulates the interaction between individual choices, goals and situational forces (Soane & Nicholson, 2008).

In line with the findings on image theory about screening (Beach & Connolly, 2005), Sutcliff and Weick (2008) suggest that interpretation and sensemaking are more important in decision action than choice. Karmanov (2009) found that aesthetic experience of landscape design is enhanced by knowledge such as the recognition of plants, the perceived meanings and narratives through which the physical properties of the environment are interpreted and evaluated, evoked associations and memories and emotional reactions. Le Dantec and Y-Luen Do (2009) also found that narratives and expressions of values helped to convey how requirements and values in design are situated together. Information overload occurs when a person receives too much information to process, compute, and interpret it and the balance between demand and capacity is disturbed. Symptoms of information overload are for example general lack of perspective, an inability to select out relevant information and increasing distraction by irrelevant and interfering cues which lead to stress and a feeling of loss of control (Sutcliffe & Weick, 2008). If information is of low quality, low value, highly ambiguous or has a short period of relevance, too much effort and time would be needed to make sense of the information. Overload can occur when input increases, but also when capacity is limited. Capacity can be limited by increasing complexity and task demands but also when time is limited or pressure is high. Distractions and interruptions indirectly also increase the chance of overload. Research shows that under certain conditions and up to a certain point, time pressure could also improve decision time and decision accuracy because it helps decision makers to narrow down their attention and to focus on the most relevant information (Hahn, Lawson, & Lee, 1992; Speier, Valacich, & Vessey, 1999).

3.4.4 Affect, mood and emotion

Emotions can influence all aspects of decision making. Kahneman (2003) calls the recognition of the affect heuristic the most important development in the study of judgement heuristics in the past few decades, which reflects a change in the general climate of psychological opinion. According to Robbins & Judge (2008) affect is defined as a broad range of feelings that people experience in the form of emotion or moods. The construct of affect has two independent dimensions for positive and negative feelings experienced by the individual at the same time (Betsch, 2005). Emotions are brief, specific and intense feelings driven by someone's concerns that could direct certain behaviour (Frijda, 1986; Gigerenzer & Selten, 2001). Emotion researchers distinguish between task-related and anticipated integral affect (emotional responses elicited by the decision situation itself or its potential consequences) and background or incidental affect (unconnected emotions brought in by an individual to the decision situation) (Gigerenzer & Selten, 2001; Mosier & Fischer, 2009). The impact of incidental emotions seems to be more than previously thought, and concerns both unconscious as well as reactions that are difficult to put aside. Research has shown that incidental emotion impacts decision making more in ambiguous situations than in unambiguous situations (Lerner & Keltner, 2000). Mate choice is an important decision based on social processes affected by emotions that could also be explained from a cultural perspective (Gigerenzer & Selten, 2001).

Appraisal theory aims to explain the impact of emotion on decision making (Ellsworth & Scherer, 2003; Lazarus, 1990). It sees emotions as specific appraisal patterns that motivate people towards specific goals and behaviours. Fear could for example lead to uncertainty reduction, self-protection and preventing harm (Nabi, 2003) while anger could induce risk-taking behaviour and stimulate actions to change the situation (Lerner & Tiedens, 2006). The work on product experience (e.g. Desmet, 2002, see also Chapter 2.4) is based on appraisal theory and therefore concerns integral affect. Regret theory and disappointment theory are examples of risky choice and modification of utilities by emotions (Gigerenzer & Selten, 2001). Emotions are also important for learning processes – a reward inspires people to keep up their good work while a punishment causes painful feelings.

Findings by Damasio (1994, 1999) on brain damage indicate that emotions are critical to rational thinking because they provide important information about how people are aware and understand the world in advance of conscious awareness. Emotions could contribute to a reduction of the options people consider and therefore makes decisions more manageable. As a result of unconscious information processing, moods and emotions people may know more than they can express, and may be unable to verbalise how they arrived at a particular judgement (Bowers, et al., 1990). Because the unconscious system is associated with affect and a lack of awareness how to be controlled, the unconscious system likely to be more compelling and difficult to control (Epstein, 1994). Epstein therefore claims that all processes in the non-conscious system are emotionally driven. Langley et al. (1995) state that emotion, imagination and memories drive decision making by forming insights. Intuitive judgements may be triggered by emotions and affect (Weiss & Cropanzano, 1996). Emotional intelligence indicates one's ability to detect and manage emotional cues and information and is composed of the dimensions of self-awareness, self-management, self-motivation, empathy and social skills (e.g. Davies, Stankov, & Roberts, 1998).

Affect could influence information processing strategies or processing styles (Muramatsu & Hanoch, 2005; Peters, Vastfjall, Garling, & Slovic, 2006). Gigerenzer and Selten (2001) suggest that emotion and social norms can be an effective stopping rule as part of their adaptive toolkit. Research findings suggest that anger could for example lead to more stereotypic judgements, less attention to the quality of arguments and more attention to superficial cues, whereas fear or anxiety could lead to systematic and comprehensive information processes (Mosier & Fischer, 2009). Positive affect could foster quicker, more superficial and less effortful strategies using little information, but could also lead to more creative, open and inclusive thoughts (Sadler-Smith & Sparrow, 2008). These findings are in line with the results of Amabile (1996), which indicate a linear relationship between positive affect and creative thoughts, cognitive variation and creative events and therefore a possible increase of organisational productivity and job satisfaction. Creativity could also evoke short term emotional reactions such as joy.

Mosier & Fischer (2009) argue that affect could lead to a limited information search and a biased integration and interpretation of information and cues in organisational decision making also. In an operational context anger could for instance encourage a 'blame' culture in which decision makers focus on responsibility and retribution rather than problem solving, while fear or anxiety may elicit a great concern for self-protection and safety. Based on research results of other scholars, Mosier & Fischer (2009, p. 101) suggest that "professionals are sensitive to task-integral affect and incorporate it into their decision process". This implies that even experts sometimes select and interpret the information which makes them feel more comfortable as input for decision making. According to Weick et al. (2005, p. 418) "expectations hold people hostage to their relationships in the sense that each expectancy can be violated, and generates a discrepancy, an emotion, and a violated interpretation". More research in natural settings is needed to further explore the effects of both integral and incidental affect on the search and interpretation of information, sensemaking, and the quality of the decision (Amabile, 1996; Mosier & Fischer, 2009; Weick, et al., 2005).

3.5 Decision making in groups

Many decisions in organisations are not made by individuals but by groups or teams of people. This section provides a short overview of group decision making and some effects that could take place within a group. Because a jury panel or steering committee can be considered as a group of experts, the concept of expert teams is also addressed.

3.5.1 Cognitive and social processes

According to Foote, Matson, Weiss and Wenger (2002), the belief drawn from the Western legal system that two heads are better than one is the reason why in many organisational today decisions are made by groups, teams or committees. A tremendous amount of literature exists on teams. This section reflects only the basic insights. A group is defined as two or more individuals, interacting and interdependent, who have come together to achieve particular objectives (Robbins & Judge, 2008). A group can be either formal defined by the organisation's structure or informal. A command group is a formal group composed of individuals who directly report to a given manager. Task groups are formal groups of employees working together to perform a task. A team is a group which generates positive synergy through coordinated effort which results in a level of performance that is greater than the sum of individual inputs (Robbins & Judge, 2008). Groups have properties, such as roles, norms, status, group size and group cohesiveness, that shape the behaviours of group members and make it possible to predict a large part of individual behaviour in a group as well as the performance of the group as a whole.

People align their frames with others in two ways: by discussion and by a shared set of beliefs and values, called an organisational culture (Beach & Connolly, 2005). If frames are similar, people are more likely to understand other people's intentions and goals. This means that conflicts tend to be about the solution of the problem, not about the interpretation of the problem. Variation in cultures tends to be less within organisations than between countries. Etzioni (1998) argues that in many areas collective thinking and decision-making are more rational than that of individuals because of the explicit openness to evidence and reasoning in the process of group decision making. In organisations shared understanding, power struggle, ill-defined problems, and unclear decision options increase the complexity of decision making. There are various models of organisational decision making: the rational model (everybody strives to achieve the same goal), the information model (decisions are taken when a satisfying options are encountered), the structural model (differentiation of capacity of organisational units), the garbage can model (an organisation as an unpredictable mix of participants, solutions, problems and choice opportunities), and the participation model (involvement of members in group decision making contributes to better decisions, greater satisfaction and greater confidence in decisions) (Beach & Connolly, 2005).

Weick, Sutcliffe & Obstfeld (1999) found that the coherence of sensemaking and sharing of perceptions and expectations enable teams to act coherently during crises and hazardous situations. People then develop collective 'mindfulness', "the combination of ongoing scrutiny of existing expectations, continuous refinement and differentiation of expectations based on newer experiences, willingness and capability to invent new expectations that make sense of unprecedented events, a more nuanced appreciation of context and ways to deal with it, and identification of new dimensions of context that improve foresight and current functioning" (Weick & Sutcliffe, 2001, p. 42). Still a lot of research is needed drawing on anthropology, economics, management, psychology, and sociology to enrich our understanding of the dynamic interplay between micro-processes and practices of actors, and the sociological and macro-economic contexts of those actors and their practices (Hodgkinson & Wright, 2006).

3.5.2 Effects of group decision making

George & Chattopadhyay (2008) argue that groups are considered to be effective for decision making tasks because they can bring many pieces of information together, they can analyse information critically and they can generate commitment in their members. According to Robbins and Judge (2008) the weaknesses of groups could be the time-consuming character, possible conformity pressure, possible domination of one or a few members, and ambiguous responsibilities. One might assume that groups are more accurate, more creative with a higher degree of decision acceptance and work satisfaction, but less efficient than individuals. However, the findings from brainstorming research show that a group is seldom more effective and often less efficient than collecting ideas of independently working individuals (Paulus & Dzindolet, 1993; Stroebe & Diehl, 1994).

Robbins and Judge (2008) summarize the factors that are known to be of influence on team effectiveness in groups: 1) context (such as trust, feedback and leadership), 2) composition (such as diversity, member flexibility and abilities), 3) work design (such as task identity and task significance), and 4) group processes (for example specific goals, common purpose and conflict levels). The effects that Rosen et al. (2008) describe for enabling team decision making include all these factors: shared mental models of the task, team and environment; learning and adaptation; clear roles and responsibilities; a clear, valued and shared vision; a cycle of pre-brief, performance and debrief; strong team leaders; a strong sense of collective, trust, teamness (the feeling of being part of an entity) and confidence; and cooperation and coordination. In heterogeneous teams each member processes different aspects of the entire mental model of the task or system. By sharing their mental models team members are able to form complementary or congruent explanations of environmental cues and implicitly coordinate their responses (Orasanu & Strauch, 1994).

Teams can consist of employees of an organisation, but in architect selection external advisors are often involved as well. Mieg (2006) introduces two types of experts that are important in transdisciplinary projects: system experts and decision making expert. System experts are individuals with experience-based local knowledge of the system of which they are part of. Employees and politicians can be considered as system experts but architects can also acquire local knowledge through involvement in projects. Consultants are often hired as decision making experts to support the process of decision making. According to Jackson (1997) the most important reasons for the growing consultancy market is an increase of complexity and dynamics of the environment and opportunity to regain the perception of control. Managers hire consultants for a strategic reason (Kieser & Wellstein, 2008): they know that those who have a chance to define a problem can influence the search for solutions and the direction of ensuing decisions. But managers, consultants and management researchers are members of different social systems in which they pursue different goals, apply different criteria for success, and speak different languages (Kieser & Wellstein, 2008). These differences seem to hamper communication between them. "People from outside the organisation can only influence organisational decision makers: they cannot make organisational decisions in lieu of the managers" (Kieser & Wellstein, 2008, p. 497).

Consultants or experts should be selected carefully (Mieg, 2001). Through their activities, consultants tend to change the conditions under which managers make decisions and, in the absence of control on the reactions of managers, focus on the impression that one is better off engaging them (Clark, 1995). Because the consultant is supposed to deliver something the client does not (yet) know, there is an information asymmetry at the very heart of any consultant-client relationship (Ernst & Kieser, 2002). Typically the activities of a consultant involve developing a vision for change, screening the exiting situation, developing alternative solutions, evaluating and ranking the alternatives and setting up an action plan (Kubr, 2002). Expertise is needed for these activities, which is often built in projects in other organisations. Often consultants produce and exploit knowledge on methods. Consultants also offer the functions of legitimization, temporary management capacity, external review, communication and acceptance of decisions, weapons for politics and interpretation and reassurance of experiences (Kieser & Wellstein, 2008). However, somehow the experience that managers, users and consultants acquire has to be applied in order to make sustainable decisions in organisation. "It is in the transformation of information to knowledge and in the location of power with regard to ultimate decision making that real influence is found" (Cairns & Beech, 1999, p. 1).

Information processing in a group could be effected by demographic characteristics of group members, the cohesiveness of group members and the numerical strength and status of diverse subgroups within a team (George & Chattopadhyay, 2008). However, research showed that group discussions tend to focus on what is known by everybody and uniquely held information tends to be ignored (Sniezek, Paese, & Switzer III, 1990; Stasser & Titus, 1985). Robbins and Judge (2008) mention communication barriers of filtering, selective perception, information overload, emotions, language and communication apprehension in a group.

People may not always share information in groups because they sometimes feel it would threaten the cohesiveness and the balance in the group or their own status (George & Chattopadhyay, 2008). These arguments have to do with salient social categorization of the group members. Philips, Mannix, Neale and Gruenfeld (2004) found that groups that could be divided into two equal sized subgroups with common information within the subgroups but unique information across subgroups did share information. This effect would suggest that information sharing in groups would benefit from equal sized subgroups. Next to that, norms that promote critical thinking, members with a counterfactual mindset and framing the problem as having a definite solution seem to increase the sharing of information (George & Chattopadhyay, 2008).

The use of explicit agendas, rules for speaking, voting procedures, and criteria for arriving at decisions can influence the outcome of group decision making. Group interaction and group consensus may lead to overconfidence in group decisions. 'Groupthink' and 'groupshift' are two phenomena of group decision making that have the potential to affect a group's ability to appraise alternatives objectively and to arrive at quality decisions (Plous, 1993; Robbins & Judge, 2008). Groupthink describes deterioration in an individual's mental efficiency, reality testing, and moral judgement as a result of group pressure (Janis, 1982). It can occur if all members of a coherent group overlook alternative frames. Novel viewpoint and fresh ideas can also be ignored or misjudged by groups with shared frames. Groupthink seems to occur most when a clear group identity exists, members hold a positive image of their group, and the group perceives a collective threat to their image (Robbins & Judge, 2008). Groupshift relates to the effects of a discussion among group members towards a more extreme position towards risk. Groups tend to take more risky decisions than individuals, probably due to diffused responsibility, social value of risk taking positions or majority influences of dominant views (Beach & Connolly, 2005). George & Chattopadhyay (2008, p. 371) suggest that "in general, diverse groups will be less cohesive than homogeneous groups, diverse groups that are dominated by high status individuals are more likely to be cohesive, and thus to suffer from groupthink, than diverse groups that are dominated by low status individuals". Coalition is a proven method to gain power in groups. Balogun et al. (2008, p. 245) sense "a difference between 'naïve' practitioners who does not understand what image they project on to others and skilled practitioners who do not only understand but are able to manipulate such images as and when required". Their work suggests that decision makers who are more cognitively skilled appear to be better in shaping power relations and a net of meanings by negotiation and therefore get more support for action or a particular point of view.

As a result of her doctoral thesis Gehner (2008) presents a strategy table (a framework to structure the formulation of a strategy) that distinguishes information sharing between decision shapers, decision takers and decision approvers in investment decision making in real estate. The actors apply organisational or project issue related qualitative and qualitative decision criteria during different decision phases of preparation (provision of information to general management), submission (transfer of information to decision takers and approvers), and approval (actual decision making and authorization). Seven decision activities are proposed as relevant for investment decision making via three indicators of timeliness (timely recognition, limiting duration), justifiability (determination of decision criteria, search for information, identification and analysis of course of action, analytical evaluation) and accountability (authorization). These different groups, power levels, decision phases and indicators show the complexity and interrelations of the organisational structure in which decisions are made in organisations that knowingly take risks. On the other hand there are the information processing concepts of sensemaking, framing, sensegiving, sensereading and sensewriting are all related to the resource, process and meaning of power effects in group and organisational decision making (Balogun, et al., 2008). A frame constitutes the relations between elements that are of interest to a decision maker and tell a decision maker what to expect in certain situations. By gathering information the frame can be tested and developed into one that fits the facts. This means that both structure and interpretation are important elements of group decision making.

3.5.3 Expert teams

Expertise appears to have a lot of influence on decision making. In general is seems that framing allows experts to be informed by richer information than a novice. Experts also seem to have more flexibility and adaptability in the use of frames and they seem to recognise meaningful patterns earlier, which gives them the opportunity to perform on a higher level (Chase & Simon, 1973; de Groot, 1946). Research shows that expert decision makers spend more time evaluating the situation while novice decision makers spend more time generating and evaluating courses of action because representation of the situation forms the basis for pattern recognition and mental simulation processes (Hutton & Klein, 1999; Sánchez, Prats, Agell, & Ormazabal, 2005; Schön, 1991). Experts can make judgements about the consistency, reliability and completeness of their information and they know when to stop analyzing and when to search for additional information (Cohen, Freeman, & Wolf, 1996; Hutton & Klein, 1999). Rosen et al. (2008) present nine mechanisms how experts achieve superior performance: 1) a tight coupling of cues and contextual features of the environment, 2) a large knowledge base differently organised from non-experts, 3) engagement in pattern recognition, 4) engagement in deliberate and guided practice, 5) a seek for diagnostic feedback, 6) better situation assessment and problem representations, 7) specialised memory skills, 8) automated small steps in decision making, and 9) self-regulation and monitoring of their processes.

In organisations teams are often assigned to perform tasks that exceed the capacity of one person. However, a group of experts is not an expert team. Salas, Burke and Stagl (2006, p. 440) define an expert team as "a set of interdependent team members, each of whom possesses unique and expert level knowledge, skills, and experience related to task performance, and who adapt, coordinate, and cooperate as a team, thereby producing sustainable, and repeatable team functioning at superior or at least near-optimal levels". Burke et al. (2006) suggest that an expert team adapts to changes in four phases of situation assessment: 1) building common understanding, 2) plan formulation by decided on the most effective course of action, 3) plan execution through coordinated team performance, and 4) team learning by evaluation. If roles and responsibilities are clear, expert team members better accept boundaries and are better able to anticipate the actions and needs of other team members. Clear values and shared visions contribute to generating goals and finding appropriate methods for reaching these goals. In group decision making, the perception of the level of expertise of a group member seems important. Bunderson (2003) found that especially in shorter tenure teams (such as project teams) group members tend to judge each others expertise based on categorization cues, such as gender or race instead of task performance cues, because there is only limited time and information for other kinds of judgements.

Feedback is critical in the development of expertise, and therefore expert teams incorporate feedback mechanisms. Salas et al. (2004; 2006) found that expert teams have leaders who solicit ideas and observations of team members, explain (when possible) why team input is rejected, seek out opportunities to reinforce effective teamwork, are receptive to and request feedback on their own performance, provide behavioural and specific solution oriented feedback, restate feedback from others to make it constructive, express satisfaction when improvements are noted, and give situational updates. Expert teams have members that trust the abilities and intentions of their fellow team members, believe in the importance of team work, are confident of the ability of the team to reach its goals and are motivated to learn how to work as a team more effectively (Rosen, et al., 2008). To complete the task successfully and in a timely manner, a team must coordinate and allocate the information and expertise of the individual members which requires different mechanism than in individual decision making.

3.6 Conclusion

The field of decision making has a long research tradition with a diverse range of research methods including experiments, surveys, observations, interviews, and is based on multiple disciplines such as psychology, administrative behaviour, economics, sociology, and anthropology. This chapter describes three generations of decision theories that present a diversity of perspectives on the individual decision maker as well as decision making in groups or organisations. The theories often differ on the proposed steps in which decisions are made and the way that people deal with their limited capacities. Research on the performance differences of decision theories appears difficult because outcomes are not easily measured and compared in a direct sense. Every generation of decision theory offers some or more concepts that can be applied on the situation in which clients select an architect. However, the concept of managerial decision making shows the most resemblance with the structure current tender regulations. Yet, the second generation theories of naturalistic decision making display most resemblance with judgement processes of jury members. Therefore the current practice of architect selections appears to be based on two conflicting models about decision making. The legal model assumes a rational and sequential decision process in which alternatives are compared based on pre-announced criteria. The naturalistic decision model attributes an important role to the use of intuition and affect. The origin of the current problems in practice could consequently be found in these different rationalities.

In general decision making can be considered as a process of making judgements that depend on the decision tasks, the level of expertise of the decision maker, and the organisational context in which decisions are made. Intuition, emotion and affect are to a larger or lesser extent part of the process of decision making. Assessing design quality can be considered an ill-structured task with a lot of ambiguity. Experts appear to be better able to make decisions in uncertain environments but expertise is often domain specific. The use of expert teams often leads to better decisions because they can complement each other in information, knowledge and critical feedback. Therefore they seem to be more able to make sense of the large amount of information that has to be processed during tender procedures. In organisations not only the input for a decision is important but also the processes that decision makers go through in order to create support for a decision. The concept of sensemaking offers a perspective that fits the situation of architect selections well. During the process of sensemaking a decision receives meaning for the members of a group, which increases the chance that a decision is implemented into action. In current practice clients appear to make sense of the process as a project unfolds, which is also an element of the sensemaking process (Weick, et al., 2005). There are also a lot of ambiguities, connections with past experiences and dialogues with stakeholders. In Chapter 5 the insights of this chapter are used to develop a theoretical research framework for the context of tender decisions of public clients in the context of architectural design as described in Chapter 4.

Chapter 4

THE CONTEXT OF ARCHITECT SELECTIONS

4.1 Introduction

An architect selection is not an isolated event. Public clients operate in a context of governance and have to consider this organisational structure in their decisions. The Netherlands is a parliamentary democracy and a member of the European Union. This affects the legal and social responsibilities of Dutch clients. Participation and stakeholder involvement are important characteristics of a democracy. The competition tradition in architecture supports this element of public debate, but also contributes significantly to the development of the professional field. The selection of architects differs from the selection of contractors in construction, but in both situations procurement law applies. Local authorities commission two to three times the volume of construction than central governmental authorities do (Weijnen & Berdowski, 2009). Public clients in the European Union have to comply with the European tender regulations, which offer various procedures to select a designer. This is accompanied by considerable costs for both the demand and supply side of the market.

This chapter addresses four contextual elements that in my view are essential to understand the environment in which architect selections take place: the political, cultural, legal, and economical context. The perspective taken in this chapter is that of the client or commissioning body rather than that of architects, contractors, citizens, legal advisors or other players in the field of construction, law, project management or public administration. Where possible and/or relevant the Dutch situation is compared to literature from other countries. The final section of this chapter addresses the perceptions and expectations of the parties involved in architect selections by building on the information provided in the previous sections. It also includes an analysis of the existing guidelines to support the architect selection process.

4.2 The political context

This section concentrates on the responsibilities of the main actors in the public sector and some basic characteristics of participation and stakeholder involvement in the area of architectural design. The cases that are included in this thesis are all situated in the Netherlands. This section therefore also addresses the architecture policy of the Dutch government and the role of the Chief Government Architect in promoting and guiding the development of architectural quality. Box 1 provides a short summary of the Dutch governance structure for those who are interested in or unfamiliar with public administration in the Netherlands.

4.2.1 Decision structure of public commissioning bodies

This research focuses on actors in the field of governance from a psychological perspective, rather than public administration which studies governance in order to improve it. So apart from the fact that the executive board is periodically chosen on democratic principles and they are part of a public organisation, contracting authorities of departmental, provincial and municipal organisations have in my view comparable roles, tasks and responsibilities to decision makers in private organisations. A public organisation is also structured as departments and led by directors (in this situation civil servants). Administrators have to run a town, a province or other kinds of governmental authorities. They are accountable for their actions and have to take the perspectives of different stakeholders into account. Legal rules, administrative rules and procedures, and external review decisions are complementary, enhancing the quality of public administration both with organisations and in dealing between administrators and members of the public (Feldman, 2003). This picture may look different from the perspective of the organisation and that of individual decision makers within it. Administrators "are likely to view appeals to courts and judicial review as a threat rather than a useful safety net in hard cases" (Feldman, 2003, p. 286). Administration and politics cannot be clearly delineated: "political choices are endemic for administration and public bureaucracies need to be understood as nested within a network of political actors" (Bryner, 2007, p. 189).

Brunsson (1989) stated that in political contexts decision making is a process of dialogue that participants engage in as a means of building rationales for action, creating visions of future states and mobilizing resources. "Because organisations have different goals and stakeholders that cannot all be satisfied simultaneously, organisational leaders have to espouse different visions at different times and support mutually inconsistent actions. Decision processes also create responsibility in that people hold to account those whom they perceive to have advocated actions or made decisions. The ways in which decision processes unfold create external perceptions of about the legitimacy of the decisions, the ensuing actions and the deciding organisation." (Hodgkinson & Starbuck, 2008a, p. 11).

Public organisations face tremendous expectations as well as challenges to satisfy competing and often contradictory values. Those values clash within bureaucratic organisations as they interact with other political institutions and as they operate within the broader distribution of economic and political power aimed at their organisations (Bryner, 2007). An important distinction needs to be made between corporate, hierarchical, collective and individual liability, between liability and responsible behaviour, and between active and passive responsibility (Bovens, 1990). "Because in complex organisations many different officials contribute in many ways to decisions and policies, it is difficult even in principle to determine who is responsible for outcomes" (Bovens, 1990, p. 324). This is called the problem of many hands (Thompson, 1980). Cohen, March and Olson (1972) described decision making in public organisations as 'garbage cans' in which people dump their preferences, alternatives, solutions and participants. Openness and publicity are important values that determine the operational environment of an administrator. Too much or too little openness could lead for an administrator to have to resign from their job.

Difficulties in aligning frames of different cultures are common in local politics (Beach & Connolly, 2005). In architect selections staff members are usually assigned to carry out a tender project. Because of the governance structure of election periods, changes in the political climate could occur during the realisation of a project. In practice this means that in general the members of a project team may remain the same but their directives may change. Governmental authorities in the Netherlands have had to cut back their expenditures for several years now and many tasks have been outsourced to private organisations or terminated. Also, turnover of staff is relatively high in local governments. A tender for the selection of an architect requires very specific legal and real estate knowledge. The Dutch architectural profession often complains about the lack of qualified personnel in the field of urban planning and design in local government (Architectuur Lokaal, 2009a). These changes, in combination with the diversity and required level of expertise, have created a situation in the Netherlands in which consultants are often hired for specific projects such as architect selections (van der Pol, et al., 2009). The periodical elections and a large amount of externally hired advisors thus created a kind of discontinuity in public administration and a conflict between the continuity of civil service staff and periodic change of administrators and consultants.

The Dutch government has been implementing architectural policy since 1991 (Atelier Rijksbouwmeester, 2008). Several ministries are involved in the development of architectural policy papers, such as Housing, Spatial Planning and the Environment (in Dutch 'VROM'), Education, Culture and Science (in Dutch 'OCW'), Agriculture, Nature and Food Quality (in Dutch 'LNV'), Transport, Public Works and Water Management (in Dutch 'VenW') and Economic Affairs (in Dutch 'EZ'). In May 2005, the latest Action Programme on Spatial Planning and Culture was presented. This programme combines architectural policy with the so-called Belvedere policy (aimed at strengthening the influence cultural history has on spatial planning), creating a framework for diverse policy proposals and concrete actions. In addition to the role the national government plays in setting an example, stimulating and supporting other parties and developing knowledge by means of policies on good contracting, practices and architectural institutions remain key principles of the programme. The Ministry of OCW bears the primary political responsibility for architectural policy. As adviser of the State Secretary of OCW the Chief Government Architect (Rijksbouwmeester) provides the Minister of VROM and the other Ministers with solicited and unsolicited advice on matters of policy and strategic developments, on architecture, urban and rural planning, infrastructure, landscape development and guaranteeing quality in legislation and regulations and in education (Atelier Rijksbouwmeester, 2008).

Local authorities, provincial authorities and other autonomous administrative authorities, such as the police regions and chambers of commerce have to manage their own housing needs. Their real estate portfolios include school buildings, sport facilities, utility buildings, housing, cultural facilities, governmental offices and infrastructure. The Rijksgebouwendienst (Dutch Government

Building Agency) deals with housing for Dutch government ministries and national agencies, the High Councils of State, independent administrative bodies and international organisations. It is part of the Ministry of Housing, Spatial Planning and the Environment. The Chief Government Architect is the primary advisor to the Director-General of the Rijksgebouwendienst on promoting and monitoring the architectonic quality, harmonising architecture with urban and rural planning, monument preservation and the application of visual arts (Atelier Rijksbouwmeester, 2008). The position of Chief Government Architect is intended to stimulate the quality of architecture in the Netherlands, for the central government, but also for regional authorities. In carrying out his/her task, the Chief Government Architect assumes an independent position. (S)he also makes specific recommendations upon request for other ministries involved in accommodations construction or construction financing, provides advice on design and recommends architects, and is a member of the Board of Government Advisors.

Box 1: Public administration in the Netherlands

The Netherlands is a parliamentary democracy with an elected government and a head of state, the king or queen. It is a member of the European Union but the Constitution of the Kingdom of the Netherlands regulates for a great deal how the government is structured. The government comprises over 1,600 organisations and bodies, including 13 ministries, 12 provincial authorities and, until 18 March 2010, 431 municipal authorities (Overheid.nl, 2009). It also includes autonomous administrative authorities, such as police regions and chambers of commerce, and public bodies for industry and the professions, such as the Water Commodity Board. At the same time, many organisations that one might assume form part of government are in private hands. They include health insurance funds, boards of private schools and benefit agencies such as the social services. Within the government sector a distinction is made between bodies that are directly elected and those that are not. Municipal councils, water boards and the Office of Representatives are directly elected by the people, whereas mayors, police commissioners and ministers are not. However, all government authorities are ultimately accountable to the public for what they do. There are three levels of government: central, provincial and municipal. Other authorities are classified mainly on the basis of their tasks. All these organisations have to comply with procurement law.

The Provincial Council ('Provinciale Staten' in Dutch) is the general administrative body of the province (Province of Utrecht, 2009). Provincial councils are elected by direct popular vote. The council then appoints the Provincial Executives ('Gedeputeerde Staten' in Dutch) who, with the Queen's Commissioner, then form the executive body. If the Provincial Council is dissatisfied with the work done by the provincial executive, it may dismiss the individual executive or executives responsible. At the council's recommendation, the Crown appoints a Queen's Commissioner, who chairs the council and the executive. Council members are elected every four years by residents of the province who are entitled to vote. The number of council members depends on the size of the population of the province. All council members belong to one or other political party. Those representing the same party constitute a political group or caucus. Combined groups of council members belonging to various parties are also possible. Each group selects a chairperson from among its members, who acts as leader and chief spokesperson. Being a council member is not a full-time profession, it is a part-time activity. Most members have ordinary jobs besides their council work. They receive remuneration and an allowance for expenses. Members may not benefit from their council membership in terms of their day-to-day activities.

The Provincial Executive ('College van Gedeputeerde Staten' (GS) in Dutch) is charged with the day-to-day management of the province. The executive is made up of the Queen's Commissioner and several members. The executive receives administrative support from the provincial clerk, who is also general head of the official apparatus of the province. The Commissioner is elected for a period of six years by the Crown and this term of office may be extended. Members of the executive are chosen by the provincial council for a period of four years. This appointment takes place on the basis of the results of elections for the council. The executive can be compared to the municipal executive. The executive may take decisions in several fields. However, where crucial matters such as budget or far-reaching plans and projects are concerned, the executive's task is to inform and advise the council. The executive then works out plans adopted by the council. The executive is also required to justify its policies to the council.

Municipalities apply the same principles as provinces. A municipality is run by a City Council (in Dutch: 'Gemeenteraad'). Council members are elected every four years by residents of the city. Most Councillors also have a job in addition to their Council work (City of Amsterdam, 2009). This is because council membership is voluntary work, for which Councillors merely receive an allowance. The day-to-day running of a municipality is the task of a 'college' made up of the Mayor and several Aldermen (in Dutch: College van B&W). Aldermen are elected by and from the Council. After they are appointed, the Aldermen remain members of the full Council and vote in its meetings. The Mayor and the Aldermen share their work: each has his or her own portfolio and areas of responsibility. The College prepares the resolutions to be adopted by the Council and implement these resolutions once they have been adopted. The Council may also reject a proposal from the College, as the Council has the final say. To be able to manage effectively, the Council delegates many tasks to the College. These mainly concern decisions taken on the basis of an established policy, which therefore does not need to be debated by the Council. The Mayor occupies a special position. He or she is not elected by the city's residents, but appointed by the Monarchy. The mayor chairs the City Council and the College of Alderman. (S)he can vote in the College, but not in the full Council. The Mayor also has a portfolio with various responsibilities of his own and is head of the police and responsible for maintaining public order in the city. To a large extent, (s)he also represents a municipality to the outside world.

4.2.2 Participation and stakeholder involvement

In decision making, governmental authorities have to deal with the interests of the public as their main stakeholders. Because the built environment affects everybody, the selection of an architect is typically an issue that could benefit from public debate to involve the community (see for example Jencks, 1987). Nasar (1999) states that competitions seldom lead to masterpieces because architects and juries stress appearance instead of the convenience and durability of the future building. Preferences of a jury do not always correspond with those of users and visitors (Nasar & Kang, 1989) but according to Collins (1971, p. 194) it is mainly the difficulty of "creating a realistic understanding in the lay mind of the difference between price and value". This implies that average citizens can do an effective job of decision making if they are provided with accurate and relevant information that is organised and presented in a way which is meaningful without being patronizing (Crosby, Kelley, & Schaefer, 1986; Robinson, 1972).

A participatory democracy is characterised by a highly decentralised democratic system of collective decision making in which individuals can effectively participate in various ways in the making of decisions that affect them. Participation is a general concept covering different forms of decision making by a number of involved groups. The groups can have different perspectives of what a good process is, such as legitimacy, ideological discussion, fairness, power struggle, leadership and compromise (Webler, Tuler, & Krueger, 2001). According to Sanoff (2006, p. 133) "the activity of community participation is based on the principle that the environment works better if citizens are active and involved in its creation and management instead of being treated as passive consumers". If done well, participation of the public can improve the quality, legitimacy and capacity of assessments and decisions (Dietz & Stern, 2008). It also can enhance trust and understanding among parties, increase ease of implementation, minimizing costs and delays (Sanoff, 2006). Participation should therefore be recognised as a requisite of effective action instead of a merely formal procedural requirement. According to Jansen, Gössling, Merks and Geurts (2005), decision makers perceive a decision as having more quality if several conditions are fulfilled simultaneously. These conditions relate to involving relevant and committed actors, dealing with the balance between stability and change, a balance between stakeholder interests on short and long term, dealing with deliberations about social responsibility ample proof, good reasoning and a guarantee of implementation.

Decision makers, such as council members or citizens, can take on different types of roles from passive auditing and information collection to providing information and active participation. The Participation Ladder of Arnstein (1969) distinguished three kinds of citizen participation: non participation, tokenism and citizen power in eights levels: manipulation, therapy, informing, consultation, placation, partnership, delegated power and citizen control. The Vroom and Yetton model of managerial decision making distinguishes five levels of individual or group level participation in organisations (Vroom & Jago, 1988), which correspond to the levels of informing, consulting, advising, coproducing and codeciding as used by Edelenbos & Klijn (2005) in urban planning. The problem characteristics that determine which of these levels of participation is appropriate seem to relate to numerous factors such as the importance of the quality of the decision, the level of expertise and information of the leader, the structure of the problem, the importance of members acceptance and commitment, the chance of acceptance of the decision by the members, the motivation of the goals, the probability for consensus, the amount of information available, time constraints, the geographically dispersion of the members, the speed of decision making and the importance of fostering members through participation (Beach & Connolly, 2005). The four types of participation in organisational decision making offered by the model of Vroom and Yetton (and comparable to the other classifications) could be applied to tender situations (Vroom & Jago, 1988):

- A: autocratic: the responsible officer decides for himself based on information available to him at the time. This information can originate from other individuals without active awareness or participation. This kind of decision making is hardly explicitly found during tenders or design competitions.

- C: consultative: the responsible officer shares the problem with relevant subordinates, getting ideas and suggestions with or without bringing them together as a group. Then (s)he makes the decision, which may or may not reflect the influence of the subordinates. Assessment by a user group can be considered as part of the consultation.

- G: group: the responsible officer shares the problem, analysis and alternatives with subordinates as a group. The responsible officer acts as a coordinator or equal group member. Decisions are made by mutual agreement within the group. Steering committees of projects could be set up like this.

- D: delegation: the responsible officer delegates the problem and the responsibility to a subordinate. Any solution that is reached will receive the officer's support. A steering committee delegating the selection process to a jury is a good example in the context of architect selections.

According to Crosby, Kelley and Schaefer (1986) and Arnstein (1969) a distinction should be made between successful methods of citizen participation and successful citizen lobbying and manipulation efforts. Participation is a concept that needs to be thought through in the sense of a match between the objectives, the methods and questions such as what, where, whose, how and when is participation desired (Wulz, 1986). It needs good preparation. Establishing a policy of inclusiveness, holding open meetings, making speeches to community groups, obtaining public input, voluntary community councils (small face-to-face groups of diverse citizens that convene for short periods of time to consider some public concern), making public announcements, holding face-to-face meetings and conducting progress surveys are only a few options in participation processes (Sanoff, 2006).

Several scholars (Crosby, et al., 1986; Dietz & Stern, 2008; Edelenbos & Klijn, 2005; Webler, et al., 2001) identified a number of principles that should be included in the design of the participation process in the public sector, such as representative non-manipulative participation selection, collaborative problem formulation and process design, process management, transparency, good-faith and clear communication, adequate funding and staff, and a focus on effective decision making. Drawing from the literature on group decision making, Crosby et al. (1986) list five factors for a successful method of participation: 1) sufficient amount of time for participants to learn the information and to reflect on the values and goals relevant to the decision, 2) an appropriate size of the decision group, 3) a planned agenda so that the important material is covered in an orderly fashion, 4) the person leading the group must facilitate the discussion, and 5) the views of the participants must be given adequate recognition.

A few suggestions for the design of stakeholder participation in architect selection can be drawn from the literature. Nasar (1999) pleads for a systematic inventory of visual quality for the purpose of pre-jury evaluation among all groups of people that might experience the building, especially users. This should be followed by an unbiased evaluation process that is based on insights from environmental psychology about preferences and meaning (see also Chapter 2.2) and completed by a post occupancy evaluation to see how the building actually performs. Crosby et al. (1986) introduce the concept of a 'Citizens Panel', which includes four days of regional and state level staff presentations on the topic, witness testimonies of several stakeholder groups and the preparation of a report of the panel members about the recommendations. Evaluation shows the concept is flexible, effective and fair but relatively expensive with a limited power to convince the people in charge. In architectural design and urban planning co-design provides an interesting option for the development of a design because designers and planners work with instead of against community groups (de Jonge, 2009; King, 1983). After all, "experiences in the participation process have shown that the main source of user satisfaction is not the degree to which a person's needs have been met, but the feeling of having influenced the decisions" (Sanoff, 2006, p. 140).

4.3 The context of design and construction

The real estate cycle consists of an initialization, design, construction, and occupancy phase. Clients play a key role in the initialization and the occupancy phase, while architects are the most essential players in the design phase and contractors in the construction phase. As mentioned in the first chapter the current selection process of architects originates from the tradition of the design competition, the tendering of works and services, and the selective search for a suitable architect or design partner (Strong, 1996). This section describes the origin and character of design competitions in architecture and the characteristics of partner selections in the construction industry. The legal aspects of tendering of works and services are discussed in the next section.

4.3.1 The concept of design competitions

Starting with the ancient Greeks, competitions have traditionally been a vehicle for the creation of major civic buildings and public spaces, such as government buildings, performing art centres, educational facilities, public libraries, museums and housing (Collyer, 2004; Strong, 1976). Design competitions are considered as the repositories of the architectural profession (de Haan & Haagsma, 1988). They have produced high profile projects as well as a lot of debate, dispute and controversies (Strong, 1996), and appear to be a significant part of the architectural culture. Most of the authors of publications about design competitions consider the phenomenon as a given without critically assessing the concept and its tradition as such, even though design competitions are often subject to intensive scrutiny (Spreiregen, 2008). Those publications that address the premises and weaknesses of design competitions show great similarities internationally in the problems they identify (de Haan & Haagsma, 1988; Heynen, 2001; Spreiregen, 1979; Stichting Bouwresearch, 1980; Strong, 1996; Sudjic, 2005). Problems mainly concern the transparency of the selection of the participants, the client's communication of the requirements to the participants, the composition of the jury panel, the objectivity of the jury's judgement, and the financial compensation compared to the amount of work (Hijdra, 2007). Spreiregen (1979) identified three myths about design competitions: competitions cost money, competitions cost time, and competition designs never get built. These myths are not supported by facts but remain very persistent.

According to Collyer (2004) the proliferation of competitions and the inclusion of so many building types in that process can be viewed as a post-World War II phenomenon, although it is relatively rare in the United States. The relevance of design competitions is acknowledged worldwide in the world of architecture. Yet information on past and recent design competitions is fragmented, inadequate and frequently unrecorded (Lipstadt, 2005; Spreiregen, 1979). Several publications describe the aims, procedures, potentials and pitfalls of design competitions in a historical perspective or show the relevance of competitions for the architectural profession (Collyer, 2004; Larson, 1994; Lipstadt, 2005; Stichting Bouwresearch, 1980; Strong, 1976, 1996; Sudjic, 2005). Others simply document the winning designs pictures or descriptions and report the opinion of the jury and some other assessors for inspiration purposes (e.g. de Haan & Haagsma, 1988; Glusberg, 1992). The experiences and interests of the client as future user of the building are usually neglected in the evaluation of the competition (Nasar, 1999; Nasar & Kang, 1989).

The selection of an architect is merely one of the several purposes of design competitions. Competitions can have educational purposes, (e.g. educating students, challenging 'conventional wisdom'), political reasons (e.g. enlarging support, marketing a project, running architecture politics, coordinating different fields of interests), cultural aims (e.g. creating a dialogue on design, contributing to the cultural dimension of the built environment, expanding the boundaries of design), and economical reasons (e.g. increasing competition, gaining insight in competences or assuring quality through jury assessment) (de Haan & Haagsma, 1988; Evers, 2008; Spreiregen, 1979; Svensson, 2008). "The purpose of competition in architecture is the pursuit of excellence. As in other endeavours, the goal is illusive. It has to be pursued constantly if it is to be achieved, and when on occasion it is, new and more distant horizons come into view. So the pursuit continues" (Spreiregen, 1979, p. 5). Spreiregen identified several benefits and problems with design competitions that still summarise the discussion among professionals today (see Table 4.1).

Pros	Cons
Identification of new talent	Cost to the client
Stimulation of old talent	Time required
Stimulation of public dialogue	Possibility of an excessively costly solution
Stimulation of professionals	Elimination of the program development phase
Exploration of new concepts	Absence of Client-Architect Dialogue
Bringing to bear the best design ability	Not suitable for complex buildings
Boost of office morale for architects	Possible inexperience of the selected architect
Development of new design forms	Possible impractical selection by the jury
Maintaining attention for design	Revelation of confidential security information
Show casting professional abilities	Costs to the profession
Bringing in a wide point of view	Realities and pressure of the patronage system
Freeing designers from constraints	Notion of 'a lot of trouble'
Test and challenge status quo	Notion that 'good design is expensive'
	Requirement of overly elaborate presentations
	Difficulty of scheduling public financing
	Uncertainty about enough or right designers
	Lack of information about the management of a competition

Table 4.1 Pros and cons of design competitions according to Spreiregen (1979)

Design competitions always involve the development of a design to the point where it realistically represents a realizable building. There are two main competition structures (the open competition and the invited competition), two main competition populations (national and international), and two main competition objectives (conception on its own called an ideas competition, and conception leading to construction, which is a design competition in the conventional meaning of the word) (Lipstadt, 2005). Spreiregen (1979) also distinguishes student competitions, (urban) renewal competitions, turn-key competitions and consultant-selection competitions as variations on the open and invited competition. The Royal Institute of British Architects distinguishes between feasibility study, design proposal and sketch design as three levels of thoroughness of the design activities required for the competition (Heynen, 2001). In a design proposal about 10 to 15% of the total building design is prepared for submission, which should be sufficient for a competition. A sketch design requires about 30 to 35% of the work of a detailed building design. The more work is required without any interaction or a proper dialogue with the client, the more the likelihood exists of redoing the work.

There are two principles that provided the foundation of architectural competitions world wide: a panel assesses the quality of the design against criteria established in the brief and the whole process is conducted in a fair and equitable manner (Strong, 1996). Strong (p. 43) lists four criteria that distinguish a competition from other forms of commissioning which also apply for tenders: "1) There are several entrants, 2) there is an identical problem for all entrants, 3) rules and procedures are prescribed and followed, and 4) systematic and independent assessment by a panel of assessors is used to select a winner." A jury has important formal, aesthetic and ethical responsibilities, such as the obligation to ensure compliance with regulations and the process, choosing a winner, description and presentation of the competition entries, an objective and unbiased assessment of the entries, independent identification of long-term value, motivation of the choice, loyalty towards themselves and not going out to the public with a deviating opinion without making a reservation against the jury's declarations (Kazemian & Rönn, 2009). The multiple aims, the complexity of the context and character of the competition rules result in eleven dilemmas that juries face during their decision process (Rönn, 2008):

1. Democracy versus Expert decision - Design competitions have open exterior and private interior; limited democratic contribution.

2. Anonymity versus Direct communication - Final product is more important than the person; limited communication between organizing body and participants.

3. Project versus Architect - Dual function of competition system; relevance for the clients and for the profession.

4. Security versus Innovation - Participants long for something new, clients require well-proven construction, efficiency, durability etc.

5. Precision versus Latitude - Degree of steering of the clients and the need for latitude of the jury.

6. Requirements versus Feedback - How to foresee the potential created by the competition and ensure quality of future buildings.

7. Minimizing faults versus Maximizing quality - The approach on how to rank quality according to the amount of shortcoming in relation to the positive qualities.

8. Letter of intent versus Educational development - When the organizing body comes in contact with the proposals they acquire a deeper understanding of the assignment.

9. Objective versus Process - The entries and organisation determine the jury process but the process aims at selecting a winner.

10. Present versus Future - The present is the point of departure for an estimation of future qualities; equally difficult for jury and participants.

11. Professional versus Community approval - Influence of multiple parties on the competition rules; composition of the jury.

According to Kazemian and Rönn (2009) jury members must be highly experienced, independent design thinkers and critics with good visionary judgemental ability in design to secure the essentials of competence and consensus. At the same time an important ongoing dispute regarding juries is whether they should be made up solely of architectural professionals or should also include non-architects closely involved with the project (users, decision makers, politicians, etc.). This dispute seems even more relevant today in relation to the current practice of architect selections. Next to the composition of the jury panel, a high-quality brief and a professional organisation are important aspects of a competition (Patijn, 2000; Sudjic, 2005). Spreiregen (2008) describes a competition process as consisting of seven components: planning, competition announcement, design, receiving and processing, jury, design announcement, and post-competition phase. The competition conditions include three basic sections of general conditions, instructions and a brief (Strong, 1976). According to Kreiner (2006) a competition brief reads as half instruction, half inspiration and should be both unambiguous and non-constraining. He discusses the inherent tensions and ambiguities of 'form versus function', 'tradition versus change' and 'requirement versus suggestion' in a brief. In every kind of competition trust must be established between the authority and the designer (Fisher, Robson, & Todd, 2007). According to Heynen (2001) two communication filters exist in communication between the client and the participant: from ideas of the client into the brief, and from the ideas of the participant to the jury. A verbal explanation of the written documents is therefore very much appreciated by the participants. This could contribute to the relationship built on trust between the client and architect. A jury report serves this purpose, just as it serves "to inform the sponsor of the reasons for the selection of the winner, to clarify the general objectives being sought in the particular design experience at had and as a record and reference of a particular moment of design thought and awareness" (Spreiregen, 1979, p. 234).

The general belief among architects is that the chances of winning a competition are very hard to predict (Kreiner, 2007b). It is said that the composition of a jury in fact determines the outcome of the competition (de Haan & Haagsma, 1988; Larson, 1994) but the strategies of participating architectural firms can influence the direction of the designs that win (Kreiner, 2008; Manzoni, et al., 2009). Kreiner (2007a, p. 1) describes the competition process as follows: "If architectural teams were to describe the competition as a dance, they would described it as a peculiar form of dance in which they are dancing with an absent partner, fancying him or her and responding to his or her imaginary movements and gestures. The absent partner is the client (and in some respects the jury which will appoint the winner of the competition). Such shadow dancing is performed currently by a small number of teams of architects, each representing on of the selected firms taking part in the competition". Strong (1976) found that although only 10% of the participants in competitions ask questions during preparations, 50% of the winners raised questions which indicates a different strategy of the potential winners. In general the economic conditions, the challenging character of the brief and overall attractiveness of a competition influence the response and therefore the accomplishment of a competition. The success of a competition however also depends on the level of the participants, and participants cannot be forced to join.

The competition principles are usually incorporated in national regulations, standardised formats and model competition conditions. Increasingly, these are being replaced by guidelines setting out good practice procedures. In the Nordic European countries (Sweden, Norway, Denmark and Finland) about 100 architectural competitions take place annually spread roughly equally over seven areas (town & urban planning, schools, culture & leisure, housing, health & social welfare, offices, and others) (Kazemian & Rönn, 2009; Rönn, 2008). About 60 to 75% of these competitions take four years to complete and 15 to 30% are cancelled. In recent years invited competitions have also become popular in the Scandinavian countries, and these competitions have become more like a form of a public negotiation (Rönn, 2008).

In the Netherlands, the number of design competitions is limited compared to other ways of commissioning jobs, but also limited compared to other European countries, such as France, Germany, Switzerland and the Scandinavian countries (Evers, 2008). Most of these counties have established official bodies to support and regulate the organisation of design competitions. Although the countries with official regulations appear to experience fewer problems with design competitions, the problematic nature of selecting architects by design competitions seems to be recognised internationally. In the Netherlands not much action has been taken to resolve the existing tension between the field of architecture and clients on matters that relate to the organisation of a design competitions (Evers, 1995). The Netherlands does not have an official body apart from an office for procurement and design competitions called the 'Steunpunt Architectuuropdrachten en Ontwerpwedstrijden'. This Office was established in 1997 by Architectuur Lokaal, an independent national centre of expertise and information devoted to commissioning building development, with the aim to advise public and private clients about selecting designers and property developers. Since the end of the 19th century Dutch regulations exist (such as the 'Algemene Nederlandse Prijsvraagregelingen') but these were, for unknown reasons, not adequately applied by clients. Because the EU Directives of 1994 refer to national regulations, representatives of Dutch governmental authorities and professionals decided to develop the 'Kompas' (van Campen & Hendrikse, 1997) that includes guidelines and models for competitions in architecture. Changes in regulations and an increasing need for support during the organisation of procedures made several organisations decide to develop new guidelines, models and other kind of publications that apply to the Dutch situation (e.g. Architectuur Lokaal, 2009b; Chao-Duivis, et al., 2007; Jansen, 2009; Patijn, 2000).

Organizing a design competition requires a substantial amount of money and energy. According to Kazemian and Rönn (2009) it can take four to ten months to carry out an invited competition in which a jury decides on three to five proposals. Because an open competition attracts on average 80 entries, implementations could take between seven and eleven months. Strong (1976) found that participants spend between 160 to 500 hours on the submissions for an open competition. A two-stage invited competition could take up to 1860 hours. These numbers and current experiences indicate that in the Scandinavian countries much more time is invested in evaluating the entries than presently in the Netherlands. The money that an average architectural design firm spends per design competition was estimated by the Royal Institute of Dutch Architects (BNA) at € 19.000 in 2008 and € 31.000 in 2009 (BNA, 2009; van den Hurk, 2008). The total costs for participation in design competitions were estimated by the BNA at €

10 million per year for the whole sector. On average architectural firms earned € 128.000 per year in competitions.

4.3.2 Partner selection in construction

The second root of architect selection refers to a selective search for a partner in construction. Architects are not the only players in the field of construction. Engineers, consultants, developers, management consultants and contractors also participate in construction tenders. Most tenders for works in construction are similar to how jobs were commissioned before the introduction of the EU Directive in 2004. These transactions were mainly based on the lowest price, assuming that the contractors would all offer comparable quality levels. Research shows an increase of the preferences to award contracts for works on the 'most economically advantageous tender' (MEAT) instead of lowest price (Dreschler, 2009). Private clients appear to be a bit more enthusiastic about the MEAT principle than public clients, which can be attributed to the accountability of the decision (Wong, Holt, & Cooper, 2000). Pongpeng and Liston (2003) describe the development of a tender evaluation tool that accumulates input from multiple decision makers, incorporates risk and uncertainty and offers flexibility to changes in the situation during design tenders. In their tool the relative importance of the bid price, in comparison to utility and social welfare determines which contractor offers the best bid and should win the tender. The methodology for best value decision making as developed by Phillips, Martin, Dainty and Price (2007) also acknowledges the uncertainty and imperfection of tender decisions. Their tool contributes to the transparency of decision making by using the analystic hierarchy process method, pair wise comparison, multiple-attribute utility theory and benchmark scores from KPI's.

Several authors (e.g. Watt, Kayis, & Willey, 2010; Wong, et al., 2000) found that for public buildings and engineering works the most important project specific criteria seem to be the actual work quality achieved on similar works (technical expertise and past performance). It appears that public clients are more focused on price than private clients, even though they should be aware of the fact that the lowest bid does not always lead to the best quality, lowest price at completion or highest client satisfaction (Sporrong & Bröchner, 2009). Large Dutch clients in construction appear to prefer the price and quality of the contractor over the financial stability, reliability and experience of the contractor (Regieraad-Bouw, 2005), while small clients think reliability and quality of the contractor is more important than client focus, price and experience. Based on a survey Phillips, Martin, Dainty and Price (2008) extracted ten factors which could be used to differentiate best value based bids in the context of social housing: understanding of clients objectives, innovative managemet, successful track record, construction practices, quality management procedures, transparency of cost data, understanding of partnering, established policy, understanding of best value, and technical ability. These factors show both the social aspects and the product related aspects of partner selection in procurement.

The preference for value based award mechanisms is not the only current trend in Dutch construction. In order to reduce the extensive costs of errors during construction, government and other institutions in the field are encouraging the use of integrated design and construction contracts. In 2007 a sample of 3200 Dutch clients indicated that in the near future they would award about 35% of the contracts as integrated contracts (Design-and-Build, Design-Build-Maintain etc.) (Regieraad-Bouw, 2005). Research conducted in Australia suggests that the culture of risk avoidance still leads clients to choose traditional procurement methods, even if alternative forms could improve the project outcome (Love, Davis, Edwards, & Baccarini, 2008). Private organisations such as housing corporations and large corporations prefer trust and reciprocal expertise, just as public clients twenty years ago, and often collaborate with the same contractors because of previous good experience (Regieraad-Bouw, 2005). This is not possible for public clients any more because of current tender obligations. The findings of Doree (1996) suggest that for municipalities the possibility of building relationships with contractors is a way to safeguard and to ensure the control over quality.

Recent numbers show that the number of innovative tender procedures in Dutch construction is still very limited (Stichting Aanbestedingsinstituut Bouwend Nederland, 2009). This could indicate hesitation in the field to prefer integrated contracts over traditional procurement methods. Phillips et al. (2008) summarises the challenges that public clients face at the moment: too much procurement is undertaken without professional support or by designated procurement staff, there is too much focus on copying best practices, problems occur in creating consensus among stakeholders, and the multiplicity of values, tools, disciplines and stakeholders creates too many options. Yet, a shift occurs in the procurement function from its traditional adminstrative and transactional role to a more strategic one (Sporrong & Bröchner, 2009). Therefore skills and professionalism of clients are becoming more important and procurement strategies should be included in policies of governmental authorities. Sporrong & Bröchner (2009) found that clients experience a high level of uncertainty about how tenders should be evaluated. In this sense we should be aware of consequences of putting a strong focus on the role of the client in changing culture in construction. Ivory (2005) indicates that clients might even suppress innovation, mainly because of their desire to avoid both short and long term risks. Project management tends to take over the perspective of the project outcome. This also shows in the kind of selection and award criteria that are currently used by clients during tenders.

Hardly any research addresses the decision criteria for architects in the context of procurement regulations. Day and Barksdale (1992) analysed how business clients select and evaluate a professional service firm in design and found four underlying dimensions: 1) perceived expertise, experience and competence of the provider, 2) the provider's understanding of the clients needs and interests, 3) the provider's relationship and communication skills, and 4) the likelihood of the provider conforming to contractual and administrative requirements. Together with the actual performance these factors also contributed to the satisfaction or dissatisfaction of the client about the services that were delivered. Nowee (2008) indentified people, process and profession related competences in his exploration of a past performance system for architects. Cheung, Kuen and Skitmore (2002) found that in Hong Kong professional qualification and experience, availability of qualified personnel, present workload, quality of work and consultant fee are important criteria for the selection of the services of architects. The research was based on four categories to measure the suitability of architects, namely the background of the firm, past performance, capacity to accomplish the work and project approach. These categories appear to relate to the financial and technical abilities as mentioned in EU procurement regulations (see section 4.4). Cheung et al. (2002) used the results of their study to provide input for an 'architectural consultant selection system' to support clients in selecting an architect. Unfortunately the response to the survey was not very high. It would be interesting to explore the criteria used by Dutch clients in such a matter as is already common in the selection of contractors.

4.4 The legal context

Awarding contracts is the third origin of current processes of architect selection. Because the Netherlands is part of the European Union, public clients wanting to award a contract also have to comply with (European) procurement law. This section summarises the most important directives applicable to architect selections and uses of the official terms as much as possible. In the rest of the section the Dutch situation is shortly explained following by a discussion of some case law.

4.4.1 EU Procurement law

"Procurement refers to the function of purchasing goods or services from an outside body" (Arrowsmith, 2005, p. 1). Procurement regulations are directed at safeguarding business connections between government and market parties. According to the UK perspective authorities should act carefully towards citizens and entrepreneurs and create "best value for taxpayers' money" (Arrowsmith, 2005, p. 4). This means that public clients always have to comply with procurement law but that only in specific cases they have to comply with European procurement law and organise a public tender. This research focuses on these kinds of situations from a Dutch perspective. On 13 March 2004 two European tender directives were passed: the Directive 2004/17/EC coordinating the procurement procedures of entities operating in the water, energy, transport and postal services sectors, and the Directive 2004/18/EC on the coordination of procedures for the award of public works contracts, public supply contracts and public service contracts.

The principles for awarding contracts are stated in article 2 of Directive 2004/18/EC as "Contracting authorities shall treat economic operators equally and non-discriminatorily and shall act in a transparent way" (European Parliament & Council of the European Union, 2004). Architectural services, engineering services and integrated engineering services, urban planning and landscape engineering services, related scientific and technical consulting services, technical testing and analysis services all belong to category number 12 or Common Procurement Category reference 867 of the Directive 2004/18/EC. Every European public commissioning client is obliged to hold a European tender procedure for services above certain threshold amounts. Until the end of 2009, these amounts were € 133.000 for services for central government authorities (€ 125.000 in 2010), and € 206.000 for services for other government bodies such as provinces, municipalities and other public institutions (€ 193.000 in 2010). Below these thresholds the principles of the European Treaty (equal treatment, transparency, proportionality, mutual recognition and confidentiality) will have to be taken into consideration. Articles 28-34 describe the procurement procedures that a contracting authority can choose from. In general cases authorities can choose between an open and a restricted procedure. The negotiated procedure or competitive dialogue are only allowed in specific circumstances. Articles 66 through 74 enable the organisation of a design contest for the award of a service contract, such as the services of an architect.

"A tender is a procedure in which several parties are invited to apply for a contract" (Essers, 2009, p. 17). In Article 1 of the Directive the procedures are defined as:

a. 'Open procedures' (in Dutch: 'openbare aanbesteding') means those procedures whereby any interested economic operator may submit a tender.

b. 'Restricted procedures' (in Dutch: 'niet openbare aanbesteding') means those procedures in which any economic operator may request to participate and whereby only those economic operators invited by the contracting authority may submit a tender.

c. 'Competitive dialogue' (in Dutch: 'concurentiegerichte dialoog') is a procedure in which any economic operator may request to participate and whereby the contracting authority conducts a dialogue with the candidates admitted to that procedure, with the aim of developing one or more suitable alternatives capable of meeting its requirements, and on the basis of which the candidates chosen are invited to tender.

d. 'Negotiated procedures' (in Dutch: 'gunning via onderhandeling met of zonder aankondiging') means those procedures whereby the contracting authorities consult the economic operators of their choice and negotiate the terms of contract with one or more of these.

e. 'Design contests' (in practice often referred to as design competition; in Dutch: 'prijsvraag' of 'ontwerpwedstrijd') means those procedures that enable the contracting authority, mainly in the fields of town and country planning, architecture and engineering or data processing, to acquire a plan or design selected by a jury after being announced as competition with or without the award of prizes.

Chapter 6 of the Directive describes rules for the publication of the notices for tender, time limits, informing candidates and tenderers, communication and reports of decisions. The contracting authority is required to place a notice in the Official Journal of the European Union. The website http://ted.europe.eu acts as a digital supplement to this journal, www.aanbestedingskalender.nl provides this function for the Netherlands. The invitation to submit a tender must contain at least an invitation to participate, a reference to the contract notice published, the deadline for the receipt of the tenders, the address to which the tenders must be sent, the language or languages in which the tenders must be drawn up, a reference to any possible adjoining documents to be submitted, the relative weighting of criteria for the award of the contract or, where appropriate, the order of importance for such criteria, and, if they are not given in the contract notice, the specifications or the descriptive (explanatory) document.

In Chapter 7 of the Directive the conduct of the procedure is described. A restricted procedure consists of two phases: a selection phase and an award phase (see Figure 4.1). In between the selection phase and the award phase tenderers have to prepare their tenders. This is sometimes called the tender phase (Architectuur Lokaal, 2009a). In the selection phase all suppliers can submit their request to be invited for a tender. In the contract notice, the contracting authorities shall indicate the objective and non-discriminatory criteria or rules they intend to apply, the minimum number of candidates they intend to invite and, where appropriate, the maximum number. In the restricted procedure the minimum amount of candidates to be invited for the award phase shall be five. During the selection phase authorities first have to verify the suitability of participants. Then they have to select the candidates qualitatively based on selection criteria. These criteria can relate to the:

- Personal situation of the candidate or tenderer.
- Suitability to pursue the professional activity.
- Economic and financial standing.
- Technical and/or professional ability.
- Quality assurance standards.
- Environmental management standards.
- Additional documentation and information.

During the award phase the contracting authority selects a winner among the candidates that were considered suitable to award the contract to. These candidates are called tenderers. In an open procedure the selection and award phase are combined. This means that the submitted tenders are evaluated on their suitability and their quality in the same deliberation.

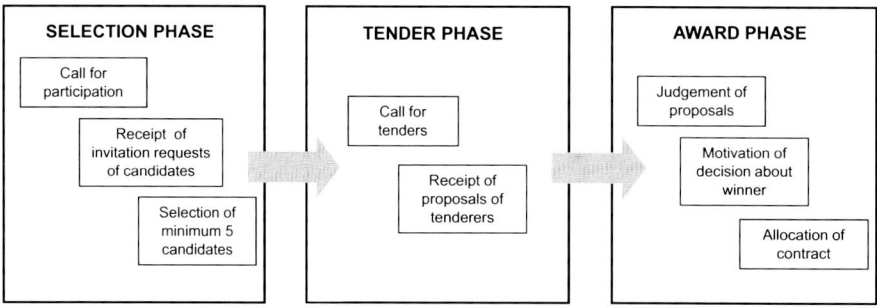

Figure 4.1 Phases and activities of a restricted tender procedure

The contract can be awarded based on 1) the most economically advantageous tender (MEAT) from the point of view of the contracting authority, and various criteria linked to the subject-matter of the public contract in question, for example, quality, price, technical merit, aesthetic and functional characteristics, environmental characteristics, or (2) the lowest price only (article 53). The contracting authority has to specify the relative weighting which it gives to each of the criteria chosen to determine the most economically advantageous tender. Those weightings can be expressed by providing for a range with an appropriate maximum spread. Where in the opinion of the contracting authority, weighting is not possible for demonstrable reasons, the contracting authority shall indicate in the contract notice or contract documents or, in the case of a competitive dialogue, in the descriptive document, the criteria in descending order of importance.

For a design contest similar rules apply for communication, notices and transparency as for the other procedures. Design contests can be organised as part of a procedure leading to the award of a public service contract, or as contests with prizes and/or payments to participants (article 67). The contracting authority also has to lay down clear and non-discriminatory selection criteria if participation is restricted to a limited number of participants. In any event, the number of candidates invited to participate shall be sufficient to ensure genuine competition (article 72). The jury has to be composed exclusively of natural persons who are independent of participants in the contest (article 73). Where a particular professional qualification is required from participants in a contest, at least a third of the members of the jury shall have that or an equivalent qualification. Article 74 describes the decisions of the jury:

1. The jury shall be autonomous in its decisions or opinions.

2. It shall examine the plans and projects submitted by the candidates anonymously and solely on the basis of the criteria indicated in the contest notice.

3. It shall record its ranking of projects in a report, signed by its members, made according to the merits of each project, together with its remarks and any points which may need clarification.

4. Anonymity must be observed until the jury has reached its opinion or decision.

5. Candidates may be invited, if need be, to answer questions which the jury has recorded in the minutes to clarify any aspects of the projects.

6. Complete minutes shall be drawn up of the dialogue between jury members and candidates.

The main differences between a design contest and an open or restricted procedure are the anonymous examination of plans and the autonomy of the jury panel. Organizing a design contest can be considered as conferring a favour upon the field of architecture (Heynen, 2001). The plans are examined by acknowledged experts but because of anonymity there is no real dialogue with the jury. The open and restricted procedures seem to require less formality than the design contest. In situations with less expected opposition, the open procedure is sufficient because there are not so many suppliers available and the time span is relatively short (Kennisportal Europese Aanbesteding, 2009). A procedure without any restrictions offers young and less established firms the opportunity to join. A client can explore concepts within a limited amount of time and money in such a procedure (Heynen, 2001). A restricted procedure gives the client the opportunity to ask a selected number of suppliers for a more detailed proposal based on more specific or sensitive information and to interact with the candidates. In case of a detailed proposal the client should offer financial compensation for the work of the candidates. Because of this and the longer time span of the procedure, the costs for the organisation of the restricted procedure are often higher than for an open procedure. The restricted tender and the restricted design contest are seen by the architectural profession as most suitable to select a design service. Both these procedures include a pre-selection of the candidates, submitting a plan by the candidates, the opportunity to invite a selected number of candidates to inform them about the brief.

Contact moments between client and designer have to meet the EU ground principles of objectivity, transparency and non-discrimination. An independent supervisor of the communication process could support correct implementation of procurement law in situations of personal interaction (Chao-Duivis, 2008). In particular the competitive dialogue procedure opens up possibilities for interaction between the client and the tenderer. However, this procedure is only allowed in specific cases of complex or unique projects. Most of the architectural design projects do not fulfil these requirements. Results of an explorative study show that procurement law provides enough opportunities for contact between the client and the potential service provider in almost all procedures but that implementation of these opportunities requires professionalism (Chao-Duivis, 2008). Especially tender candidates are interested in contact with the contracting authority to understand the intention of the client (Heynen, 2001). Contracting authorities perceive these moments of interaction mostly as threats for a successful tender because a lot of imponderables exist (Chao-Duivis, 2008). A distinction needs to be made between static interaction to improve understanding by raising questions and providing answers, and a dynamic dialogue in which ideas are exchanged. Interaction during a restricted or open tender procedure can take place by:

- Written questions and answers.
- A joint information meeting.
- Presentation of a vision.
- Visit to a reference project.

In current tender practice the written questions and answers and the presentation of a vision (or sketch design) are the most popular. Sometimes information meetings are organised by clients but visits to reference projects hardly ever take place. The reasons for this remain unclear. It is self-evident that in situations of anonymous assessment of the tenders the interaction possibilities have to be adjusted to the chosen procedure, for example after the selection phase.

Table 4.2 provides an overview of project characteristics and the tender procedures for general service contracts, so without the competitive dialogue and negotiated procedures. A distinction is made between procedures with design components and with design ideas that include a vision about the future building only.

* assumes anonymous examination of the plans # assumes autonomy of the jury panel	Open procedure	Restricted procedure with presentation of vision	Restricted procedure with design proposal	Open design contest*#	Restricted design contest#	Ideas competition*#
Exemplary function with stakeholder interaction			+	+	+	+
Stimulation young talent	+			+		+
Complex and/or important location			+		+	
Exploration of concepts and possibilities				+	+	+
Specific project requirements needed		+	+		+	
Specified project definition available			+		+	
Interaction with participants desired		+	+		+	
Limited time and money available	+	+				

Table 4.2 Overview of project ambitions and characteristics in relation to possible tender procedures for general projects

This distinction is based on experiences from practice, options in procurement law and the publication of Heynen (2001).

4.4.2 The Dutch interpretation of procurement law

The Netherlands has a long history in the field of rules to invite market parties to compete. Chao-Duivis (2009, p. 21) describes that already in 1376 a tender was used to "select a carpenter for a bell cage in Gent", a former part of the Netherlands. From the second half of the 18th century onwards competitions were used to stimulate sciences (Evers, 1995). In 1815 the first procurement regulation was announced for the realm government at Royal Decree (van Romburgh, 2005). The aim of these rules was, just as they are now, an efficient management of resources and fighting corruption of civil servants. The first European rules for invitations to tender arose in the beginning of the seventies with the aim of reaching free, honest competition within the European Economic Community. The World Trade Organisation was launched in 1995 to officially liberalise trade, negotiate trade agreements and settle trade disputes, but the trade system is about half a century older (World Trade Organization, 2009).

In the Netherlands the mixture of rules and regulations originating from the 1980s and 1990s were in 1993 brought together in a national legislative framework. As a result of the parliamentary enquiry in the construction industry in 2001, the Dutch government decided to develop a common tendering regulation called 'het Aanbestedingreglement Werken' (ARW 2004, later replaced by ARW 2005). In reaction to the introduction of the two renewed European tender directives 2004/17/EC and 2004/18/EC of 13 March 2004 the Dutch directives were on 16 July 2005 further simplified to the one directive for the supply of product, services and works: the BAO ('Besluit Aanbestedingsregels voor Overheidsopdrachten'). The Utilities sector has its own directive, the BASS ('Besluit Aanbestedingen Speciale Sectoren'). The BOA and BASS are still leading for procurement in the Netherlands. In the Netherlands a need exists for a clear and uniform frame for Dutch contracting authorities in the public sector in order to stimulate innovation, integrity, open access for small and medium sized companies, and sustainability and reduce overhead. The bill that would provide this new uniform frame (in Dutch 'Aanbestedingswet') was rejected by the Dutch Upper Chamber in July 2008, mainly due to the strict frame and low level of the provisions in the proposed bill. In April 2009 a new concept of the law was presented for consultation to a number of parties and an advisory council. The current proposal of the 'Aanbestedingswet' has been approved by the Cabinet in November 2009 and is now (midst 2010) to be approved by the Council of State.

The relatively scare number of case law on tender regulations got a 'boost' when the EU came into being. For the understanding of procurement law court decisions are therefore very instructive, not to say essential. The decisions of judges shape the interpretation of the official regulations and provide input for debate. Based on a recent case about a library Evers (2010) for example questions in the context of architect selections the position and decision task of a professional jury panel and the actual meaning of a vision compared to a sketch design. Collins (1971) links architectural judgements to legal judgements because the principles of architecture are as meaningful and genuine as the principles of law. He (Collins, 1971, p.48) states that "in law and architecture any valid decision must depend on wider contexts: the context of history (which provides precedents), the context of society (which provides safeguards for the public with regard to the possible effect of any decision on those not immediately involved), and the context of the physical environment (which provides both a sense of place and the judicial guidelines of customary law)". Therefore history, society and the physical environment must be involved in the process of reasoning and evaluation about architecture. "And when an architect can enunciate his reasoning with the same clarity and precision as a High Court judge, he may feel assured that his judgement is professional in the noblest and most apt sense of the term." (Collins, 1971, p. 48). He further states that lawyers as well as architects are increasingly influenced by the growth of political sciences as an academic discipline. "By the study of legislative processes the lawyers try to control politics" (Collins, 1971, p. 64). Collins suggests that in architecture the concept of debate could perform this role of controlling politics and policy.

Until a few years ago Dutch architects did not, as contractors did do, go to court when decisions during the tendering phase were not acceptable. This practice has changed and more and more case law develops on tender procedures in which architects are the suing party (van Wijngaarden & Chao-Duivis, 2010a). Apart from this new body of case law there are also cases in other fields that are of use for tenders concerning architectural services. These cases show that in general judges are willing to take the specific nature of the work of the architect into consideration. The starting point is that in using the MEAT award criterion (the most relevant criterion with these contracts) contracting agencies have some room for discretionary decisions. Judges tend to easily accept these decisions. To give the reader an impression of the situation in the Netherlands several more or less representative cases will be described.

The District Court of 's Hertogenbosch stated on 8 February 2008 (168487/ KG ZA 07-822, LJN: BC3956) that in cases of intellectual qualities candidates and tenderers should accept a considerable amount of subjectivity, provided that chances for all candidates remain equal. On 2 April 2009 the District Court of Haarlem (154394 / KG ZA 09-79, LNJ: BH9497) acknowledged that awarding a contract of taxi services based solely on the criterion of quality is not easy to reconcile with the basic principles of objectivity, transparency and equal treatment, because of the subjective judgement required. Therefore the assessment system should be as objective as possible. The contracting agency is not required to describe in full detail how he expects the tenderers to show how they prove that they meet the required standards. The system should leave room for tenderers to ensure some level of competition and innovation. The tenderer has to show not only that, but also how a maximum score will be reached. According to the judge the contracting authority secured the basic principles of procurement law by the use of an independent committee of experts in the award phase. This committee first judged the proposals individually and then discussed the results in order to determine the final score.

The District Court of 's Gravenhage (332764/ KG ZA 09-336, LJN: BI8767) argued on 29 May 2009 that even if the motivation of a decision is not fully clear, the use of the award panel, consisting of nine people judging independently from another, implied a careful judgement. By supplying the scores on the award criteria and further explanation about the argumentations the contracting authority provides enough insight in the way the assessment was done during the award phase. At 16 February 2010 the District Court of Rotterdam (345682/ KG ZA 09-1364, LJN: BL4031) pronounced a claim for further explanation in an award matrix even not admissible because the jury had reported their motivation in a jury report and had acted according to the call for proposals. The element of independent members of the jury is considered an important one by Dutch judges, see e.g. the Appellate Court of The Hague, 8 February 2007, 06/1421 KG en 06/1430 KG, LJN AZ8670. In this case the jury consisted of several disciplines and the proposals were anonymous: these two aspects formed a sufficient guarantee that the jury would act in a non discriminatory manner. The District Court of Middelburg (13 february 2009, 65920/ KG ZA 08-244, LJN: BJ1373) confirms that announcing the weight of the sub award criteria is enough. A further breakdown of the criteria used is not required by the Dutch procurement regulation. In the same case the use of a presentation as one of the award criteria was deemed to be allowed. A presentation offers the opportunity to judge the level of performance in relationship to the requirements for the assignment that is being tendered.

It is essential that contracting agencies stick to the way they announced they would judge the offers. A mistake is easily made here, as is shown by the case judged by the District Court of Utrecht, 1 August 2006, nr 214609 (to be known from van Wijngaarden & Chao-Duivis, 2010a). In this case the contracting agency had announced that points would be awarded for specific qualities in the award phase. The actual awarding of the points however was based on the relationship of the presentations from the different tenderers and not on the qualities itself. This meant a forbidden change of award criteria. Another mistake which could haven been avoided played in the following case, decided by the Appellate Court of Amsterdam, 200.026.280, 4 August 2009, LJN: BK8538. In this case the members of the jury had judged the tenderers on the basis of eleven points with either 'positive' or 'negative' followed by a motivation. An employee of the contracting agency was supposed to translate these judgements into numbers after which the jury would check this. This last check had not taken place. The contracting agency therefore was not allowed to award the contract to the winning party, but was ordered to let the jury finish its work. In this same judgement the Court allowed the contracting agency to use wishes on the one hand and requirements on the other hand. Not living up to the wishes does not automatically mean that the tenderer is excluded from further participation in the procedure. The Court judged this to be acceptable in the award phase. However, this judgement has also been criticised.

From the perspective of decision making some characteristics of legal judgements can be read between the lines. In each of the above cases the judges argued that the value of the proposals can be seen in comparison to the other proposals. It also suggests that a proposal could be evaluated against other projects or the current context on a holistic level if the aims and context of these projects support each other in the implementation of the project. Most cases acknowledge that subjectivity exists, especially in the award phase when the economical most advantageous tender includes judgements about quality. The principle of transparency requires an objective assessment system that can be checked afterwards to ensure objectivity, not an objective topic of assessment (PIANOo-vakgroep Aanbestedingsrecht, 2009). The commitment of an independent committee of professional staff members or external experts will support the transparency of the assessment process, especially if the members first evaluate the proposals individually and then discuss about the average outcome. Van Wijngaarden and Chao-Duivis (2010a, 2010b) suggest a distinction between a consensus model and individual independent jurors in situations in which a jury panel decides about the winning tenderer. Justification of such a decision can be done by giving insight in the assessment process and the scores of the proposal in question. A matrix sheet with aggregated scores on the decision criteria is not required. Further explanation of the sub criteria and scores of the other proposals is according to case law not necessary as long as they can prove that they meet the required standards.

4.5 The economical context

From a legal perspective, the process of architect selection is considered as the purchase of a service for which procurement law applies. So far there has been no systematic research in the field of design tenders in the Netherlands. It is therefore difficult to draw a comprehensive picture of the economical context of architect selections which could indicate the amount and kind of tenders for the selection of architects. A number of documents published by different authorities and consultancy firms show divergent facts and figures about procurement in general. Without the financial characteristics of architectural design firms, it is hard to draw a picture about the relation of the organisational structure and the way their business is affected by the tender industry. In this section several sources are compared in the context of architect selections to provide a picture of the Dutch economic situation of architect selections.

4.5.1 Dutch market potential

According to the latest estimates the total purchasing volume of the Dutch Government was € 57.4 billion in 2007, of which € 38 billion was spent on supplies and services (Weijnen & Berdowski, 2009). A volume of € 10.3 billion of services had to be publicly tendered because it exceeded the monetary thresholds of € 2.9 billion by central governmental authorities and € 7.4 billion by local authorities. This means that more than 70% of service related tenders are organised by local contracting authorities, such as the provinces, municipalities, district water boards, or authorities other than the state. According to Dreschler (2009) the total turnover of architects and engineering firms was € 5.4 billion in 2005. About € 2.0 billion was spent on residential and commercial building projects, such as public administration offices (5.2%), education (9.9%) and health and welfare work (12.4%), while € 0.85 billion was spent on services for town and traffic planning. The Royal Institute of Dutch Architects estimated a total net turnover of the architecture branch of € 1.3 billion in 2006 (Vogels, Mooibroek, & de Vries, 2008).

A report by the consultancy firm Berenschot under the authority of the Ministry of Economic Affairs shows that there were about 6,000 national and European tenders in the Netherlands between June 2008 and June 2009 (Ruiter, et al., 2009). About half of these tenders were above the thresholds and one third included a tender for services. In their sample of 703 national and European tenders, 12.1% used the restricted tender procedure; 27% of the sample was tendered by large municipalities (over 100,000 inhabitants), 18% by small municipalities, 22% by central governments and 15% by provincial governments. According to Essers (2009) the central government complies with procurement law better than the municipalities, and large municipalities comply better than smaller ones. He also mentions that the allocation of contracts for works and supplies complies more with procurement law that the allocation of service contracts.

Tenders for building projects show better results than tenders for infrastructural works because contracts are most often awarded based on the economically most advantageous offer and a restricted procedure. Clients in such tenders prepare themselves better, and with more transparent requirements fewer lawsuits are needed (Koenen, 2009). There is a noticeable need for clear procurement policies and guidelines for implementation in the Netherlands. Many mistakes are made and Dutch clients often do not fulfil their legal obligations. About 56% of the contracting authorities did not publish their short listing system in a restricted tender and about 48% did not announce the weight of the criteria (Ruiter, et al., 2009). The same report gives an indication of the requirements that clients ask of the tender candidates and the relation to the final award decision. It shows for example that in 89% of the service tenders on average 2.5 reference projects are asked to select tenderers. The final decision for selection appears to be determined for 68% by the assessment of these reference projects. Other important requirements were related to quality measurement (11%), personnel (8%) and technical capabilities (7%).

4.5.2 Tenders for architect selection in the Netherlands

Most Dutch clients require candidates for a tender in architectural design to be a registered architect. Compared to doctors and lawyers who developed their profession in the late Middle Ages, architects have a relatively recently established profession (Mieg, 2006). A growing body of abstract knowledge, a justifiable reference to a social core value, formal academic training and a national professional association are phenomena that are connected to professionalization. In the Netherlands the title of Architect is protected. In 2009, approximately 13,000 people were registered as architects, urban planners, interior designers or landscape architects in the Dutch Architect Register SBA, but not all these people practice their profession (Stichting Bureau Architectenregister, 2009). Approximately 10,000 of them are architects. The BNA is the professional organisation of architects in The Netherlands and has about 3,000 individual members and 1,500 firm members. There is an increase of the number of architects in the Netherlands: in 2007 about 850 new persons registered in the register while around 300 had their names removed (Stichting Bureau Architectenregister, 2007). In 2007 about 2,850 architectural firms were registered with the Dutch Chamber of Commerce (Vogels, et al., 2008). About 1,500 of these firms were one-man businesses. A report of Senter Novem shows that in the Netherlands about 60 architectural firms have 50 employees or more (Senter Novem, 2009). According to the same report worldwide there are only 100 architectural design firms that have more than 100 employees, of which 5 are Dutch.

The Royal Institute of Dutch Architects annually surveys Dutch architectural firms on their organisational characteristics. About 170 Dutch architectural firms participated in a periodical comparison about the year of 2007 (van den Hurk, 2008), 443 participated in the same kind of research about 2008 (BNA, 2009). This is only a small portion of the total number of their firm members but not every firm is willing to participate in these kinds of research. Results show an average yearly net turnover of € 512.417 in 2007 with a net profit of 9.2% and € 610.000 in 2008 with a net profit of 14.7% per firm. Only 67% of all Dutch architectural firms is profitable. Especially large firms (over 40 FTE, average net turnover € 7.4 million) and the very small firms (0-2 FTE, average net turnover € 59.000) were profitable in 2008. The average rate per hour in 2007 was € 72 for design work and the average net turnover per FTE was € 81.142, which seems to increase with the size of the firm. In general the number of employees of an architectural firm increased from an average of 4,5 FTE in 2004 to 6,6 FTE in 2007 and 7,1 FTE in 2008. This means that the size of the average architectural firm increased over the past years but not every employee of an architectural design firm is an architect. The clients of architectural design firms mainly consist of private persons, project developers, business relations and housing corporations.

Governmental authorities were responsible for only 8% of the turnover (van den Hurk, 2008). About half of the turnover in 2007 was earned in the housing segment. Offices, health care and education create respectively 11%, 7%, and 4% of the yearly turnover. This image was about the same for 2008, but for 2009 a major decrease of turnover is expected (van den Hurk, 2009).

The portfolio of a firm shows the kind of projects they are involved in. For many years half the portfolio of architectural firms has consisted of complete commissions (from initialization to realization) and half of partial assignments (Teunissen, 2009) but the amount of partial assignments seems to be increasing (Senter Novem, 2009). The portfolio of architects still consists mainly of traditional design-bid-build contracts. Small firms generally do not participate in the public tender market as they are usually not able to fulfil the financial selection requirements (Vogels, et al., 2008). In 2006 22% of the commissions were awarded in competition. Only a very small amount of the income in 2007 seems to be obtained by design competitions (2%) and European tenders (0,5%) (Vogels, et al., 2008). According to the results of the periodical BNA research, this increased in 2008 into 5% of the commissions from governmental tenders and 3% from design competitions by governmental authorities. Architectural firms have many small commissions. Almost half of the new commissions in 2006 were lower than € 5.000, while only 4% exceeded € 250.000. Especially project developers, housing corporations and governmental authorities allocate contracts above € 250.000. In 2007 61% of the contracts were based on a fixed price agreement, 17% based on a percentage of the building costs and 19% on an hourly rate (Vogels, et al., 2008). Only a few Dutch architectural firms operate internationally (Senter Novem, 2009).

To draw a picture of the total costs of tendering in relation to the volume of trade, the costs for the organisation and joining of a tender project needs to be brought in mind. Based on data of Sira Consulting, the costs to an organisation for a restricted procedure are estimated at € 17.500 and the open procedure at € 15.000 (Essers, 2009). According to the same data a restricted procedure will cost about € 47.000 for a tenderer, an open procedure costs € 38.500. Estimated total costs in 2008 for participation in an EU tender were about € 25.000 according to the BNA (Vogels, et al., 2008), but their members estimated the costs in 2009 at € 45.000, which implies a significant increase in the costs for tenderers. These data do not include the costs of candidates that are only involved in the selection phase, which could be estimated several thousand Euros per candidate with about 50 to 100 candidates per tender. The BNA estimated the total costs for tenders in their sector at € 6.5 million (based on 260 tenders in 2006) with an estimated total sum of commissions of € 99 million. This implies that costs for tenders are about 6% of the potential income by tenders. Taking into account the net turnover of the architecture branch (€ 1.3 billion as estimated by the BNA in 2006) the costs are about 0,5% to 8% of the branch turnover.

There are two main sources about the amount of tenders for architect selection since the introduction of the Dutch procurement directives BAO and BASS in 2005: a report of Kroese, Meijer and Visscher in commission of the Chief Government Architect in 2008, and the numbers of the 'Steunpunt

Architectuuropdrachten en Ontwerpwedstrijden', abbreviated as the 'Steunpunt' that supports public clients in organizing tenders and competitions in the sector of architectural design. According to Kroese et al. (2008) 204 tenders and 7 design contests were announced for architectural design contracts in the period of January 2006 till October 2008. About 50% were from municipalities, 20% from educational institutions and 1% from the ministries. Most of the tenders contained educational buildings and governmental offices. 68% of the tenders were guided by consultants. Only 111 notifications were made on the award decisions. The contracts were awarded to 74 different architectural firms of which 61 with more than 10 employees and an average of 39 employees (Kroese, et al., 2008). Tenders are thus awarded by competition to a limited share of the market (maximum 5%).

The Steunpunt analysed the publication of all Dutch tenders with a design component in the period of 16 July 2005 till 1 November 2009 (Geertse, et al., 2009). In total there were 388 tenders for architectural services and 183 for development competitions announced in this period. 89% of these tenders were EU restricted tender procedures (see Table 4.3). The numbers of restricted tenders continue to increase; until 1 November 2009 there were already 118, compared to 80 in 2006. The latest insight for 2009 show a total amount of 140 tenders for architectural services (88% restricted tender procedure) and 48 for integrated contracts and development competitions which include design (65% restricted tender procedure) (Geertse, 2010).

A local government acted as contracting authority for about half of the tenders, the central government for an average of 7% (Geertse, et al., 2009). 55% of all contracting authorities were municipalities, 6% central governments, 3% provincial government, 15% educational institutions, 4% housing corporations. The distribution of projects over the different building segments is fairly consistent. Most of the projects (average 65%) concerned social real estate (school buildings, sports facilities). City halls belong to the category of governmental offices (12%). Figure 4.2 shows the tenders for architects per building sector for 2009.

Large commissions are important for firms because they contribute to the stability of the firm. Larger projects in architecture, especially buildings that are used by a lot of people or situated at a central location, add important value to the portfolio and reputation of a firm. Although project developers, housing corporations and private business also commission these kinds of buildings, governmental authorities represent an important group of clients. Taking into account that most

	2005	2006	2007	2008	2009	Average %
Restricted EU procedure	19	73	60	85	104	89%
Open EU procedure	1	4	2	9	3	5%
Restricted NL procedure	0	2	3	1	1	2%
Open NL procedure	0	1	8	2	10	4%
Total	20	80	73	97	118	388

Table 4.3 Numbers of architectural design services tenders from 16-07-2005 till 01-11-2009

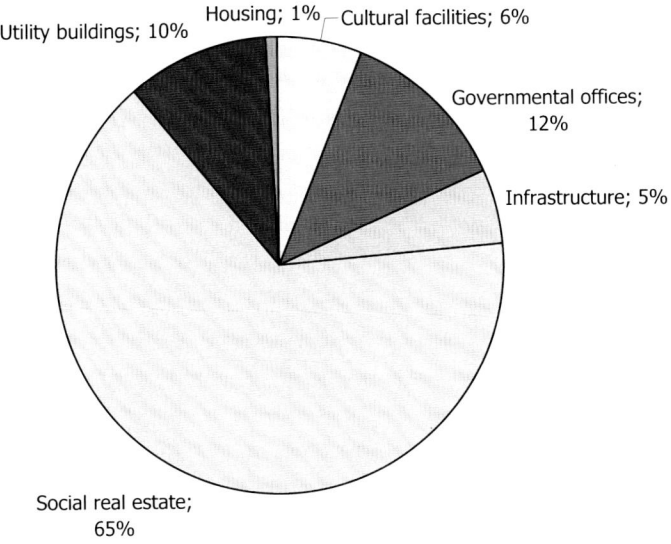

Figure 4.2 Distribution of public tenders for architects in
different building sectors in 2009 (based on Geertse, 2010)

of the contracting authorities for these projects are local, the monetary threshold
for a tender was until December 2009 € 216.000. This is about half of the annual
turnover of an average architectural firm and excludes about 70% of the firms.
Participating in the award phase of a tender procedure costs an architectural firm
about 5 to 10% of their annual net turnover. For large firms this is only 0.01% on
their net turnover. These numbers demonstrate that tenders are presumably more
feasible for large firms than for the average architectural firm. Because at least five
firms participate in a restricted tender the total costs (of clients and architects)
probably are between € 100.000 and € 250.000 per full tender. Recent numbers
show around 190 restricted tenders a year for buildings and area developments
(Geertse, 2010), which implies a total of € 18 to 45 million per year spent in
Dutch society on architect selection tenders.

   In 2009 about 60% of the tenders were organised with support of consultants
who acted as contact persons. In total 103 different consultancy firms were active
in the field of tenders to select architects (Geertse, et al., 2009). Two firms were
identified as market leaders with 24 and 31 tenders each over a period of 4 years,
ten of them organised at least 3 tenders a year. 275 of a total of 388 announced
the winning firm of the contract. In total 180 architectural firms were awarded
a contract through a tender in the past five years. This is about 6% of the total
population of about 3,000 firms as registered at the Chamber of Commerce. In
the past 5 years 10 architectural firms got awarded 6 or more tenders. These 10
firms won about one fifth of all tenders in the past 5 years.

   The average requirement for net turnover of the firms to participate in the restricted tenders was € 862.500 for the period of 2006 to 2008. This is about 1.5
times as much the average net turnover of all architectural firms in the Netherlands
in this period. This can be considered as relatively high compared to the annual
turnover of an average architectural design firm and excludes about 85% of the

market. The lowest requirement was zero, the highest € 2 million. The requirements appear to relate to the size of the project in gross square meters. In about 70% of the tenders required reference projects that were realised more than three years ago while the Directive talks about a maximum of three years ago. Only 25% allowed projects that have not yet been built. Approximately 70% of the tenders required 3 or fewer reference projects.

The data of the Steunpunt are based on publicly available information. This means that not all of the 388 tenders that were identified by the Steunpunt were included in the analysis because of a lack of information. Other analyses, such the ratio between fee and net turnover, the number of participants per tender, the kind of firms that were selected for the award phase or the award criteria, were not possible either because of missing data. At the moment it is therefore not possible for any institute or governmental agency to provide evidence based recommendations for the level of requirements of other kinds of standards that need sufficient benchmarking.

4.6 Current practice in the Netherlands

The previous sections described the administrative, cultural, legal, and economical context of architect selections. The current practice displays a scattered and dissatisfying image of how architects are currently selected in the Netherlands. There is clearly a substantial need for change. The different players in the field all have their own interpretation of the fundamental weaknesses of the tender system and the implementation capacities of public commissioning bodies. Solutions are sought in different directions but no real action has been undertaken. In this section the perceptions and expectations of the players in the field are addressed and existing supporting tools are briefly analysed to create an overview of current practice.

4.6.1 Perceptions and expectations

Despite the financial problems at the moment in the real estate market recent numbers show an increase in the number of tenders for architectural services in the Netherlands (Geertse, 2010; Geertse, et al., 2009). This increase cannot be explained only in terms of increased purchasing volume of governmental authorities, but also shows the impact of the implementation of the new regulations starting since the introduction in 2005. Increasingly the misuse of tender regulations was exposed in the media. Because architects had previously not felt the need to go to court, little case law was available for the interpretation of EU procurement law in specific cases of architect selections. In December 2008 a first meeting about tendering processes was organised in the Netherlands by Architectuur Lokaal in collaboration with the Ministry of Housing, Spatial Planning and the Environment (VROM), the Royal Institute of Dutch Architects (BNA), the Association of Dutch Project developers (NEPROM), and the Association of Dutch Municipalities (VNG) to discuss the problems around architect selection in the context of procurement. They searched for solutions but did not fully agree about the problem or solution space yet.

At the follow-up meeting in December 2009, the director of Architectuur Lokaal Cilly Jansen described the essence of the problem as "a confrontation between the need for certainty and the desire for creativity" (van der Pol, et al., 2009, p. 10). This certainty versus innovation dilemma was also stated by Rönn (2008) for design competition the European Nordic countries. In the early days clients selected a firm that they could trust. In procurement situations they do not know what and who they can expect to react on their request for invitations. In order to reduce the number of potential participants and decrease uncertainty they issue higher financial and technical requirements. However the selection of an architect is based on judgements about intellectual capacities and creativity, which is not easily assessed by financial and legal criteria.

The financial thresholds for projects to be publicly tendered are relatively low so almost all of the interesting projects for the sector require official tendering. One of the key problems with the EU procedures is the cost and work involved in administering the invitation (Strong, 1996). Response rates can be high. A client organisation may be required to send out a few hundred selection manuals even when it intends to select only a few service providers to tender. Even when this process of invitation and documentation is digitised, the amount of administrative work is high. The candidates need to submit substantial paperwork to show their capabilities. If they are asked to submit a design proposal in the award phase this means even more workload and investment (see section 4.5.2 for an estimation of the costs involved in tendering). An overview of the most pressing issues is given in Table 4.4, which is based on several publications (Architectuur Lokaal,

Selection phase
- Jumble of guidelines
- Unclear selection criteria
- Too high qualification requirements
- Too many suitability requirements
- Design activities during selection
- Too many candidates are selected for the award phase
- Careless requests to participate from the candidates

Tender phase
- Missing, too ambiguous, or too strict briefs and ambition documents
- Little or no financial compensation for design activities
- Unrealistic building budgets from clients
- Little or no interaction between tenderer and client
- Lack of professionalism during interaction with other tenderers
- Delay during the procedure

Award phase
- Unclear or not well-considered award criteria
- Indistinct user and citizen participation
- Lack of political support
- Conditional offers
- Too much or too little work done by the tenderers
- Incomplete cost calculations
- Mixture of politics and procurement
- Ambiguous or incomplete motivation of the award decision
- No official announcement of the winner
- Negotiations after announcement of the winner

Table 4.4 Overview of most pressing issues during tenders in architecture

2009a; Atelier Kempe Thill, 2008; Kroese, et al., 2008; Postel, 2001; van der Pol, et al., 2009).

The selection of architects via a tender procedure still poses difficulties for contracting agencies and more case law can be expected. Based on such case law (see also section 4.4.2) contracting agencies should be able to limit the risk of a dispute by: 1) using independent jury members and jury members who are properly qualified, 2) following the procedure the way it was announced, 3) using unambiguous language, and 4) motivating their judgements. Current experiences in practice as well as judgement and decision making theory suggest that in all four possibilities to limit the risk of potential conflicts arise in the differences between expectations from a legal perspective and a behavioural perspective. It is, for example, very hard to find independent jury members in a relatively small professional community such as architecture. The concept of sensemaking suggests that decision processes do not always go as one expects and that priorities may shift during the process. The subjective and intangible characteristics of architectural quality make the services of architects ambiguous by nature with multiple perspectives and interpretations, and motivating decisions is usually difficult when intuition is used in making judgements about the quality level of designs.

The underlying difficulty of architect selections appears to be the ambiguity in the actual aim: Is it the selection of a product (a design), the selection of a service (design activities) or the selection of a partner (a designer or a firm)? The unconscious mixture of these different aims causes inconsistencies in the design and execution a tender, as also addressed by Evers (2010). Current discussion in architectural practice mainly focuses on the level of requirements for selection and judgement about the qualification of tender candidates. An appropriate answer to the question of suitability of professionals has not been found. Some suggest using other procedures more often, such as the design contest or competitive dialogue. Others talk about lowering or balancing the financial requirements, or adjusting the way reference projects are used (Architectuur Lokaal, 2009a; van der Pol, et al., 2009). The majority of tenders ask for reference projects, which shows the emphasis on experience. However, often these projects need to be realised in the last three to five years. Because an architect only delivers the design and not the construction, a gap exists between the date the design service was delivered and the building was established. Next to that, designs are sometimes not realised or delayed for external reasons, especially if architects participate in an ideas or design competition.

Another important issue during selection appears to be the organisational structure and size of architectural design firms. In current practice clients appear to perceive that the suitability of tender candidates increases with the structure and size of a firm. An architectural design firm can be considered as a professional service firm. Architectural firms often choose to remain a small or medium sized enterprise to preserve creativity. This means that the character and structure of a professional service firm do not match the economic principles of size and large turnovers to assure the reliability of a tender candidate. The style of an architect tends to be derived from the level and kind of design quality established in previously delivered buildings (van Eldonk, 2008). On the one hand, large firms with established designers have typically acquired a great amount of experience and therefore suitability to deal with any accommodation problem in a public context. Experience on the other hand is not the same as the ability to design and offering of the required expertise to deal with specific complexities in design. Architects are proud of their capabilities to analyse complexities on an abstract level in order to find solutions beyond the obvious. They deliver intellectual services. In this sense expertise matters. However, it seems that expertise can be built by experience but is not guaranteed by it. Design excellence and innovation is therefore not always directly related to experience. Some architects might even suggest the opposite. The suitability of candidates should therefore not always be related to experience or size of the company. In order to change the current practice one of the suggestions for clients is to make a distinction between innovative projects that require 'naïve' entrepreneurship and fresh creativity, and classical projects that need certainty in order to distinct the kind of suitability a design project requires (van der Pol, et al., 2009).

The report by Kroese et al. (2008) and report of the tender meeting in December 2008 (Architectuur Lokaal, 2009a) show different solutions and perceptions of the problems that are currently experienced. The architects, clients and consultants do not seem to trust each other and often reproach each other for the current course of events. (Former) Chief Government Architects mainly blame the consultants for increasing the complexity of the procedures. Consultants point their fingers towards the clients and talk about cultural differences between the disciplines and the lack of spirit and innovation of the clients they represent. According to the consultants and architects clients should prepare themselves better, both in relation to their ambitions and with respect to their procedures (Architectuur Lokaal, 2009a; Hijdra, 2007). The use of a generally acknowledged model to structure the tender procedure might even be required. The general quality of the selection and awards committees has been criticised and architects are also complaining about the fact that hardly any feedback is given to the tenderers about their decisions. High levels of requirements in the selection phase might decrease risk for the client but prevent new parties from entering the market. In the end this could lead to a poor architectural culture. Clients suggest that participants contribute to the lack of trust themselves by communicating about their displeasure through the court instead of with the authorities themselves (Architectuur Lokaal, 2009a). Clients suggest that architects could also generate ideas for selection procedures that would lighten the burden in running tender projects instead of blaming them for avoiding risks (van der Pol, et al., 2009). All actors aim at simplified procedures with mutual trust and room to realise shared dreams. Good practices should show the potential of tenders instead of the negative consequences and pitfalls.

The current procurement regulations provide enough room to enable clients to fulfil the aims of their projects (Kroese, et al., 2008). A more qualitative procedure is surely legally possible, just as interaction to create better understanding between the actors (Chao-Duivis, 2008). An advantage of the fact that clients have to comply with procurement law is the increased transparency of decision making and the decrease in importance of the old-boys network, which should create chances for new firms and ensure a fair level of playing field. Several suggestions have been made to improve current practice in the selection phase of the tender, such as only asking for proportional and reasonable requirements, allowing unrealised projects or projects of a different sector as reference projects, a databank with references and exhibits, minimizing the number of required documents, and the use of personal statements ('eigen verklaringen') about the financial and organisational requirements (Atelier Kempe Thill, 2008; Kroese, et al., 2008). Other suggestions relate to the inclusion of experts in the jury panel or award committee, the role of citizens and users in the tender phase, the use of digital means for communication and evaluation of the requests for invitation and tender proposals, and wild cards for 'young' firms (Atelier Kempe Thill, 2008; van der Pol, et al., 2009).

Several authors suggest establishing an independent tender authority to change the current practice. The Chief Government Architect could play an important role in this authority because (s)he and the staff of the Atelier Rijksbouwmeester already support a number of central governmental bodies in architect selections. The Steunpunt of Architectuur Lokaal is also often mentioned as candidate for a national expertise centre. Examples are taken from neighbouring countries such as Belgium, Germany and France. During 2009 several complaints and service desks were (re)established by the Dutch professional organisations, such the BNA, Architectuur Lokaal and NEPROM. These initiatives are in addition to the existing Dutch Public Procurement Expertise Center PIANOo and the Procurement Institute for Construction and Infrastructure called 'Het Aanbestedingsinstituut', which mainly focus on tenders for works, supplies and other services rather than architect selections. In December 2009 the Atelier Rijksbouwmeester, BNA, NEPROM and VNG announced their support for the use of a new tender model for architect selections, the 'Kompas Light' (Architectuur Lokaal, 2009b), which would replace all the previous guidelines. Their aim is to stimulate all other kinds of institutions, consultants and public clients to use the same model and build a new standard.

4.6.2 Models and guidelines

In an attempt to support public commissioning clients several models, guidelines, regulations and project management guides have been developed. Table 4.5 provides an overview of the steps that need to be taken in the process of the preparation and realization of a restricted tender (e.g. Arrowsmith, 2005; Chao-Duivis, et al., 2007; Essers, 2009; Heynen, 2001; Pijnacker Hordijk, van der Bend, & van Nouhuys, 2009; Spreiregen, 1979; Strong, 1976).

Further analysis of the publications about tenders in the field of architecture shows that the publications can be divided into five categories:
1. Procurement methods and models.
2. Competition models and guidelines.
3. Decision support and navigation systems.
4. Project management tools and methods.
5. Tips & recommendations (often in combination with other categories).

Preparation
a. Form a project team
b. Explore the market
c. Determine if procurement law applies
d. Decide about the tender procedure and planning
e. Formulate a tender intention announcement
f. Decide the object of the tender and contract
g. Determine the ambition and brief
h. Determine the decision makers
i. Decide the criteria and method for the selection
j. Decide the criteria and method for awarding the contract
k. Decide the rules of the tender

Realisation
i. Announce the tender
ii. Send out call for participation
iii. Receive and process candidates' credentials
iv. Assess and select the candidates
v. Issue the call for tenders
vi. Receive and structure the tender proposals
vii. Structure and assess the proposals
viii. Judge the proposals
ix. Allocate the contract preliminary
x. Justify the decision
xi. Final allocation of the contract
xii. Publish the contract allocation
xiii. Follow up and quality control

Table 4.5 Overview of steps to take during a restricted tender in architecture

These categories of support tools for the organisation of a tender in architectural design show the multi-faceted nature of a tender project. Categories are often combined in the publications, for example the models with the tips and recommendations, or the competition models with the procurement models.

The publications on procurement methods and models are written from a legal perspective. They describe the possibilities as well as the impossibilities that procurement law offers and sometimes also some tips and tricks. Most of these models arise from the publication of the EU Directive in 1994; some were made especially for architect selections. The most important Dutch examples at the moment are the 'Leidraad Aanbesteden' (Jansen, 2009) for tenders in construction, and 'Kompas Light' (Architectuur Lokaal, 2009b). The Kompas Light is an appropriate example of a combination of a tender model with a competition model for architect selection.

The competition models are often written by architects or other professionals in the field of design and developed by governmental bodies. These guidelines and models often derive from competition regulations in 19th and 20th century and include recommendations to support a decision. International examples are 'Participating in Architectural Competition (Strong, 1976), 'Design Competitions' (Spreiregen, 1979), 'Model for Running Design Competitions' (Nasar, 1999), 'Qualifications–Based Selection' (Ontario Association of Architects, 2008) and 'Overheidsopdrachten architectuur' (Heynen, 2001). Examples for the Dutch context are the 'Kompas' for design competitions (van Campen & Hendrikse, 1997), and 'Reiswijzer Gebiedsontwikkeling 2009' for development competitions (Kersten, Wolting, ter Bekke, & Bregman, 2009). The '10 tips of the Chief Government Architect' (Patijn, 2000) shows a pragmatic approach to stimulate professionals in the field.

Most of the decision support systems focus on selection of contractors or procurement methods for integrated construction projects. Decision support systems are dynamic systems which include functions to compare procurement methods or develop tender documents, such as the LEA DSS (CROW & Balance and Results, 2009) or systems such as those described by Love et al (2008). The Dutch LEA DSS system is based on the 'Leidraad Aanbesteden' (Jansen, 2009) under the authority of Regieraad Bouw (Dutch Council for Innovation in Building and Construction). Other Dutch systems support specific elements of the tender process, such as past performance of contractors (Koolwijk, Geraedts, & Chao-Duivis, 2005) or analysing and comparing design alternatives (Bittermann, 2009; van Loon, et al., 2008). The Archiselect system offers a digital portal to communicate with tender candidates and tenderers and collect the requests for participation and proposals (ICOP, 2006). Internationally several other decision support systems are described by researchers for the selection of architects (Cheung, et al., 2002) or the evaluation of tenders with multiple decision makers (Pongpeng & Liston, 2003). Still more research is needed to validate the criteria that form the basis of these comparison systems. But one must bare in mind that although a plethora of tools and techniques are available, "no specific techniques have gained widespread acceptance, particularly by the public sector" (Love, et al., 2008, p. 773). Therefore a framework or process navigator is preferred over a prescriptive solution.

The Office of Government Commerce (OGC) in the UK offers several procurement documents that provide a project management context for the organisation of tenders in construction. Examples are PRINCE2, the Best Practice Guidance and Collaborative Procurement. In the Netherlands this kind of support does not seem to be offered by comparable institutes such as PIANOo. Several knowledge centres, such as SBR (Knowledge Platform for Construction) and CROW (National Information and Technology Platform for Transport, Infrastructure a nd Public space) do provide support and knowledge on this matter. Dutch consultancy firms seem to fill this gap in project management. The Regieraad Bouw published several reports about innovation, collaboration, trust and integrity in the construction sector (Boudewijn & Broekhuizen, 2007, 2009; Glunk & Olie, 2008).

Based on the previous sections several issues can be distinguished that seem of specific interest for the selection of an architect:

• Skills and means required to organise a tender.
• Formulation of the competition brief.
• Composition of the jury.
• Techniques and methods to present the proposals.
• Assessment, evaluation and judgement of the proposals.
• Participation of users, visitors and citizens.

These issues are all part of the analysis framework that was used in the empirical cases of this thesis that led to a categorisation of actor identification, analysis of the project characteristics, ambition & brief, tender procedure, stakeholder involvement and decision process (see Chapter 6). A comparison of the publication categories to these issues shows that none of the publications connects the required activities of preparation and realization to the output of selecting an architect (see Table 4.6).

Although most of the publications mention aspects of the issues, the focus remains scattered. Some publications focus on construction in general while others are specific for the situation of selecting an architect. The models focus on the tender documents as a result of the project, while project management tools focus on the process of preparation and realization. This implies that in order to make a tender in architecture successful, several kinds of publications are needed. The existing publications all show options and underlying considerations, often based on implicit knowledge, but do not tell exemplary stories that illustrate what actually happens in the complexity of organizing a whole tender process. The models do not provide choice outside the options provided in the regulations and existing models and do not trigger action. They do not consider the multi-disciplinary and potentially conflicting nature of the problem of assuring quality during the selection of an architect. In my opinion this is due to the fact that they all try to solve the problem from within their own rationality and not focus on the underlying principles of human behaviour.

4.7 Conclusion

In this chapter I addressed the political, cultural, legal and economical context of architect selection, and showed some current suggestions for improvement. Yet these suggestions do not appear to solve the complete set of problems occurring in current Dutch practice today. The need to implement existing guidelines or a single model seems more urgent than the development of new models and guidelines. Most of the commotion about European tenders in architecture is caused by architects. Public clients are remarkably absent in the discussion, just as academics. Systematic data collection is missing. Without benchmark data no recommendations can be made on the level of requirement, the chances of winning or the size of financial compensation in the award phase, and neither can we build policy that is based on facts and knowledge. Many questions remain unanswered. Why for example do clients prefer the restricted procedure over other procedures, such as a design competition? What are the true costs of tenders, and are these costs in balance with the benefits? And why do architectural firms still participate in so many tenders, even though their chances do not appear realistic? How should a tender procedure be designed differently?

Architects often seem to forget that before the introduction of the procurement regulations, the patronage system did not make it easy for young talent or for less noted firms to enter the market either. European tenders are a step in the right direction to open up the public market and in selecting architects in the fairest possible manner. Although the commissions of public clients are often challenging projects, the recent growth of public-private initiatives creates new

	Procurement models	Competition models	Decision support systems	Project management tools
Actor identification		x		x
Project characteristics			x	x
Ambition and Brief	x	x		
Procedure	x		x	
Stakeholders		x		x
Decision process	x			

Table 4.6 Focus of publications about architect selections

possibilities for architects as part of consortia as well. This requires a different way of organizing a design project (Renier & Volker, 2008). Still, about 95% of projects are commissioned without a public tender because they concern contracts beneath the monetary thresholds or are for private clients. Although most complaints about architect selection concern public clients, it is not clear that the commissioning system of private clients is any better. What exactly are the differences between the system(s) that private clients use to select an architect and those used by public clients? And why do architects not complain about them as much? Will the selection of an architect ever satisfy both sides of the market? After all, a competition per definition creates more losers than winners. I feel that without a change in culture and in the level of professionalism the current problems will continue. A better spread of knowledge and policy making could contribute to this change but it requires also a change of attitudes and behaviour of the actors themselves to realise it.

A client fulfils a very important role in a building project, especially during the preparations of a tender competition. The choices made during the preparation phase determine to a considerable extent the results and appropriateness of the tender, as well as the style of the architectural design. Existing knowledge remains scattered and is not used adequately by the contracting authorities. Only the procurement models have an obligatory nature and could actually be enforced. However, it appears that it is the perception of these legal obligations rather than the actual procurement law that prevent a selection process based on open dialogue between the client and the architects about design quality. There are no open discussions about the difficulties experienced by clients as well as architects. Professions tend to search for solutions within their own domains while an architect selection is in fact a multi-disciplinary phenomenon by nature. A gap exists between the existing structures provided to support decision making and actual decision making of public clients. Clients do not seem able to make enough sense out of the existing structures. This research aims at contributing to bridge this gap from a psychological perspective by describing and explaining how selection decisions were taken in the past and how they could be affected in the future. The aim of this research is therefore to expose underlying behavioural phenomena of clients selecting an architect for a building in the public domain.

Chapter 5

THEORETICAL FRAMEWORK

5.1 Introduction

The research questions in this study are:

1. How do public commissioning bodies decide on the selection of an architect in the context of EU procurement law?

2. Which situational characteristics influence the process of decision making of public commissioning bodies in this context?

3. What are the implications for the design of procedures for the selection of architects?

The previous chapters introduced a number of theoretical and practical insights about the selection process of an architect in the context of EU tendering regulations. EU procurement law sets the requirements for the selection process. However, public clients are finding it difficult to comply with these rules and regulations. The previous chapters indicate that tendering involves a process of conscious and unconscious decision making in which conflicts can occur between legal obligations, governance responsibilities and community related expectations. I propose that decision making in such a specific, elusive and dynamic context requires expertise in different domains to be able to interpret and make sense of the decision task. A consequence of seeing architect selection as such a dynamic process of sensemaking is that it is not realistic to expect clients to apply pre-announced decision criteria and behave according to a pre-designed procedure. In this chapter the most important insights of the previous chapters are integrated in order to define possible implications for the design of an architect selection process in the context of European procurement law. These implications take form of a series of proposed success factors.

The framework provided in this chapter can be considered a bridge between the theoretical and the empirical part of the research because it identifies the characteristics found in the theories about value judgement and decision making in the context of architectural design that have to be considered in a tender design. These characteristics lead to five sensemaking processes of which I assume to play a role in the decision making processes of public clients for the selection of an architect. In the following section of this chapter I argue what kind of processes can be expected in the decision behaviour of public clients during a tender and propose underlying success factors for that specific sensemaking process. The chapter ends with the explanation of the research design and methodology for empirically validating the proposed processes and success factors. Based on the empirical research in Chapters 6 and 7, conclusions will be drawn in Chapter 8 on these sensemaking processes and the proposed success factors, for the design of future tenders.

5.2 Proposed success factors for a tender design

This section explores the five sensemaking processes I identified from the literature that characterise the decision process of public clients during architect selections: 1) reading the decision task, 2) searching for a match, 3) writing the decision process, 4) aggregating value judgements, and 5) justifying the decision. For each sensemaking process the theoretical concepts and findings of the previous chapters lead to several possible success factors for the design of an architect from a clients' perspective. In total I propose fifteen success factors that relate to five sensemaking processes. These processes and factors are only somewhat chronological and interrelate. The diverse roots and conceptions of architect selection appear to cause a considerable amount of ambiguity in the expectations about the selection process of an architect. In my opinion clients therefore have to identify their ambitions and aims first before they start designing a tender. However, the aims and ambition are subject to the interaction that takes place during a tender between the supply and demand sides of the market. The decision process during the implementation of the tender is based on the assessment of the proposals in order to judge design quality. In this process the different kinds of value judgements made of the proposals have to be aggregated. Expertise plays an important role in defining and shaping the structure of the procedure that should lead to a decision about the winner of a tender project. At the end of a decision process the decision needs to be justified to the stakeholders.

5.2.1 Reading the decision task

The first sensemaking process I identified as part of the decision process of public clients is based on the sensereading concept as described by Balogun, Pye and Hodgkinson (2008). During the selection process clients have to interpret the circumstances in which they have to make their decision and shape the meaning of a subject. Shaping meaning and judging its character and significance is part of a framing process that is required to make decisions (Beach & Connolly, 2005; Tversky & Kahneman, 1981). The reading of the decision task is also evident in the fact that a public commissioning client acts as a client instead of a customer, and has to try to analyse distinctive dimensions of architectural language and competition to know what to expect (Jones & Livne-Tarandach, 2008; Strong, 1996).

The perspectives on design quality as discussed in Chapter 2 - the architectural design approach (2.2), the cognitive approach (2.3), the interaction approach (2.4), and the process approach (2.5) - led to a definition that defines design quality as a value judgement that results from the interaction between an individual and a design object. Because of this judgements about design quality are always made in relation to the existing values, structures, ambitions and needs of an individual stakeholder and their potential for the future. The decision task of a client during the selection process of an architect requires several judgements about the

```
┌─────────────────────────────────────────────────────────────┐
│                    PERSPECTIVE/ INTERESTS                      │
│                                                                │
│        artistic expression vs. professional interest vs. user service │
│                        experts vs. laymen                      │
│                       process vs. product                      │
│                         costs vs. quality                      │
│       aesthetics vs. functionality vs. construction vs. durability │
│                     purchase vs. maintenance                   │
│                         goals vs. means                        │
│                     differences vs. similarities               │
│                   active juror vs. passive perceiver           │
│                      debate vs. decision making                │
│                                                                │
└─────────────────────────────────────────────────────────────┘
```

Figure 5.1 Overview of perspectives in assessing quality and value in design

potential design quality. In order to make a judgement, the meaning of a subject has to be framed. After further analysis of the perspectives on design quality I propose that the difficulties about judging design quality relate to: 1) the perspective by which the assessment is made (see Figure 5.1), and 2) the configuration of the assessment (see Figure 5.2).

The assessment perspective depends on the interests and values of the person who judges the quality of the design. According to Bucciarelli (2003, p. 99) "the object can mean, can be, different things to different people". In the same line of thinking I suppose that different participants in design see the object of design differently depending upon their competencies, responsibilities and their technical interests. Because of this multifaceted character of the object, a design is one object with multiple object-worlds. Further analysis of the literature led to a list of potentially conflicting perspectives and interests that could occur during a process of architect selection, either from within the perspectives themselves or between the interests and values that caused the value judgement (see Figure 5.1). For example, an architect might be more concerned with the reactions of the architectural community than that of the potential users of a building, users might be more interested in the functionality and aesthetics than in construction or durability, and a project leader might focus more on the process efficiency and effectiveness than on the product quality. But as a user, a project leader or an architect might also be interested in the functionality and aesthetics, and a project leader might also be concerned about the reactions of the architectural community. These perspectives and values need to be balanced within an individual and within a group of individuals in order to reach a final judgement about design quality.

The object of measurement and the nature of the measurement are also very relevant in comparing the levels of design quality. As displayed in Figure 5.2, measurement levels could for example differ between the object and the context, the individual versus a group, or aspects versus the totality. In Chapter 2.2 I found that literature in architecture usually does not differentiate carefully between measurement levels, but does distinguish different characteristics of the design. The nature of the measurements determines the structure of a measurement. Design objects can have characteristics that can be dynamic, static, tangible, intangible, perceived, sensed etc. (see Figure 5.2). These diverse characteristics influence the ability to quantify or qualify the assessment of the design qualities. A building

```
┌─────────────────────────────────────────────────────────────┐
│                  MOMENT OF MEASUREMENT                        │
│                                                               │
│        representations of future objects vs. existing physical objects │
│                  pre-purchase vs. post-purchase               │
│              initialization vs. realization vs. use           │
└─────────────────────────────────────────────────────────────┘

┌─────────────────────────────────────────────────────────────┐
│                  OBJECT OF MEASUREMENT                        │
│                                                               │
│                 built objects vs. built environment          │
│             individual vs. organisation vs. society level     │
│                 overall value vs. costs/quality balance       │
│                      groups vs. individuals                   │
│                       aspects vs. totality                    │
│         separate aspects vs. the relative importance of the aspects │
│               tangible aspects vs. intangible aspects         │
│                    satisfaction vs. dissatisfaction           │
└─────────────────────────────────────────────────────────────┘

┌─────────────────────────────────────────────────────────────┐
│                  NATURE OF MEASUREMENT                        │
│                                                               │
│                  qualification vs. quantification             │
│                      values vs. value system                  │
│                object as such vs. object as perceived         │
│                      simulation vs. real life                 │
│                        static vs. dynamic                     │
│                      measurement vs. meaning                  │
└─────────────────────────────────────────────────────────────┘
```

Figure 5.2 Overview of configuration options for comparing quality and value in design

can be seen as consisting of features or parts that can be measured and sensed physically, such as its indoor climate or height. It also contains many aspects that provide meaning, such as the impact on the environment and the sense of home. These differences relate to the tangibility of a feature. Design quality relates to the sum of these different qualities. In the situations of architect selection it seems likely that judgements will always produce conflicts about the potential design quality that the candidates offer. An assessment about design quality is very complex and diverse and cannot be defined in a static manner. Some means of arriving at and publishing the judgements will inflame conflicts, while others may allay them. The assessment system of design quality should address both tangible and intangible characteristics but also channel a decision. Because of this complexity I presume that design quality is eventually perceived on a holistic level during decision making. In order to successfully read the decision task I assume that:

Factor 1

A tender will produce fewer conflicts and will be more successful if the tender design allows for a holistic judgement that incorporates potentially conflicting judgements within itself.

The myriad of theoretical and practical insights in the previous chapters about value judgements and decisions showed that decision making with the aim of selecting an architect addresses a lot of potentially conflicting characteristics that originate from the diverse roots of the phenomenon: the tendering for works and services, the search for a partner, and the design competition (Strong, 1996). During architect selections assumptions are made about future functioning of a building and certain alternatives need to be compared. At this stage in the design project a valid and reliable estimation of the construction cost in relation to the potential quality is much more complicated than the relationship between price and quality at the start of a construction project. This appears to be reflected in the number of tenders awarded on the lowest price (preferred in construction) and the most economically advantageous tender (preferred in architect selections). Yet a supposedly irreconcilable phenomenon takes place during architect selections. This has to do with the fact that the final aim of the client is to buy a work (a new building), but the client must first select an architect based on a representation of the potential work. The process of architect selection only allows an indirect link with the final aim of the client: a new building. The products of the architect and of the contractor are often separated during the selections processes but at the same time also they are connected by in the minds of the decision makers. The mixture of aims and means causes that the selection process of an architect is a complicated process with a somewhat unclear aim at the beginning.

Figure 5.3 presents the tensions that appear often from the different interests at play in architect selections. The left side of the figure shows the architectural competition tradition, which is based on the client's intention to acquire a design product as a patron. This tradition acknowledges the artistic characteristics of an architectural design. When clients look for an architect for the design of a public building, architects are often asked to show their competences to the public and to the client anonymously. This public debate is part of the competition tradition. During a competition the focus lies on the design of, or vision for, the future building. The architectural community is represented in a jury committee that has the authority to appoint a winner based on anonymous evaluation of the design proposals. This tradition is based on the assumption that a partner in architectural design can be chosen by judging the competences of a person by (anonymously) evaluating his or her physical work or a preview of this product.

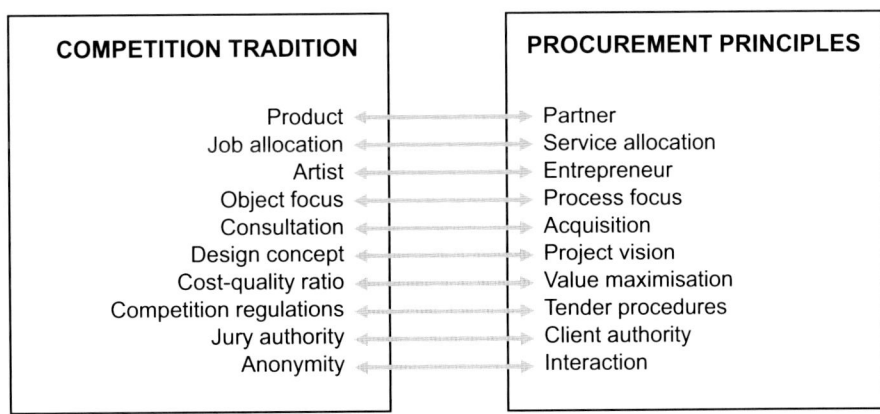

Figure 5.3 Tensions originating from the competing conceptions of the architect selection process

The right side of the figure shows the procurement principles and their managerial processes that apply to the situation of architect selection. In the procurement approach architects are considered as entrepreneurial service providers competing for contract. Underlying the EU procedures is the idea of selecting a partner for a building project who is capable of designing the future building. Such a process focuses on maximum value for the client and therefore the client has the final decision authority – 'he who pays the piper calls the tune'. In order to know with whom the client will be doing business with, interaction is an important element in the selection process.

Both approaches in Figure 5.3 display a view of the search for a partner in design in line with their own rationality and traditions. In my opinion both approaches are appropriate. However, the exact blend in any particular selection procedure must reflect as long as they fit the aims of the client. Both the competition and the tender tradition require intuitive judgements as well as rational analytical thinking to judge the competences of the tenderer and the quality of the proposal. But what exactly is the service that a design firm delivers and how can it be assessed? A client therefore has to analyse their aims of the selection process first and then start to develop an assessment process for selection. Day and Barksdale (1992) and Cheung, Kuen and Skitmore (2002) found that architects are assessed on their perceived expertise, experience and competences, the background of the firm, their understanding and relationship with the clients, and the likelihood that they will conform to the contractual and administrative requirements. These reasons reflect the characteristics of the nature of the service and the practice of assessment in design. Architectural services address unique circumstances and the results are not easily foreseen – one cannot expect an architect's next building to look like his/her last – so that the judgement must be made not only on previous designs, but also in part on the character an skills the architect can demonstrate or describe in his/her proposal. I therefore conclude that the selection of an architect relates to the combination of the architect as a person with certain competences, the quality of the potential product, and the characteristics of the firm they represent. This 'package' cannot be taken apart during decision making. The priorities and aims of a client determine the relative balance between the competition tradition and procurement principles. I therefore presume that a possible success factor in reading the decision task will be to acknowledge that:

Factor 2

A tender design will only be successful if it addresses the characteristics of the architect as a person, the proposed design, as well as the firm that they represent.

5.2.2 Searching for a match between aims, ambitions, needs, and opportunities

The second process of sensemaking I distinguish in architect selections is based on the interaction principle of market parties underlying procurement law in combination of the image theory of Beach (1990) and the matching principles of the adaptive toolkit with heuristics (Gigerenzer & Selten, 2001; Gigerenzer, et al., 1999). In the architect selection process the governance structure of the client needs to be taken into account and connected to the economical and legal context of architectural design. According to image theory decision makers match their own values and beliefs with specific goals and operational plans. The first process of reading the decision task is an important step in defining the goals. This process strongly relates to the operational plans. In an architect selection process clients have to find a way to match their ambitions and needs with the opportunities that are offered by the architects that join a tender, a process comparable to the findings of Elsbach and Kramer (2003).

In absolute numbers the number of tenders in which architects are selected for a building project in the public sector is limited; recent numbers show for 2009 about 140 restricted tenders in the Netherlands (Geertse, 2010). However it regards a significant portion of all buildings with a public character built in any given year. The implications, and therefore also the complexity of the assignment of those projects that have to be tendered, have a high impact on society. In developing buildings public clients make decisions that could have a tremendous impact on the wellbeing of many people or groups of people. For architects these kinds of projects are of major interest because of the opportunity to build up the portfolio, the publicity that comes with the project, and attractiveness of the assignment to design a public building. Once the architect is chosen, (s)he will become an important member of the project team as the building will greatly depend on his or her capabilities to visualise and build the 'dream' of the client.

For public commissioning clients the proposals of the candidates in the selection phase and tenderers in the award phase are essential input for decision making. Decision makers experience all kinds of dilemmas about the aims, actions and actors of a decision (Etzioni, 1988). A decision about the winner of the tender has to be supported by the organisation of the commissioning body, the market parties in question as well as the community in general (Feldman, 2003). A tender will therefore probably not only consist of a tender brief and a procedure to make the final decision, but also include a procedural process in order to collect the proposals from the market and a concept to involve stakeholders in decision making.

I suggest applying the term 'design' and the activity of designing on decisions that are made to shape the tender procedure; a client has to design a tender in order to select an architect. The design of the tender procedure should specify the format of the information used during the assessment and judgement of the proposals and determine the actors in decision making. Based on the aims of the client, a perspective has to be chosen to design the tender regulations that set the direction of the tender (see Figure 5.3). The direction and aim of the tender should be reflected in the decision criteria and the procedure that are used. In terms of this research this means that the term 'design' sometimes refers to the services and products offers by the architects (the supply side of the market) and sometimes by the commissioning public clients (the demand or purchasing side of the market. I assume that it is the potential fit between the design of the decision process and the characteristics of the tender project that eventually determines the quality of the match between the client and the architect. This leads to the following potential success factor in searching for a match between the aims, needs, ambitions and opportunities of an architect selection process:

Factor 3

The fit between the aims and the design of the tender will contribute to the success of the tender project.

Citizens, politicians, and the architectural community have great interest in these kinds of tender projects. It usually depends on the impact of the future building if public clients involve the citizens, experts in architecture and city planning or other stakeholders in the process (van der Pol, et al., 2009). Some of the problems in current practice indicate that the noise that is created around architect selections partly relates to the involvement of stakeholders, especially citizens, and their potential influence on the final results. This could be caused by the lack of a clear distinction between participation aims and lobbying purposes of clients (Crosby, et al., 1986). For the selection process of an architect I identified five actor groups that could be involved in decision making: stakeholders that originate from the governance structure (e.g. citizens and professionals), stakeholders that are part of the client organisation (e.g. employees, political parties), staff members that are part of the project team, members of the jury panel acting as a selection and/or award committee, and the steering committee of the public commissioning client. The stakeholders all have a different role and interest in the building project and the selection process of the architect. For all actor groups the advisory or decisive rights needs to be properly addressed (see for example Edelenbos & Klijn, 2005; Vroom & Jago, 1988).

Table 5.1 shows four strategies I distinguished in participation roles of actors during competitions and tenders: the tradition competition, the participatory competition, the compact tender, and the participatory tender. The model of Vroom and Yetton (1988) offers 4 types of participation in organisational decision making, which could be applied on tender situations: autocratic (A), consultative (C), group (G) and delegation (D) (see Chapter 4.2.2). Applying these notions to tender situations seems promising because it distinguishes different 'design' types of participation in selection processes and differentiates the actual actors in the selection process from the other stakeholders. Assessment by staff as part of the project and analysis team can be part of the consultation, jury processes can be considered as group decision making processes and a steering committee appointing a jury panel to select the winners seems to be a good example of delegation

A = autocratic, C = consultative, G = group, and D = delegation	Traditional competition	Participatory competition	Compact tender	Participatory tender
Stakeholders originating from the governance structure (visitors, citizens, professional field)		C		C
Stakeholders as part of the client organisation (employees, political parties)		C		C
Staff members - Project team	(C)	C	(C)	C
Selection & Award committee - Jury panel	G	G	C	C
Commissioning client body - Steering committee	D	D	A or G	A or G

Table 5.1 Typology of participation options in decision making in the context of architect selections

in the context of the tender. In my opinion a client has to design a participation process that fits the level of expertise of the stakeholders. In Chapter 4.2.2 two methods of stakeholder participation were introduced that are suitable for citizens in the context of architect selections. In case of a design competition or restricted tender with design proposals, the proposals can be evaluated on their preferences in the tradition of environmental psychology (Nasar, 1999). This requires a research approach with a Likert-scale on certain variable and statistical analysis and a proper randomised sample and not a superficial 'beauty contest' as currently used in practice. The other option would be to assign a citizen panel by on the ideas of Crosby, Kelley and Schaefer (1986) that is educated on the theme and supported in writing a report about their recommendations.

The main difference between a competition and a tender in current practice appears to be the role of the jury panel. Are they consultants for the steering committee, or do they have responsibility for the outcome of the decision making? Should the award committee consist of government representatives, or should it consist of experts in the field of architecture and urban planning? What should be the relationship between the steering committee and the award committee? These considerations also originate from the mixed origins of architect selection procedures as described by Strong (1996). On the organisational level a conflict could occur between the project structure of the tender or competition and the organisational structure of the commissioning body. Responsibilities are delegated to actors that are not part of the organisational structure and therefore have different responsibilities. This leads to the following possible success factor for participation of stakeholders in architect selections:

Factor 4

The fit between the position and type of the stakeholders and their role in the decision process will contribute to the success of the participation strategy of the tender design.

5.2.3 Writing the decision process

The third sensemaking process that evolved from the previous chapters is the process of writing the decision process. This sensemaking process is based on the concept of 'sensewrighting' and 'sensegiving' of Balogun et al. (2008) in which the decision task is inherited and the process of decision making is influenced by the decision makers. I combine this sensemaking process with the concept of sharing frames among decision makers (Beach & Connolly, 2005). In my opinion this process can be compared to the activity of 'writing' a process because a decision process is shaped and thus written during the duration of the preparations and the implementations of the tender project.

The literature review in Chapter 3 showed that decision making is a process of goal setting, perception, information processing, framing, comparison, evaluation, deciding on action and finding decision support which occurs at individual as well as on the level of the team (Beach & Connolly, 2005; Hodgkinson & Starbuck, 2008b). Therefore goal setting should be the first step in a tender process. This relates closely to reading the decision task (section 5.2.1). I consider the preparations of a call for proposals as a process of building frames according to which the proposals of the architects will be judged. To ensure that the frame of goal setting is similar to the frame of evaluation and action decision, preparations should be conducted in collaboration with the jurors. Aligning a decision frame is not the same as copying a frame so the actors are still able to pursue their own goals. In case a client decides to hire consultants for the management and legal implications of the tender and invite external experts as members of their jury panel, they have to be aware of the differences in goals and criteria for success (Kieser & Wellstein, 2008). Because of the specific character of an architect selection an information asymmetry and communication filters are expected between the consultants, jury members and the client representatives.

Numerous decisions about the procedure, brief and stakeholder participation have to be taken before the evaluative work of the jury process starts. A few examples of such decisions are the size, type and content of the contract, the kind of tendering procedure, the decision criteria or the format of the proposals. By creating mixed teams with staff and external advisors, information about these issues can be exchanged and critically assessed by a diverse group of people in the preparation and the realisation phase (George & Chattopadhyay, 2008). If these actors would share their decision frames, often originating from different fields of expertise and affiliations, this would increase the processes of sensemaking between the decision actors. I therefore suggest the following success factor in writing the decision process:

> **Factor 5**
>
> *All actors in the decision process need to align their frame of references during preparation of the tender process in order to reach a decision at the end of the process that fits the ambitions and aims of the client organisation.*

A brief is an ill-structured problem with no 'right' solution. The judgement of the quality of a design proposal can therefore be characterised as a judgemental task about several tangible and intangible quality aspects. This requires domain specific knowledge. The combination of different kinds of judgemental tasks implies that two separate processes of rational analysis and experiential intuition (the dual-process theory, see for example Chaiken & Trope, 1999; Kahneman, 2003) are involved in making decisions about the best architect. The value judgements ask for both conscious, analytical assessment processes as well as intuitive, unconscious assessment processes. During these processes people use heuristics to save time and energy, but by doing so systematic errors are made (Plous, 1993; Simon, 1997). Therefore measurements should be taken to prevent biases.

The process of decision making about design quality during a restricted tender officially requires two phases. In the first phase a selection is made between the architects that showed interest in the contract, the tender candidates. According to procurement law this decision should relate to the technical, financial and organisational qualities of the firms. In the second phase a decision is made which results in awarding the contract to one of the tenderers. Legally this decision should relate to the offers or proposals made by the competing firms. The check on financial or legal requirements of the firms in the selection phase can be considered as a well-structured task with objective criteria for success. Either a firm fulfils the requirements or it does not, which suggests a nominal measurement scale. I presume that such a task can be done with limited domain specific expertise. The outcomes could be valid input for a decision support system to create an overview of the information of the candidates. The evaluation of a reference project or Curriculum Vitae to judge the competences of an architect however has a less tangible character. In my opinion this can be considered as a judgemental task on an ordinal scale that can only be completed by someone with domain specific knowledge. During the award phase the architects usually have to present a design proposal or vision on the building project. This requires a large amount of domain specific expertise and should thus be performed by domain specific experts. This leads to the following assumed success factor for the process of writing a decision process in architect selections:

Factor 6

The type of expertise needed for the various decision tasks during the selection process of an architect needs to be aligned with the nature and content of the decision task.

The fact that levels of design quality can be distinguished (see Chapter 2.6) implies an unconscious or conscious comparison of the perception of the design object with expectations and perceived values from a certain anchor point for assessment of design quality. This anchor point provides a point of reference that could be based on previous similar experiences but could also be determined by the use of heuristics that compare this situation to a previous one with a different nature (Plous, 1993). The context and project definition are both starting point and end, initiative and touchstone, of the assessment process (Heynen, 2001). Because experts are able to recognise patterns, they use different reference points than lay people which provide a standard for comparison (de Groot, 1946). This allows them to be informed by richer information than novices which decreases the uncertainty of the decision task (Mieg, 2001). In architecture this means that experts are able to use knowledge about previous designs to assess the quality of an architect for a particular project. They know the market and players in the field. They can distinguish design styles, working methods, cost-quality ratios, specialities and more aspects that determine if the designer would suit a project or not. Laymen do not share this knowledge and can therefore only use their own experiences. In the context of architect selections this means that laymen probably have to build their frames of reference from their current and limited experience of a single case, while experts can use their relatively broad experience to build a comprehensive frame of reference for the tender. Because of the shared frame of references, judgements of experts and members of the same culture or profession are usually more consistent than judgements of lay people or individuals.

Connected to the different perspectives of individual judgements are the issues of communicating preferences and opinions. Users, designers, clients and other stakeholders speak different languages (Bucciarelli, 2003). Because the built environment is surrounding us all, everybody has some experience in making judgements about physical building. Also in architectural tendering processes the quality of an answer depends on the quality of the inquiry (Heynen, 2001). From my perspective there are three phases in an architect selection process in which a message of one stakeholder group is communicated to and interpreted by another group: 1) from ideas of a client into a tender brief, 2) from a tender brief of the client into a design proposal of the tenderer, and 3) from a design proposal of a tenderer into the interpretation of ideas by a jury panel. All these moments of transference include one or more communication filters. It requires expertise to make these translations as fluently as possible. Although Jones and Levine-Tarandach (2008) suggest that architects address all these stakeholders by using multivalent keywords, my assumption is that architects often expect others, including lay persons, to understand and use their architectural language.

A question that remains unanswered is the level of expertise needed to make valid judgements from a client's perspective. Experts are usually more experienced in reading designs that represent (future) buildings and they are able to incorporate experiences about building features in all phases of the life cycle of a building (initialization, design, construction and use) (based on Cohen, et al., 1996). Experts are better at seeing significance of information, identifying important cues for risks, and estimating consequences (Hutton & Klein, 1999). They can judge autonomously and feel the need to discuss and harmonise with the professional field. Experts are generally better at expressing their needs and share the same language and profession in which judgements and knowledge are communicated (Brenner, et al., 1996; Mieg, 2006). However, the perspective and level of the expert appears to influence the value of its judgement. The findings of Brown and Gifford (2001) even suggest that less experienced architects are better at predicting the preferences of the general public than the experts. In this perspective Vollaard (2009a) suggests a course for professional architects who want to participate in a jury panel because a client's perspective requires different skills than being an architect. For the a successful writing process I suggest taking the potential differences and limitations of different levels and kinds of expertise into account when selecting decision makers for the judgemental processes about design quality. This could mean that people would have to be trained in assessing design quality or take a 'language course' in order to participate in architect selections. In this line of reasoning the following possible success factor can be phrased:

The use of intuition seems to increase the performance of judgemental tasks

---

Factor 7

Educating decision makers in reading architectural designs from a client perspective will be beneficial for the process of decision making during architect selections.

(Dane & Pratt, 2007). However, intuition and unconscious decision making cannot be forced. It could however be stimulated by a large amount of domain specific knowledge, experience in decision making and more time during decision making. Kazemian & Rönn (2009) describe how Finnish juries always meet several times instead of one long session. In this way they create room for intuition and reflection in their process of decision making. In uncertain and ambiguous situations could cause emotions that do not relate to the situation influence the process of decision making. Fear and anxiety could lead to more systematic and comprehensive information processing, but could also influence the quality of decision making negatively by producing a blame culture or self-protection (Mosier & Fischer, 2009). Positive affect can stimulate creative thought and an open atmosphere, but could also lead to more superficial and less thorough information strategies. Even for experts it is hard not to be influenced by personal interests and preferences. The definitions of intuition suggest that affect and emotion are an integral component of intuitive judgements. In my opinion the literature is inconclusive about the relation between product emotion, intuition, integral affect, and their potential influence on the selection process. It also remains to be seen if emotional responses from clients can be steered by a conscious design of the architects. According to Kreiner (2007b, 2008) the chances of winning a competition cannot be influenced by the participating architects because the evaluation criteria are developed by the decision makers during the decision process. His findings suggest that an architectural firm can steer their portfolio by strategically submitting only proposals with a 'thrill' factor.

Several scholars (e.g. Fawcett, et al., 2008; Gifford, 2002) have shown that professionals and daily users clearly differ in their preferences about buildings, but not as extremely as professionals in architecture sometimes suspect. I assume that this could be explained by the biological origin of aesthetic pleasure which relates to a human need for structure. Contrary to the other elements of product experience, the attribution of meaning and the emotional response (see Chapter 2.4), aesthetic pleasure appears to relate to cognitive perception and not specifically to expertise (e.g. Gifford, et al., 2002). Expertise can influence the attribution of meaning because it is influences by cultural and personal experiences. Both the perceived as well as the potential qualities of a proposal (Heynen, 2001; Kazemian & Rönn, 2009) appear to be part of the attribution of meaning in a process of recognition, interpretation, association and assignment. Every product experience includes an emotional response that gives the person an indication about the potential harm or benefits for their product concerns (Desmet, 2002; Schifferstein & Hekkert, 2008). The level of design excellence often includes a wow-experience or affirmative surprise, which appears to be comparable to the 'thrill' factor of Kreiner (2007b). This emotional response is mainly positive and seems to focus on the moment of purchase. I assume that experts are better at dealing with (product) emotions and intuition because their frames of reference are more stable and they are better able to filter non-relevant or disturbing information. This does not mean the emotions do not play a role in expert decision making at all, but to a lesser extent than with novices. Therefore I consider the involvement of experts as a possible success factor in writing the decision process of architect selections:

In both the competition and the procurement tradition (see section 5.2.1) a dialogue between the client and the architect is sought to gain information about a potential partner. The more architects show interest in a tender and the more

---

Factor 8

The involvement of (external) experts will be beneficial for the process of decision making about design quality because they have domain specific expertise and are better at controlling product emotions and using intuition than novices.

---

extensive (design) proposals are requested, the more information is gathered. All this information needs to be assessed and evaluated in order to make sense out of the data. Therefore the tender format should provide a clear information structure without surplus requirements. At the same time uncertainty is reduced by information. The literature about design competitions suggest that clients face the dilemmas of security versus innovation, precision versus latitude, present versus future, and requirements versus feedback during decision making (Rönn, 2008). These dilemmas contribute to the degree of uncertainty that accompanies a selection process. There are different forms of uncertainty: inadequate understanding, lack of information and conflicting alternatives (Lipshitz, et al., 2001). Emotions have a positive as well as a negative dimension and uncertainty can thus result in fear but also in excitement (Betsch, 2005; Mosier & Fischer, 2009). Linking back to the literature on product emotion (Chapter 2.4), clients will probably expect to be surprised by the proposals of the architects. At the same time they also fear the possible consequences of the decision outcome. Information overload could occur, especially in situations in which a decision is based on information which is low in value, quality and relevance, or highly ambiguous (Sutcliffe & Weick, 2008). Narratives and background information can support sensemaking and help to create an overview of the information (e.g. Karmanov, 2009; Le Dantec & Y-Luen Do, 2009). This implies that a personal explanation of the architect is beneficial for the process of decision making. This leads to the following possible success factor for writing an architect selection process:

Factor 9

A personal explanation by the architect will improve the clients' understanding of the proposal.

Svensson (2008) found that a consensus among jury members is not the same as an average of their opinions, but rather the result of a negotiation process. Jury members usually reach a consensus through several rounds of discussion or voting. During these rounds adjustments are made to judgements due to external influences, such as opinions of other members of a group, changes in the context, or internal personal factors, such as moods or other kind of emotions. The findings of Kreiner (2006), Svensson (2008) and Kazemain & Rönn (2009) show that decision criteria are developed during the evaluation process in a design competition. Therefore I presume that decision making in the context of architect selections can be conceptualised as an incremental and cyclic process based on a dynamic interplay of judgements. In this interplay, discussion about important issues is probably even more important for sensemaking than the actual choice. Taking all contextual and characteristics of architect selections into account, I consider negotiation as an essential element for support of the decision and the process of sensemaking. This leads to the final possible success factor in writing the decision process:

Factor 10

Discussion and negotiation among the decision makers will positively influence the decision making process of selecting an architect.

5.2.4 Aggregating different kinds of value judgements

The fourth process of sensemaking relates to the potential clash between the legal procurement rationality and its underlying rational decision model, and the complex character of making judgements about design quality when choosing an architect. As shown in Chapter 2.6 and section 5.2.1 a judgement about design quality have multiple layers and perspectives. Simon (1997), Etzioni (1988) and Snellen (1987) describe the different rationalities in which decisions are made. Especially during the aggregation of the different kinds of value judgements problems could occur that endanger the result of a tender project.

Procurement law does not distinguish between a natural person and a legal entity. It also does not specify who and how many persons should take decisions or how many characteristics should be judged. In my opinion a distinction should be made between an individual judgement, a judgement of a group, and a decision about design quality. Unlike value judgements, decisions are intentions for action that include an element of choice (Hodgkinson & Starbuck, 2008a). I consider decisions about design quality as the final result of a judgemental process. In the context of architect selections one full proposal is chosen as a winner of the tender competition. This is in line with the assumption in section 5.2.1 about reading the decision task that a decision about design quality is always taken at a holistic level (Factor 1). Both judgement and decisions can include comparisons of alternatives at holistic or aggregated levels. Individual judgements must be aggregated to form group judgements and to arrive from fragmented design qualities to holistic design quality.

Figure 5.4 shows four points of departure for an assessment of design quality: a single judgement about separate qualities, a single judgement about holistic quality, multiple judgements about design qualities, and multiple judgements about holistic quality. Linking these perspectives provides six relations between the four different assessment structures. Depending on the aim and starting point of an assessment, one or more steps are needed to make an assessment. The legal structure of an architect selection requires the distinction of criteria for several separate design qualities, but one tenderer to win the tender. The regulations for a design contest suggest that a group of jury members evaluate the proposals on different aspects and then reach a decision about the winner, which can be positioned in the lower segment of the scheme. For the other procedures the number of decision makers is not mentioned, which means that theoretically relation 1 could be applicable: from a single judgement of separate qualities to a single holistic judgement. In this approach the aspect of negotiation and discussion is missing, which is conflicting with proposed success factor 10 of the previous section about writing the decision process.

If a judgement about design quality is made by more than one individual, there are different ways to come to a common judgement for the group. The first approach is to aggregate the individual judgements without interaction in a system and average the scores (relation 2 and relation 6). This method shows weaknesses in the measurement scales of the intangibles (see Chapter 2.2) but shuts out social influences. It can therefore be perceived by the outside world as more accurate. Disadvantages of this method are that insights of other decision makers are not shared and decisions are less easily accepted (e.g. George & Chattopadhyay, 2008). In the legal system this is referred to as the independent expert model (see Chapter 4.4.2). Examples of methods that are based on this principle are the Delphi method, which is based on a ranking of individual judgements of several

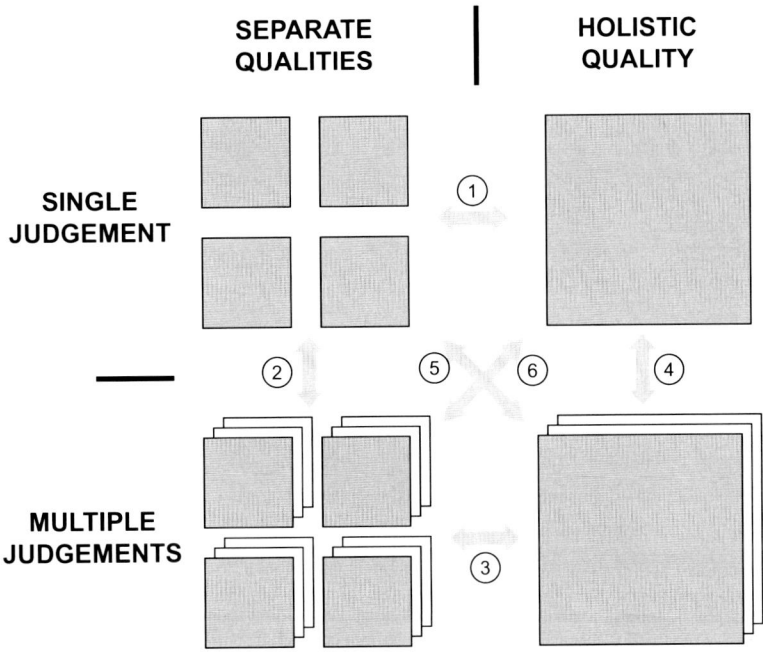

Figure 5.4 Four points of departure and interrelations for quality assessment

design qualities without social interaction of the decision jurors (relation 6) or the Song festival method (countries independently express their grades) is based on holistic individual judgements about design quality (relation 4). The Olympic scoring system for sports such as gymnastics or figure-skating is based on individual judgements about design qualities (relation 2) that are expressed in holistic judgements and related to other jury members (relation 3) to present a ranking that shows the final winner (relation 4).

Another approach is to discuss the differences between the individual judgements on a holistic level and define one judgement for the group (relation 4), discuss the separate qualities with the other group members and then reach a decision (relation 6), or also discuss the proposals on a holistic level in between (relation 3). The more differences in perspectives, the more difficult it is to discuss issues but in every situation a consensus must be reached. The consensus model is also acknowledged in case law. As discussed earlier the nature of the object affects the ease of the discussion because intangible characteristics elicit more diverse judgements than tangibles. Because of the consistency of the judgement, in practice tangible aspects are more often seen as a solid base for discussion (Macmillan, 2005). Yet, leaving out or quantifying the intangible characteristics does not benefit the validity of the judgement. The fact that during discussion more information can be put on the table and discussions contribute to decision acceptance can be considered as a benefit. At the same time there is more pressure to conform, possible domination of one or two members and chances of groupthink and groupshift as explained in Chapter 3.5. The results of both approaches can be regarded as inter-subjective and a consensus that is based on trust in the other group members. A decision about design quality can therefore be considered as an inter-subjective consensus among the members of a group of decision makers (Kazemian & Rönn, 2009).

Case law shows that the court acknowledges the subjective character of services (see Chapter 4.4.2). Based on the case law and theories of value judgements I conclude that objectivity should be sought in the assessment system, not in the assessment itself. Every judgement contains an assessment of a design object that depends on the actors, the means and the context. Because of the complexity of an building (design), the question should not be if the assessment is objective or subjective, but if it is valid – if it actually measures what it should be measuring, if the judgement is not biased or inconsistent. This means that a 'judgement based approach' should not be perceived by definition as more subjective than the 'manage and measure approach' (Gann & Whyte, 2003). In section 5.2.1 I described the characteristics of the decision task in relation to assessment of design quality. The nature of the measurement and the perspective of the stakeholder cannot be changed, but the object and moment of measurement can be structured in order to improve the comparability of the alternatives (see Figure 5.1 & 5.2). Bucciarelli (2003, p. 40) states that we should know better than "to think in two worlds: the social - where subjectivity, opinion and values matter, and life depends on needs, desires and interests of people - and the technical/scientific other - where objectivity, uniformity, scientific law and cold value-free instrumental reasoning matter, where life becomes banal, mundane, autonomous, purely instrumental". Taking the characteristics of design quality into account, I suggest that structuring the assessments and making the aggregation system explicit can contribute to a successful aggregation system of value judgements and valid comparison of the alternatives that are offered by the architects. This leads to the following possible success factor for aggregating the different value judgements that are made in architect selection processes:

Factor 11

Making the assessment structure and aggregation system of value judgements explicit in the preparation process will contribute to the validity of the comparison of alternatives in the actual tender process.

Both the literature about design quality assessment and that about decision making provide little insight on the aggregation principles of the value judgements. Based on the review in Chapter 2 I found four basic levels of design quality: under-performance, basic performance, added value, and design excellence. These four levels form an ordinal scale from under-performance to excellence. Although this is often ignored in current practice, an ordinal scale can not be used as a ratio scale because there is no natural and fixed zero-point and the intervals between the levels are not equal. This means that scores should not be added, subtracted or averaged and only non-parametric analytical statistics can be performed on these data (de Keyzer, 1998). Only the median should be used as an indication of average preferences. In practice many examples can be found in which ordinal scales are misused for calculations. Satisfaction is for example often measured on a 5-point scale or by grading on a scale from 1 to 10. These scores are then summed up per category or item and averaged to make a statement such as 'The average is 3,9 on scale from 1 to 5, which implies a satisfied population'. This way of dealing with ordinal data is very appealing because it offers a sufficient system to compare individual judgements. Still without the context and the original formulation of the scales in which the measurements were taken, an average number is methodologically invalid. In addition items and individuals are treated as equal, even if this is not the case.

Some non-parametric statistical analysis techniques take this inadequacy into account but cannot 'cure' the problem. Heynen (2001) mentions the Argus method to be used in architect selections, which is an abbreviation for 'Achieving Respect for Grades by Using Ordinal Scales only'. Another approach to improve the power of the outcome is to include the weight of the items in the average. This is the method as used in a multi-criteria analysis (MCA) or in prospect theory (Kahneman & Tversky, 1979), one of the first generation decision theories. Still, if a measurement on an ordinal scale is taken as basis for the calculations, the outcome has limited power. Yet calculative systems with criteria and weighted interests are often applied by clients during architect selections to create a ranking among the tenderers. This means that criteria with different measurement scales are treated as equal input for the ranking. In current construction the issue of monetizing value is considered to open up the difficulties of measuring characteristics in design (see for example Dreschler, 2009). Procurement law even suggests the use of a system that is based on these principles to safeguard transparency. In my opinion clients should be aware that although this multi-criteria system can support the analysis of the data, the scientific value is limited without awareness of the inadequacies.

The aggregation of judgements raises another issue that should be made more explicit, namely the compensating or overlapping character of design qualities. The framework of the Design Quality Indicator (see Chapter 2.2) suggests that qualities can overlap, but does not include this potential overlap in the assessment system. At the same time several authors (e.g. Gerritse, 2008) assert that in architecture the value of the whole exceeds the sum of the parts. The theories and frameworks discussed in Chapter 3 do not explicitly address the compensating and conflicting character in relation to value judgement. Procurement law assumes that selection and award criteria are mutually exclusive with clear boundaries and that they together fully cover the award decision. From this perceptive neither a potential overlap nor additional criteria are taken into account either. Hogarth (1988) and Mieg (2001) raise the issue of the fundamental difference between compensating and non-compensating judgements in decision making. A compensatory strategy avoids conflict, while a non-compensatory strategy confronts conflicts. The concept of the 'wow'- factor (Desmet, et al., 2007), the 'surprises' in the model of Kano (1984) and the 'thrill' factor of Kreiner (2007b) suggests that an affective response can compensate deficiencies and ease choice. For the validity of the value judgement, an assessment system for design quality should therefore also take possible compensation, increasing insights, and possible overlapping qualities into account. This leads to the assumption that allowing compensation in the aggregation system of value judgements would contribute to the success of an architect selection process. This leads to the following proposed success factor:

Factor 12

Allowing for compensation in aggregating value judgements about design quality will increase the validity of the assessment.

5.2.5 Justifying against different rationalities

The final process of sensemaking I identified as part of the decision process of a public client selecting an architect deals with the justification of the final decision about the winner of the tender. This process originates from the potential conflict between the legal, political, economical and the social rationality as described by Snellen (1987) and the characteristics of organisational decision making as described by several authors in the book about organisational decision making of Hodgkinson and Starbuck (2008b).

In justifying a decision, clients have to evaluate their intentions in relation to their responsibilities. Public clients represent their citizens and employees and can be held morally, personally, collectively or hierarchically accountable for their decisions (Bovens, 1990). In the context of architect selections there is a lot of uncertainty. Additionally the information provided in the proposals is not always comparable and could lack information on crucial points. As stated above, design problems are ill-defined with no 'right' solution and alternatives could be conflicting. This creates responsibilities that are multifaceted and conflictive by nature and makes a valid comparison of alternatives and a risk implication very difficult and subject to error. The early decision theories address the perception of risks and outcome value more explicitly than the second generation theories (Beach & Connolly, 2005). The architectural community perceives Dutch public clients as risk avoiding decision makers (van der Pol, et al., 2009). Apart from the fact that risk avoidance does not have to be negative in a public function, this aversion to risk could be explained by a lack of expertise among public clients about architecture. Expertise could support proper risk estimation, control over the process and therefore reduce the fear of failure (Mosier & Fischer, 2009). Fear could also be reduced by trust in other members of the group that may or may not have domain specific expertise (Salas, et al., 2006). According to Bucciarelli (2003) trust is a social matter that binds beliefs and people together. Knowledge presumes belief and belief rests upon trust. If public clients involve numerous parties, consultants and stakeholders, it could be hard to establish trust (Kieser & Wellstein, 2008). Thinking about and defining roles, responsibilities and authorities of the decision shapers, decision takers and decision approvers beforehand (see Gehner, 2008) could improve the level of trust within the group of decision makers and between decision makers and stakeholders of the project. Assigning decision tasks to people with the right level of expertise could also decrease the level of fear for decisions. In relation to this aspect of responsibility the authority of the jury panel is one of the issues that the clients needs to decide upon in their tender procedure (see also section 5.2.2). This leads to the following assumed success factor for the final justification of the decision:

Factor 13

Carefully addressing the roles and responsibilities of the decision makers in the design of the tender will contribute to the trust between the decision makers and increase support for the decision among the stakeholders.

If a jury panel consists of experts and acts as an expert team (Salas, et al., 2006), it can be used to ease the justification process of decision making. Spreiregen (2008) reported in relation to the quality of a jury decision that only an expert panel could have selected the winning design for the Vietnam Veterans Memorial in Washington DC because a lay jury would certainly have overlooked it. Jury members have certain obligations that include control over the regulations, independent assessment of the proposals, motivation for the choice of winner, and loyalty towards the rest of the jury panel (Kazemian & Rönn, 2009). They have to deal with time pressure and have responsibilities towards the client, the profession, the participants, their own organisation, and society. Jury members should be capable of fulfilling these obligations and responsibilities. Doubts of the jury about uncertainty of their judgements are reduced by their competence and consensus among them (Kazemian & Rönn, 2009).

However, a team of experts appears not to be the same as an expert team. An expert team has to build common understanding, formulate plans on the most effective course of action, plan execution by coordinated team performance and learn by evaluation (Salas, et al., 2006). The use of a team can also create commitment to each other and the groups they represent. In creating an expert team, attention should be given to the composition, the context, the work design, and the group processes. Research suggests that equally sized subgroups in teams would benefit decision making and too many high profile members could increase the chance of groupthink (George & Chattopadhyay, 2008). In case stakeholders are represented in the jury panel that do not have specific expertise in architectural design, they have to be supported by the other members of the panel in reading design information. This does not imply that their judgements are of lesser quality but they might have difficulty in creating an overview of the information and expressing their preferences. All jury members provide input for the decision process from their own expertise. For the jury to act as an expert team the chair of the jury panel has to solicit ideas and observations of team members, stimulate and enable all kind of feedback and give situational updates (Salas, et al., 2004).

The research on expert teams focuses on diverse groups that act as coherent and repeatable teams. It suggests that teams are able to improve performance on repetitive tasks (Rosen, et al., 2008). A jury or client panel as often applied in tenders nowadays does not conform to this definition of expert team because of the temporary and the incidental character of a tender. However, it raises an interesting training issue. It implies that not only the level of experience as an individual decision maker, but also experience as a member of a specific team must be taken into account. The series of meetings in Scandinavian countries imply a kind of team development which opens up the possibility for feedback, monitoring and coordination (Kazemian & Rönn, 2009). In between these meetings several informal sessions are held where the members meet in smaller groups to scrutinise the proposals, assess their qualities and clarify issues about quality. All these activities stimulate a learning process which could also be done with similar tasks and providing feedback before the actual assessment process starts. This might even enable jury panels to be deployed as expert teams in different tenders to judge similar decision tasks. The concept of the expert team leads to the following possible success factor for a justification of the final decision:

Factor 14

A carefully composed, trained and well-guided jury panel will contribute to the success of an architect selection process.

In Chapter 4.6.2 I identified different models and tools that currently exist in the context of architect selections: procurement models, competition models, decision support systems and project management tools. None of these tools cover the full playing field of a selection process in architecture. An interesting issue is the use of decision support systems and other models in relation to transparency of the process. In traditional design competitions it is accepted that only the names of the jury members are published beforehand instead of exact procedure and definite criteria of decision making. In case of an experienced and well-known jury, the names actually provide participating architects with the information they need to take their chances in the competition. The justification of the final decision was considered to be reflected in a jury report. Still the underlying reasoning and negotiations remained within the group, but participants trusted the jury members in making sound decisions. Because of the intuitive character of jury decisions it would also be to justify a decision otherwise than by a written report. Almost all decision systems used during tenders use decision criteria, in line with the legal way of thinking about decision making. Most public clients decide on the members of their jury panel themselves. Often these panels include members of the board or city council without professional design experience. Because these people are less experienced in judging architectural quality based on drawings, they may get distracted more easily by criteria that are less relevant to the assignment (Evers, 1995). This increases the chance that tenderers will not accept the justification of the decision.

A procurement procedure offers structure during the assessment process. According to current interpretation of the legal requirements, the output of a decision support system can provide a justification of a decision because the outcome of the process can be retrieved by the itemization per tenderer or per criterion. Wierzbicki (1997) however questions if the outcome of such a decision system would be accepted as the final decision by the decision makers and suggests using a decision support system only as part of the process of decision making. Examples of the Urban Decision Room (van Loon, et al., 2008) or the Design Quality Indicator (Gann, et al., 2003) show positive results in increasing the transparency of the decision process when the preferences are used as incremental input during the process. A decision support system could therefore improve the transparency of the decision process but does not automatically reflect the final decision nor lead to a valid decision.

Taking the idea of sensemaking into account, trust will probably play an important role in the acceptance of a tender decision. If decision makers are not aware of the meaning of the output of a decision system or were not attending the process of decision process, it is very hard to legitimise or trust a decision. Future research will have to show the influence of decision support systems and other widely acknowledged methods and tools on transparency, objectivity and motivation of a decision. For now I suggest that the use of the decision support system or comparable structure can improve the transparency of the decision process to the decision makers, but does not contribute to the acceptance of the decision by the tenderers and the other stakeholders. This leads to the final possible success factor in the process of architect selection:

Factor 15

The use of a decision support system during the selection process will improve the transparency of the decision process to the decision makers but cannot be used to justify a decision to the tenderers and other stakeholders.

5.3 Research design

This research focuses on the complete process of decision making from the perspective of public clients selecting an architect in the context of EU Procurement law. The aim of the research is to analyse the causes of current problems in the decision practice of public clients during architect selection and provide input for the future design of tender procedures in architecture. In the previous section the theoretical frame and potential factors of influence on the process of architect selection have been discussed. Based on these insights I proposed fifteen success factors about value judgement and decision making that could provide a solid basis for the empirical work in a structure of five sensemaking processes.

To account for the fact that the research field on architect selections is nascent and neither empirical studies nor theories exist that address processes of decision making in this context I chose the case study method to gather empirical data. Building theory from case studies is a research strategy that involves using one or more cases to create theoretical constructs, propositions and/or midrange theory from case-based empirical evidence (Eisenhardt, 1989) and that typically answers research questions that address 'how' and 'why' particularly well in unexplored research areas (Edmondson & McManus, 2007). Each case serves as a distinct experiment that stands on its own as analytic unit and theory is built based on induction and replication logic.

The research design is based on the roadmap as proposed by Eisenhardt (1989), which synthesises previous work on qualitative methods, the design of case study research and grounded theory building. This roadmap describes that after the selection of cases and crafting of the research protocols, multiple sets of data need to be collected in the field in a flexible and opportunistic manner. A within-case analysis provides familiarity with the data and forces the researcher to look at the data from different perspectives. Hypotheses and theory can be built based on the constructs found in the case analysis in combination with a reflection of conflicting and similar literature. In line with the roadmap the theoretical framework and the data collection and analysis of the cases developed through a parallel but interactive process. The structure of the success factors and the constructs used during analysis of the data are combined in the final chapter in order to draw conclusions.

The method of studying cases makes it possible to study decision making in a real life context on different levels of individual, group and organisational decision making (Hackman, 2003; Yin, 2009). According to Eisenhardt and Graebner (2007, p. 25) "theory building from multiple cases is emergent in the sense that it is situated in and developed by recognizing patterns of relationships and their underlying logical arguments among constructs within and across cases". In this research I used case studies as a rich empirical description of particular instances of a phenomenon that are typically based on a variety of data sources (Yin, 2009). Each case can be described as an instrumental case study in the sense that it is an intensive study of a single unit for the purpose of understanding a larger class of (similar) units (Flyvbjerg, 2004). A variety of different forms of data was collected for each case to allow for triangulation between self-report, observed behaviour and official justifications.

Because the research aims at developing theory instead of testing it, theoretical sampling is appropriate. In this situation cases were selected because of opportunities for unusual research access and revelatory situations (Yin, 2009). Although transparency of public governance seems to imply otherwise, gaining access to tender situations proved to be very difficult. I experienced that tender situations often have a very sensitive and delicate nature. Next to that it appeared hard to

Case	Type of client	Type of procedure	Research methods
A School with Sports facility	School board with municipal representatives	Restricted tender for traditional design contract	Non participatory observations Interviews Document analysis
A City Hall with Library	City council with library representatives	Restricted tender for traditional design contract	Interviews Document analysis Non participatory observations
A Provincial Government Office	Provincial Executives	Restricted tender for Design Build contract	Passive participatory observations Informal conversations Document analysis
A Faculty Building	Dean as representative of the Executive Board of the University	Open ideas competition with prize money	Active participatory observations Documents analysis Interviews

Table 5.2 Overview of empirical data

trace clients preparing a tender before the official announcement is made to trace their motives. Within these limitations and the available time, I conducted three instrumental cases (Stake, 1995) in the context of a restricted tendering procedure in the Netherlands: a School, a City Hall, and a Provincial Government Office. The cases differed in the scope of the brief, the type of tender and the characteristics of the selection process. Additionally I conducted one case about an ideas competition for a new Faculty Building of a Dutch university. Table 5.2 provides an overview of the cases and their main characteristics.

The cases show the process of architect selection from a psychological perspective in their full complexity including the interrelations of the phases, actors and characteristics. To explore the different research methods and develop research protocols some pilot studies were conducted (Volker & de Jonge, 2007; Volker & Heintz, 2007). Chapter 6 describes the research methodology and results of a cross-case analysis for patterns across the three tender cases. The cases were analysed as separate identities first (Volker & Chao-Duivis, 2010; Volker & Lauche, 2008; Volker, Lauche, Heintz, & de Jonge, 2008) and then systematically compared on appearing constructs. The findings are described for each of the central constructs of an overall framework of the process of tendering for an architect. The framework identified three actor groups (a steering committee, a project team and a jury panel), four elements of the project (project characteristics, client governance, stakeholders and project management), and four fundamentals of a tender design (tender brief, process procedure, stakeholder involvement and decision process). These same constructs were used in the analysis of a fourth case about an international open ideas competition. The results and research methodology of this case are reported in Chapter 7.

The characteristics and processes as distinguished in the theoretical framework and the empirical cases were validated in a workshop with ten professional clients, legal professionals, and architects. The validation workshop consisted of two validation steps: reflection on the findings and a modelling exercise. A description of the workshop and the results are integrated in Chapter 7.

In Chapter 8 the emergent framework from Chapters 6 and 7 is compared with the evidence from each case and the sensemaking processes and success factors proposed in this chapter. Then conclusions are drawn about the decision process of public clients during architect selections and the implications for the design of an architect selection procedure in the context of European tendering regulations are described. Chapter 8 concludes with a reflection on the research approach and suggestions for future research.

Chapter 6

THREE EMPIRICAL TENDER CASES – CROSS CASE ANALYSIS

6.1 Introduction and research questions

Managerial decision making is a complex phenomenon that can best be understood in the specific contexts in which it takes place. A specific context, like the selection of an architect or the evaluation of a design proposal, is very difficult to simulate or trace. Therefore case studies are the most appropriate research method for exploring the complex relationships and associated causal mechanisms (Gerring, 2004) between decision making in design and the perception of design quality. As mentioned in the first chapter very little empirical research has been done in the specific context of client decision making during European tenders or design competitions (Strong, 1996). Just recently some papers have been published about the preparations and decision process for a design competition (Kazemian & Rönn, 2009; Kreiner, 2007b, 2008; Svensson, 2008) but tender procedures are not explicitly part of these publications. According to architects and consultants in practice, the lack of professionalism of public clients is the main reason for the increasing dissatisfaction about the selection of architects in the context of EU procurement law (van der Pol, et al., 2009). Clients acknowledge the current problems with architect selections but also point to more systematic difficulties. So far tender situations have not been analysed on the underlying difficulties that clients experience. The aim of this chapter is to help remedy this situation by exploring three conceptually different tender cases in the Netherlands.

The chapter focuses on the following research questions:

- How do public commissioning bodies decide on the selection of an architect in the context of EU procurement law?

- Which situational characteristics influence the process of decision making of public clients in this context?

6.2 Research methodology

The case study method was chosen in light of the fact that the research field is nascent, lacking both empirical studies and theories that address processes of decision making in this context. An instrumental case study can be considered as a rich empirical description of particular instances of a phenomenon that are typically based on a variety of data sources (Yin, 2009) with the purpose of understanding a larger class of (similar) units (Flyvbjerg, 2004). The research design fits the idea of the roadmap as proposed by Eisenhardt (1989) which synthesises the principles of qualitative methods, case study research and grounded theory. This chapter describes the results of a cross case search for patterns across three cases. The cases were analysed as separate identities first and then systematically compared on the basis of constructs as they appeared in the cases. The findings are described for each of the central constructs of an overall framework of the process of tendering for an architect (see section 6.3). The same constructs are used in the analysis of a fourth case concerning a Faculty Building in Chapter 7. In each section conclusions are drawn that contribute to the building blocks of a tender design. The reflection provides a short summary of the findings of these three cases.

A variety of different forms of data was collected for each case to allow for triangulation between self-report, observed behaviour and official justifications (see Table 6.1). In the School and Provincial Government Office case the observations and both formal and information interviews were the main source of information, while in the City Hall case the documents and interviews were most important. Unfortunately the official notes of the observations of the School case were destroyed in the fire at our faculty building in May 2008. But because most of the analysis had already been done and the other data, including pictures and personal notes, were still available, the School case could still play an important role in the research. The cases were set up according to the principles of Yin (2009) and Stake (1995) in order to address the potential shortcomings of using a case study method in terms of limited generalisability, validity and reliability. The semi-structured interviews were recorded and transcribed. The coding and analysis of all documents was done by the author in Atlas.ti, a software package to support qualitative coding, and validated by the supervisory team of the research project. In the City Hall case, the framework of the Design Quality Indicator and its resource envelope (Gann, et al., 2003; Whyte & Gann, 2003) was used to analyse the arguments for selecting the best design. In total 388 phrases were analysed and coded to one of the fourteen design aspects in the City Hall case. A more detailed description of this analysis can be found in Volker, Lauche, De Jonge and Heintz (2008). All data were collected in Dutch. Therefore all citations used in this chapter are translations by the author.

	School and sports facility	City Hall with library	Provincial Government Office
Observations	2 meetings (14 hours in total) 6 presentations	3 meetings (7 hours in total) 10 presentations	19 meetings (53 hours general report) 1 meeting (2 hours detailed report)
Documents	8 documents (competition rules, matrix, official correspondence)	6 documents (competition rules, memos, official correspondence)	15 documents (working documents, competition rules)
Interviews	8 interviews with project team members and representatives of the stakeholders	9 interviews with project team members, representatives of the client body and participating architects	informal conversations with project team members and members of the client body
Degree of involvement	Involved as observer from official announcement until final award decision	Involved as observer after first round of selection until final award decision	Involved as participant observer from before official announcement until the process was cancelled after selection phase

Table 6.1 Overview of research methods and data per case

6.3 Framework for data analysis

In this chapter the results of all three cases are integrated to validate the propositions as formulated in the theoretical framework for client decision making in the context of EU procurement law. The interview data, observations and documents of all cases are combined and validated by triangulation as much as possible by a process of interpretation, reflection and evaluation of the findings. The within-case analysis lead to the structure of three themes: the actors, the project and the tender design (see Figure 6.1). The framework shows resemblance with the three concepts that Jansen, Gössling, Merks and Geurts (2005) used for a content analysis study of the relation between decision making and decision quality: decision process characteristics, characteristics of the decision results and context characteristics.

The principles behind Figure 6.1 can be explained as follows. The client's steering committee, the project team and the jury panel (selection and/or award committee) are the key actors of the tender project. The input of the other stakeholders, including the proposals of the architects, is part of the information that decision makers use during the process. Architects are therefore considered as suppliers of information in the decision process.

A tender project consists of certain characteristics within a context of client governance, social stakeholders and project management. During the whole tender process changes in the environment cause a constant interplay between the actors and the project characteristics. The actors and the project characteristics provide the input for the design of the tender, which consists of the tender brief, the process of the procedure, the structure for the involvement of stakeholders and the decision process. The first three elements (tender brief, process procedure

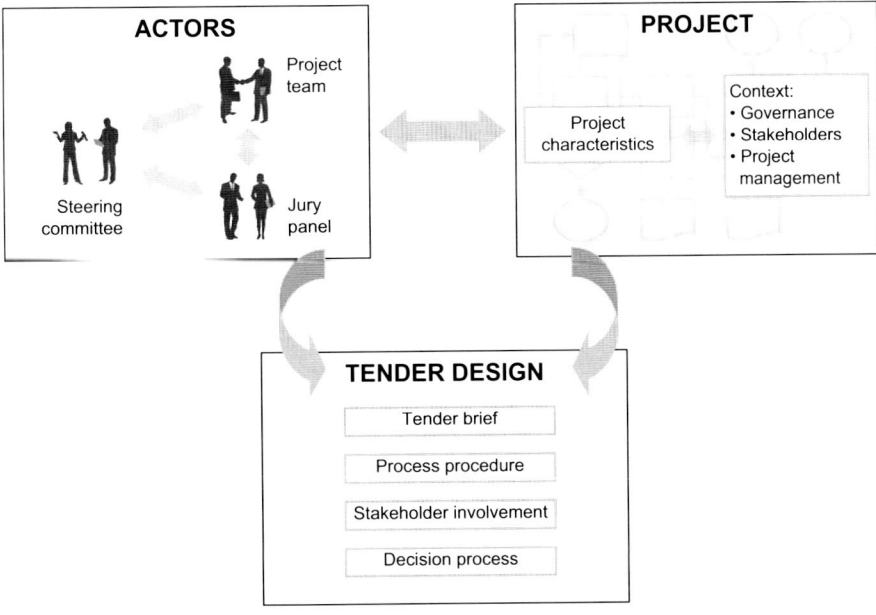

Figure 6.1 Theme structure of cross case analysis

and stakeholder involvement) provide the structure in which the decision process takes place. They therefore need to be designed before the tender can be announced. These design activities are mostly done by the project team. Therefore most of the analysis of the data is done from the perspective of the project team. The steering committee is considered to be of most influence on the involvement of stakeholders, in collaboration with the project team. The jury panel conducts most of the decisions in the actual selection process.

6.4 Case descriptions

In total three case studies were performed in a real life context in which the restricted tender procedure was applied. Table 6.2 provides an overview of the characteristics of the three cases. The School and City Hall cases were successful tenders, which led to a contract between the client and the architect. The tender of the Provincial Government Office was cancelled after the first round of selection. Pragmatism, the aim for good collaboration, and a drive for objectivity turned out to be important themes in the School case. In the case of the City Hall one of the main aims was to apply democratic principles in choosing a building that which would add value to the city. The Provincial Government Office case can be characterised as a political process in which some of the actors strove to gain control over the process and sought support within and outside the organisation. The following sections describe the cases in more detail.

6.4.1 A pragmatic process: A School with Sports facility

The first case is about the selection of an architect for a large elementary school, called 'de Twaalfruiter' in a town in the middle of the Netherlands called Vleuten. The tender took place in the period from August till November 2007. Vleuten falls under the jurisdiction of the City of Utrecht. The project also includes a sports facility with two halls and additional facilities. The School Board (of the Foundation of the Roman Catholic Elementary Schools of the places Vleuten, De Meern, Haarzuilens) and the Department of Sports of the City of Utrecht are the official awarding authorities. The main shareholders in this case are the

	School with Sports facility	City Hall with Library	Provincial Government Office
Contract	Traditional	Traditional	Design & Build
Total building costs (excl VAT)	€ 5.7 million	€ 31.8 million	€ 39 million
Gross floor area	5,700 m^2	18,300 m^2	16,000 m^2
Selection decision mainly based on:	Portfolio, reference projects, CV lead architect	Reference projects, profile of lead architect	Reference projects
Award decision mainly based on:	Presentation of design vision	Sketch design, scale model and presentation	Concept design and presentation
Tendering phases analysed	Selection and awarding phase	Award phase	Selection phase
Financial compensation for tenderers award phase	€ 0	€ 15.000	€ 150.000

Table 6.2 Overview of restricted tender cases

School Board, two departments of the City of Utrecht (Department of Sports and Department of Education) and representatives of a holding company developing the neighbourhood called the GEM Vleuterweide (Dutch abbreviation for Common Exploitation Company Vleuterweide). This case is done from the perspective of the School Board because they took the lead in the project. They chose to use the restricted tender procedure to select their architect. A consultant was hired to act as official awarding authority in terms of organising the tender, providing a brief, and managing the overall project. The consultant functioned as the contact person for this tender.

In the first round of the Vleuten case, the selection criteria for decision making stated in the competition rules were:

- Legal requirements - to be fulfilled by signing a declaration form.

- Economic and financial requirements - to be fulfilled by signing a declaration together with a signed statement of the annual turnover of the last three years.

- Professional requirements - to be fulfilled by a portfolio with three comparable projects designed by the proposed lead architect, a statement of willingness to take on the job based on total engineering (awarding the contract of all design activities (functional, technical etc.) to one party acting as contact person and coordinator), three letters of reference signed by the client in question, and short descriptions of the firm's organisational structure, the number of employees, the firm's quality management system and the experience of the lead architect proposed for this job.

The selection committee consisted of the four client parties with similar voting rights. They all assessed the portfolio, the reference projects and the CVs of each candidate during a one-day meeting. The other documents and requirements were checked beforehand by the consultant. Different weights were assigned to the decision criteria for the selection phase (Portfolio: 3, CV: 2, and Reference projects: 1). The weighting factors of the criteria were not mentioned in the official competition rules but decided on in an earlier meeting and applied in a matrix (selection criteria on the horizontal axis, participants on the vertical axis) during decision making. Selection as well as awarding would be 'based on unanimity'.

In total 35 firms applied for this tender, of which three were excluded from selection because they did not comply with the minimum requirements and exclusion grounds. The client decided to select six tenderers for the award phase. The candidates received a letter describing the selection decision, accompanied by the official minutes and an appendix with the matrix form filled with the assessment scores of all candidates.

The award committee consisted of the School Board and the Department of Sports, advised by the GEM Vleuterweide and the Department of Education. The contract would be awarded based on the most economically advantageous tender. The official instructions to the tenderers for the elementary school and complementary sports facilities read as follows (translated from Dutch by the author):

"Please provide a vision for the present construction task. Feel free to elaborate on the presentation of the vision on all these issues. The award committee expects visuals to enable assessing of the vision about the project. Please go into the spatial-functional brief (the concept of the school and the relation with the sports facilities), ambition for a sustainable and durable building (suggestions for a healthy interior climate and the process to realise this ambition), and the urban context (contribution to the architectural quality of the area and relationship with the neighbouring houses and facilities)."

The main requirements of the brief were (see also Figure 6.2):

- Integration of the school building and the sports facility, into a single design.

- The presentation of legal aspects necessary to permit the commission to be awarded as a 'total engineering' project.

- A building volume of 4,160 m² gross floor area for the school and 910 m² gross floor area for the sports facilities. A site plan is required for the area of the school building, sport facilities and outside space of about 5,370 m² with a footprint of about 2,200 m².

- A budget for the school building of about € 5.3 million including VAT, and a budget for the sport facilities of € 1.5 million including VAT.

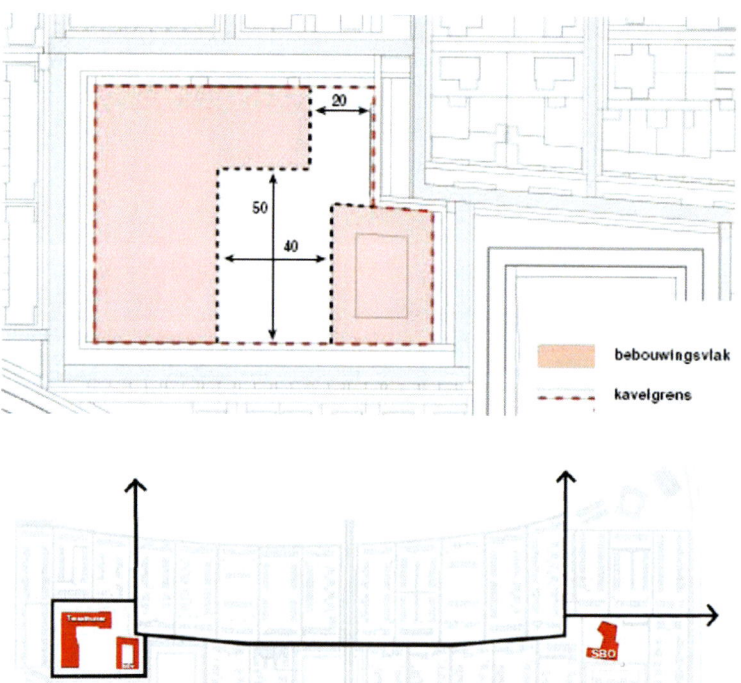

Figure 6.2 Detailed map of location and urban environment with main access routes. (source: brief in competition rules, drawing by Enno Zuidema Stedebouw)

The official award criteria as stated in the competition rules were:

- Vision for the project (the brief, ambition for sustainability and the urban planning context) – to be assessed based on information provided by the participant in a format of their choice.

- Professional competencies of the designer and the firm – to be assessed based on the presentation of their portfolio and of a presentation of their vision for the project.

- Communicative skills – to be assessed based on their ability to understand and translate the requirements of the client, their empathy, and ability to communicate in Dutch.

- Vision (or proposal) for the realisation of the project and the estimated fee of the participant – to be assessed based on a separate written document handed in at the time of presentation.

In the matrix sheet that was used during decision making these criteria appeared to have different weighting factors but these weights were not mentioned in the official competition rules.

The client requested that the architect who would be responsible for the project in case of winning the contract would perform the presentation to the jury panel. They did not require the entrants to hand in their presentations beforehand. The School Board and the Department of Sports decided to award the contract to Architecten- en Ingenieursbureau Kristinsson. The participants received a letter announcing the award decision, accompanied by the official minutes and a matrix sheet filled with the assessment scores of all six participants. As is common in the field of school design, the participants received no financial compensation for their submissions. In February 2010 the architect was still working on the detailed design. The project delivery is planned for September 2011.

6.4.2 A democratic process: A City Hall with Library

The City Hall case study concerns a restricted European tendering procedure to select an architect for the design of the new town hall and library for the historical city centre of Deventer. The tender took place in the period from February till June 2006. Deventer is a city of 100,000 inhabitants in the eastern part of the Netherlands. The design brief called for 18,300 m² gross floor area and was challenging because of the central location and narrow streets of an historical character, the parking requirements, and the presence of an office building and of a theatre considered to be of architectural and historical value. The City of Deventer wanted to utilise the professional insight of the competing architects to determine the fate of the two buildings, and therefore gave the architects as much freedom as possible in providing a design solution.

The client consisted of the City Council and the Board of the Mayor and Aldermen. The board of the Deventer library was involved as one of the future users of the building. The City Council acted as the commissioning client. A project team and steering committee were assigned to the tender project. The project team consisted of an interim manager and several staff members of the municipality. They hired several consultants to assist them on legal and financial matters. The steering committee included members of the executive board of the municipality, the chair of the Deventer library board, the head of the real estate department of the municipality and the project manager.

The official tender brief from the City of Deventer read as follows (translated from Dutch by the author):

> *"Deliver a sketch design of maximum 15 A4 pages, clarification, a plan of action and scale model for the new City Hall, the library, and an underground parking garage on the location of the former theatre in the Grote Kerkhof square, resulting in the capacity of 18,300 m2 in total."*

The main requirements in the brief were described as (see also Figure 6.3):

- The construction budget was set at € 31.8 million excluding VAT (total project costs € 55.9 million) – including machinery (climate control, lifts etc) and a parking garage.

- The council hall was to be restored and the design should fit into the existing urban development plan. No decision would be made beforehand whether the facades of the current buildings at the location should remain or be demolished.

- The main entrance was to be located at the square and the reception and the public functions at the ground floor.

- A new concept of office organisation would be applied, so a variety of workplaces had to be offered and the building should offer a high degree of flexibility in itself.

- The design concerned the urban fit and contours of the building, as well as the vision and translation of the organisational processes into the design, such as the synergy between the library and the city, the welcoming of clients and the physical translation of the workplaces.

The contract was to be awarded based on the most economically advantageous bid that fulfilled the aesthetic and functional aspects of the design. According to the competition rules, all three criteria weighed equally. The criteria were described as:

- The degree of flexibility of the programme concerning the synergy between library and City Hall, integration of front and back office, and technical and environmental durability.

- The intelligence and creativity of the solution in its historical context.

- The contribution to the diversity and restoration of split urban character of old and new.

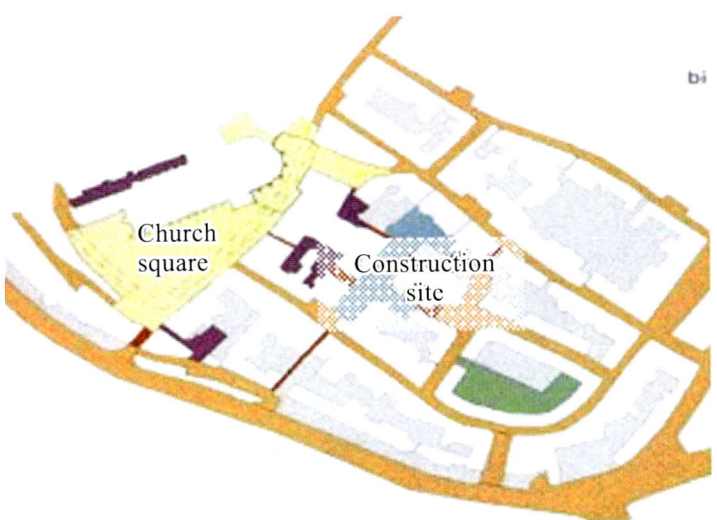

Church
square

Construction
site

Figure 6.3 Current situation and historical map of the location (source: appendices of competition rules)

In total 22 firms showed interest for this tender by submitting the documents asked for in the call for participation. A selection was made by the selection committee based on the selection criteria of the reference projects (maximum of 60 points) and the proposed lead architects (maximum 40 points). The selection committee consisted of representatives of the Board of the Mayor and Aldermen, the municipality, the Deventer library and the project team of the tender. The selection process existed of a meeting of the selection committee with discussion and voting. Because the selection phase of the tender had already taken place when the case study started, there were only limited data available about this phase (mainly interviews).

Design 1 Architecten Cie

Design 2 Architectuurstudio HH & Witteveen+Bos

Design 3 Ibos Vitart

Design 4 Kraaijvanger·Urbis

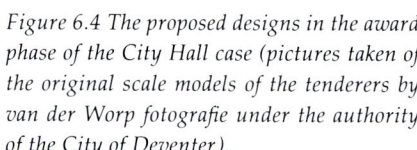

Figure 6.4 The proposed designs in the award phase of the City Hall case (pictures taken of the original scale models of the tenderers by van der Worp fotografie under the authority of the City of Deventer).

Design 5 Neutelings Riedijk Architecten

Five firms from the 22 were invited to join the award phase of the tender. Figure 6.4 represent illustrations of the scale models of the designs that were presented during the tender. The architects, except for the winner, each received € 15.000 compensation for their activities in the award phase. The award committee, who had to determine the winner of the tender, consisted of representatives of the political parties of the City Council. The award committee considered the advice of several stakeholders (library, user group of the city and the citizens) as well as the advice of an expert committee during a voting process. An advisory report by the project team provided them with an assessment of the proposals in light of the stated project requirements.

The design proposals can be described as:

- Design 1 (Architecten Cie.) consisted of three separate but connected buildings. It placed the library at the front and the city offices at the back and demolished all the existing buildings at the site.

- Design 2 (Architectuurstudio HH in combination with Witteveen+Bos) used a system of layers to fit into the surroundings and placed the library at the back of the building, creating a large internal square. Several existing buildings were included in the design.

- Design 3 (Ibos Vitart) could be described as a long glass ribbon. The entrance of the library was situated at the front. All existing buildings at the site were to be demolished.

- Design 4 (Kraaijvanger·Urbis) organically filled the available space and renovated most of the old office building. The entrance of the library was situated at the front.

- Design 5 (Neutelings Riedijk Architects) was shaped as an ellipse and provided a large winter garden at the top. Almost all existing buildings were renovated and the library entrance was placed at the front.

The firm of Neutelings Riedijk (design 5) was chosen as the winner of this tender.

In 2009 several members of the current executive Board of Mayor and Aldermen resigned over this project because the council did not approve the detailed design of Neutelings Riedijk. The main reasons were the fact that the design did not meet important local planning requirements, e.g. that the height of the building should not exceed the height of a nearby church. The project also failed to provide a convincing and supported calculation of constructions costs that was within budget, even after a second opinion by an independent engineering firm. These conflicts with the requirements were known during the tender selection process but were ignored in the decision making process. The most recent information regarding the City Hall case suggests that the project has been indefinitely postponed.

6.4.3 A political process: A Provincial Government Office

The third case concerns the selection of a Design-and-Build consortium for a new part of the governmental offices of the Province of Utrecht. This case involved the period from July 2007 till January 2008. The current offices of the Province consist of a tower, built in the 90s, and a lower block called 'The Stars', built in the 70s. This case regards the tendering procedure for a new building replacing 'The Stars'.

The Provincial Council (PS) is politically the representative body of the province and therefore the commissioning client in this case. Provincial Executives (GS) are charged with the day-to-day management of the Province. Several of the executives were part of the steering committee of the tender project, accompanied by the heads of the provincial organisations to be housed in the new building. The project team consisted of two part-time project leaders, a secretary, two staff members of the purchasing department, and several consultants. The client chose to tender their contract as a Design-and-Build contract and to use a restricted tendering procedure. Unfortunately the tendering procedure was cancelled just after the qualitative selection phase due to reconsideration of the original positions of the decision to replace the lower part of the building.

To apply for the tender interested parties were asked to form a consortium, which should include at least an architect, a contractor and a project manager. The consortium would be responsible for delivering the new building (design as well as construction) for a fixed price of € 39 million excluding VAT.

The brief for the Provincial Government Office reads as follows (translated from Dutch by the author):

> *"The new building has to form a new office concept with flexible workplaces and contemporary facilities. The building should be more efficient (technically and functionally) than the current building and requires a new central entrance. The building has to offer about 10,200 m² usable floor area (about 16,000m² gross floor area) and space for about 500 workplaces. A special point of attention is the connection / relation to the tower."*

Eight consortia applied for this tender. The minimum requirements and selection criteria were based on strict organisational and financial requirements of the consortium but the deciding factor was suitability of the architect, to be assessed based on three reference projects that were designed by the proposed lead architect. Three consortia were excluded for failing to meet the minimum requirements, resulting in five candidates. The selection committee, consisting of members of the executive board of the Province and an independent chair, determined the degree in which the reference projects fulfilled the criteria of a 'public, timeless, business-like character with a human scale' in a single meeting of three hours. They were assisted by an architectural expert. All consortia participating in the award phase would receive financial compensation of € 150.000.

The selection criteria were:

- Financial requirements – to be documented by an audit certificate with the annual turnover and liquidity of the last three years of at least one of the consortium members.

- Professional requirements of the contractor – to be documented by a certificate of a professional register, a certificate of the quality management system and a certificate for the VCA (Safety, Health and Environment for contractors) of at least one of the consortium members, and three reference projects with specific minimum requirements about the function (minimum 50% office), scale (minimum of 10,000m²) and budget (€ 14 million construction costs), that had to be delivered in the past five years, including a statement of satisfaction of the client, at least one of the project needs to be sustainable and at least one of the project needs to be a utility building realised by an integrated contract.

- Professional requirements of the architects involved in the consortium – to be assessed by reference projects of three new utility buildings that were designed by the proposed lead architect and delivered in the past five years, visualised on an A0 poster without any reference to the name of the architect nor the consortium.

If the tender had not been cancelled the participating consortia would have been judged on a detailed design in the award phase. Award criteria for the economically most advantageous tender would most likely have been: 'architectural quality', 'functional quality', 'durability', 'maintenance', 'price', and 'plan of action'.

6.5 Findings about actors

The results of the cases in relation to the actors show that a tender is considered by clients as a project and treated similarly to other projects. Due to the governance structure of public commissioning clients, decisions were usually taken in groups and not by individuals. Many people are involved in the decisions made in a tender project. Table 6.3 shows an overview of the actors in the cases and their positions. In some cases the steering committee consisted of virtually the same people as the selection committee. In the School case the selection and award committee were similar but different in authority, while in the City Hall case the membership of the award committee was completely different from that of the selection committee. The characteristics of the committee and team members influence the implementation of the tender greatly. The next sections describe the characteristics of the actors in a tender project more thoroughly.

	School with Sports facility	City Hall with Library	Provincial Government Office
Steering committee	Members of the School board, head of the municipal department of Sports, project manager	Members of the executive board, chair of the library board, department heads of the municipality, project manager	Members of the executive board, head of the departments of provincial organisation
Selection committee	Members of the School board, employees of the municipal department of Sports, employees of the municipal department of Education, representatives of the GEM Vleuterweide	Members of the executive board, chair of the library board, departments heads of the municipality	Same as steering committee: members of the executive board, department heads of Provincial organisation, with external advisor
Award committee	Members of the School board, employees of the municipal department of Sports	Representatives of all political parties, advised by citizens, experts and user groups	Probably same as selection committee but not constituted.
Project team	Two management consultants, director of the School	Interim project manager, members of executive board, head of staff departments, permanent staff members, and management consultants	Internal project managers, permanent staff members, and several management consultants
Characteristics of the tender process	Pragmatic process, drive for objectivity and collaboration	Democratic process, ambition to make a statement supported by community	Political process, fear to loose control and search for internal support

Table 6.3 Overview of actors in the cases

6.5.1 Steering committee

A public client is embedded in a governance structure which cannot be reduced to one single decision maker. Still decision makers are held accountable for the public money they spend. One of the participating architects in the City Hall case quite ironically stated in an interview that:

> "A public client is in fact not a single client but a monster with many heads, a conglomeration of clients who does not know what she wants and always tries to find a solution in the middle."

In order to realise the building project in all cases a project structure was created with a steering committee to make the most important general decisions and a project team to organise the project and prepare the decisions of the steering committee. The tender was usually a part of the total building project with the same steering committee but a separate project team. In all cases the final decision about the winner of the contract was to be approved by the official commissioning body: the School Board, the City Council including the Aldermen or the Provincial Council. The steering committee mostly consisted of a few members of the public commissioning client body, complemented with members of the permanent managerial staff with facilities or finances in their portfolio. Two of the School Board members were real estate professionals in daily life, but in the other cases none of the steering committee members were professionals in architecture or related fields. In fact, none of the commissioning clients were professional clients which meant that a building project was a rare event for most decision makers. This can be explained by the fact that a public client is part of the public administration and is therefore run by politicians instead of professionals in construction or architecture. In the cases in which (one of) the project champion(s) was part of the commissioning body, the School and the City Hall, the decision processes seemed to be more fluent that when the project champions were only part of the project team.

The building project was perceived by some members of the steering committee as a positive challenge, while for others such a project appeared to be the impetus of uncertainty and fear. This perception was partly influenced by the personality, competences, experience, ambition and expertise of the person responsible for the project. One of the managers of the Provincial Government Office described this phenomenon during a meeting of the project team:

> "Obviously there is the fear of a political Waterloo. Uncertainty about the contract dominates the process. ..... Fear is always a bad advisor."

The fact that an architect has to be involved in a project also seems to be perceived as a risk by certain clients. Architects have earned a reputation for spending too much money and being too dominant in taking design decisions. Some clients do not seem to be bothered by this reputation and acknowledge that they need an architect to transform their ambitions and requirements into a physical shape. The School Board, for example, was really looking forward to working with the architect; they actually perceived the architect as a partner in this project. Therefore they chose to organise the tender at a phase in their project in which the brief was not yet fully specified. One of the basic ideas for the City Hall was to use the creativity of the architect in shaping their ideas of 'a new home for the city'. They chose to ask for a sketch design in the award phase to see how this transformation would work out in their situation. During the preparations of the Provincial Government Office the lack of detail on certain aspects of the brief actually increased the anxiety about the possible dominance of architects in the project. This feeling was increased by the fact that they would award an integrated Design-Build contract in which the client cannot make many changes to the design before construction. The search for control over the project was shown with a comment drawn from the observations:

> "We need to find enough space between the preliminary award decision and allocation of the contract.... Everything seems to be disputable. He who pays the piper should call the tune."

6.5.2 Project team

Because of a lack of knowledge, expertise and capacity about tendering and building projects, management consultants were hired in on a project basis in all cases. All tender project teams consisted of a mixture of staff members and external advisors. Only in the case of the Provincial Government Office were the project managers permanent staff members. They were supported by a legal advisor, a management consultant and some colleagues from the purchasing department. The School Board (made up of part-time volunteers) hired a management consultant to run the whole building project, including the tender. In the case of the City Hall an interim process manager took the lead, while being advised by a legal consultancy firm and several employees of the municipality. In most cases the project manager of the (future) building project was not actively involved in the tender. This meant that the (future) project manager would not have a say in the decision about the person (s)he will be working with. The project manager of the City Hall case explained during an interview that in his case this was done on purpose to make a clean start of the next phase:

> "I have been there for two years and that is a long time for an interim manager. I played some dirty tricks to realise this tender. Then the bad guy needs to be replaced by the good guy. In the next phase someone needs to build a trustful relationship between the architect and the client."

In the City Hall case the external project manager was found to have the advantage of being able to make decisions autonomously in order to keep the process running. In the School case the consultant acted as an independent mediator between parties. Then again, the findings of both cases indicate that external members of the project team are not always aware of the sensitivities of the project compared to permanent staff members. Therefore they might be less able to translate the ambitions of the client into the design of the tender and take delicate issues into account. In the Provincial Government Office case a well reputed legal advisor became involved in the final stage of the project. The legal advisor rewrote the call for participation as taken down by the project team and threatened to withdraw their legal responsibility if the paragraph about the architectural ambition of the client and the related selection criteria was not made more operational than as written by the project team. However the subjective character of this aspect of architecture made it almost impossible for further objectification. Finally the issue was solved through the advice of a legal expert from the Dutch Government Building Agency who recently worked on a legal guideline for tenders for the central government. The Provincial Government case showed that too much involvement of a legal advisor could lead to a legalistic procedure in which complying with the law came to be more important than selecting an architect. This incident also shows the delicate and rather undeveloped status of architect selections and the lack of experience with interpretation of the regulations.

Among the issues for the steering committee to decide was the composition of the selection and award committees, the performance levels required to meet the requirements, the final approval of the selection and award criteria and the project brief. All the documents for the steering committee were prepared by the project teams. The results of the cases indicate that the strength of the link between the project team and the steering committee determines the basis for the success of decision making and the translation of the ambitions of the client into the process procedure of the tender. This can be enabled by the project manager who connects the project team, jury panel and steering committee, as in the City Hall case and the School case, or by putting members of the project team and/or steering committee in the jury panel. In the Provincial Government Office the link between the project team and the steering committee cum jury panel appeared to be very weak and presumably one of the reasons that the tender was cancelled.

6.5.3 Jury panel

All clients appointed a jury panel (in tender usually referred to as a selection and/ or award committee) to judge the proposals in the separate phases of the tender process which was just as the project team and the steering committee especially composed for the project. In all cases the data collection started after the composition of the actor teams. It is therefore hard to retrieve the exact reasons why people were involved in the jury panel or other committees. The interviews indicated that the panel members were mainly selected because of their position within the organisation and related field of expertise, or their position within the most important stakeholder organisations. None of the members in the jury panels of the cases was specifically asked because of their expertise about architect selections or architecture in general. The selection committee the School consisted of representatives of the steering committee and the core organisation of the client, completed with representative of the supervisory groups. The same committee acted as the award committee - only the rights of the supervisory groups changed from decisive into advisory. In the City Hall a similar committee acted as selection committee. The award committee however consisted of representative members of the political parties, who had voting power proportional to the size of their party. This political composition of the award committee reflects the character of the award phase in which many stakeholder groups were involved. All stakeholder groups had advisory rights to the members of the award committee. It should be noted that the Mayor and Aldermen were not part of the award committee; rather their political parties were represented in the award committee. They did have the final approval authority over the decision of the award committee. Neither the consultant nor any other member of the project team had decisive or consultative rights in any of the cases. However, they often facilitated and/or chaired the committee meetings. Only in the case of the Provincial Government Office did the steering committee also act as the selection committee.

6.6 Findings about project characteristics

The findings about the project characteristics address the relations of the tender project with the overall building project and the governance structure of the client. It shows that a tender project is embedded in a network of potential conflictive stakeholder interests and expectations. Additionally the object and character of a project – the creation of a physical building – creates issues of discussion on different levels and perspectives that have to be taken into account during the decision process. Because of the multiple perspectives that characterise public administration, private and professional interests of decision makers often interact. The following sections discuss the project characteristics and decision issues that were found in the cases in more detail.

6.6.1 Project characteristics

Tendering regulations apply when public clients want to build their own office buildings but also for other public buildings, such as educational buildings, prisons, museums and social housing. In this research one case concerned the construction of a primary school building and two cases concerned future municipal and provincial offices annex public information 'houses'. Within these projects two kinds of briefs were distinguished: 1) the brief for the tender process, and 2) the brief for the building project. The building brief is meant to be point of reference for the design and construction of the new building and includes physical, social, planning and financial requirements. The tender brief should express the ambitions of the client to support the selection process of the architect and provide a basis for the bid of the tenderers. Theoretically the tender brief could be the same as the building brief but a clear difference in aims was found. Next to that there are privacy related issues that could prevent clients from using the exact building brief in a tender procedure.

In all cases the tender brief was based on the certain specifications in the building brief and the site and budget for the project were clearly defined. In all cases the development of the building brief was however not completely finished at the time the tender was announced, which indicates the ambiguous character of the requirement in this early stage of the building project. The differences in the kinds and relative importance of the award criteria between the City Hall and the School suggest a strong connection to the kind of tender brief and the reasoning behind the award procedure (sketch design versus project vision). When asking for a sketch design in the tender procedure the tender brief needs to be on a detailed level, as in the case for the City Hall. The School attached a tender brief with less detailed functional requirements for the School Building but very specific requirement for the Sports facilities. The representative of the Sports department of the City of Utrecht explained during a retrospective interview that they used standardised briefs for the development of sport facilities, and thus also for the use in tenders. Overall it meant that the tenderers in the School case did not have enough information to make a sketch design, which suited the aim of the tender.

Because of the importance of the brief as a reference point for the tenderers in preparing their bid, these findings indicate that a tender brief should be specifically designed for a tender. Apart from the monetary thresholds, the size of the project did not seem to be important for the design of a tender. But because a tender is always based on assumptions made in an early phase of the project and the long time span of a construction project, these assumptions often have to be adjusted along the way. In all cases the requirements as stated in the competition rules already changed during the period the tender was running and decision makers had to deal with increasing insights about their projects.

Most building projects have specific constraints that create specific challenges for the designer. Some challenges relate to the characteristics of a building sector, others to the location or users. In the Netherlands, school buildings for example generally have a limited budget determined by the central government. Transference and accessibility of information is very important for a library or a City Hall. Analysis showed that in some cases the client used these challenges to create input for the selection process. They deliberately left some challenges open for the architect to provide a solution during the tender. In the case of the City Hall it was still to be determined if the office building and the theatre were to be demolished. Also the tender brief in the School case opened up several possibilities to integrate the sport facilities with the school building.

In all cases the obligation to set a good example to others played an important role in the character for the future building. In the School case and the Provincial Government Office sustainability issues were very prominent in the building brief. For the School Board it was very important to provide a safe environment for the children. The City Council and the Provincial Executive Board aimed for prestigious offices. At the same time, they all felt the pressure to spend their money as efficiently as possible and to be accountable for their actions. The data suggest that the ambition level of the client determines the impact of the exemplary role of the client on certain specifications in the tender brief; the greater the ambition, the more specific were the requirements incorporated in the tender brief. This reflects on the profile of the architect clients are hoping to find through the tendering process.

6.6.2 Client governance

From a legal perspective the main aim of a tender procedure is to select an architectural design firm. The findings of the cases showed that a tendering procedure often has multiple aims, apart from awarding the contract to an architect. These aims related to the process, the outcome and/or the project of the tender. If the aims for the process of the tender were in line with the principles of EU tendering, such as being transparent (City Hall) or being a responsible client (School), the chance of conflicts with the regulations was limited. However, in the case of the Provincial Government Office most decision makers perceived the procurement regulations as inconveniently restricting their range of action. The rigid approach implied by the legal procedure did not suit the strategic aims of the decision makers. Nor does the legal process seem to accommodate user participation. In the City Hall case and the School case the tender project was seen as an opportunity to involve stakeholders, to create support within a community and to show the competences and ambition of a client. User participation during the tender was therefore seen as a positive contribution to the building project. The members of the steering committee of the Provincial Government Office perceived user participation as a threat to their decision making latitude and to their chances of obtaining a secure position within the organisation.

A new building is often perceived by clients as an opportunity to change the virtual and social organisation as well. Clients believe that architects could contribute to this change by persuading the stakeholders to support other decisions concerning the organization made at the same time. In the City Hall case the client used the building project to promote the centralisation and synergy between the municipality and the library organisation, and the implementation of a new digital library concept, and demonstrate the potential to increase the liveliness of the location. Similarly the Province wanted to use the architect selection process to support the introduction of a new office concept of non-territorial workplaces. The School building would be the official start of a new school in a recently developed urban area and the implementation of an innovative school concept. Because of the importance of the architect in the process of a project, these complementary project aims also determined the interests of the client during the selection of an architect.

Results show that in some cases the tender project was used to strengthen collaboration within the client organisation or with parties directly related to the organisation. The School Board involved the City of Utrecht for mainly strategic reasons. Their ambition was to create a sustainable and 'healthy' building, which would require additional finances on top of the standard funds provided by the municipality. In the City Hall project, the Mayor realised that adding an architectural icon to this historical location could raise conflicts with the advocates of the historical city centre. The ambitions of the administrators of the city were therefore doomed to conflict with the concerns of the other stakeholders. In both cases user participation was used as a means to provide a fruitful foundation for the project and limit the risks of failure. In the case of the City Hall this did not prevent the project from being delayed - a conflict about the budget and the city skyline led to a managerial crisis three years after the contract was awarded.

Apart from the perception of the legislation and user participation, the governance structure itself appeared to be of direct importance for the success of a tender. Both in the case of the Provincial Government Office (cancelled after the selection phase of the tender) and the City Hall (in which the project was put on hold during the design phase) the governance structure seemed to conflict with the long term perspective of a building project and hindered the continuation of the (building) project. Because of the periodical changes in the councils, decisions taken in a previous term of office need to be executed in the following. Detail and justification of the decision is missing because employees involved in the tender have shifted department or organisation, as was the case for the City Hall and the Provincial Government Office. This effect seemed to be enhanced by the personal characteristics and leadership competences of public decision makers and the impact of these competences on the internal support of the decision made in the tender.

In the case of the Provincial Government Office a greater degree of uncertainty was found among the actors compared to the other cases. In the School and the City Hall the project champions, respectively two members of the School Board and the Mayor, created a strong framework for decision making by stating their ambitions clearly. Still, while a shared frame of ambitions might contribute to the ease of the decision, it cannot make differences in the interests disappear. The findings indicate that a high level of uncertainty can be attributed to a lack of clear project ambitions and therefore also a lack of consistent or shared interests of the decision makers. Further analysis showed that independently of the ambition level, in every case differences between the interests of decision makers could be found. In the next section about the social context of the stakeholders these differences will be addressed more specifically.

6.6.3 Social context of stakeholders

All cases illustrate that the public character of a client led to of the involvement of a large number of stakeholders in decision making (see Table 6.4). Because the built environment is something generally known and experienced by all kinds of people, a plethora of ideas, interests, ambitions and opinions could be found among the stakeholders.

In the City Hall case the decision makers mainly differed in their concern for user needs (staff versus political parties), the integration in the context (citizens versus politicians), the importance of financial means (between different political parties), and the political aims of the project (between City Council and Board of Mayor and Aldermen). During the School case differences mainly occurred between the School board and the GEM together with the Municipal school department that related to the interest level (urban level versus building level), the user levels (shareholders versus users), the input of means (realisation versus maintenance), the perspective in time (short term versus long term use), interests in the product (quality versus time and money), interests in the process (control versus risks), kinds of responsibilities (supervision versus commissioning), and need for innovation (personal use versus exemplary function). In the case of the Provincial Government Office the perception of risks, the decision supportive level, the responsibilities, and the level of involvement in decision making differed between the stakeholders of the project, especially between the Executives and the staff members of the project team. All together this led to the following overview of pairs of potential conflicting interests that need to be addressed during the selection process of an architect as displayed in Figure 6.5.

	School with Sports facility	City Hall with Library	Provincial Government Office
Commissioning client body	Municipal department of Education	City Council	Provincial Council
Representatives of the client body	School Board, Municipal department of Sports	Board of Mayor and Aldermen	Provincial Executives, Queen's commissioner
Shareholders and supervisors	Common Exploitation Company, Municipality	Province, Department of Interior and Kingdom Relations	Department of Interior and Kingdom Relations
Daily users	Teachers, Children	Local government servants and officials, Library employees	Provincial employees
Non-daily users	Parents, Neighbours	Citizens, Library users, Tourists, City centre shop keepers, Neighbours	Provincial inhabitants, Visitors, Business park companies
Representative groups of the users	Employees union, Parents' council	Employees council act, Historical interest groups, Political parties	Employees council act, Biological or environmental interest groups, Political parties

Table 6.4 Overview of most important stakeholders per case

CLIENT RELATED ISSUES

Short term	Long term
Certainty	Innovation
Political support	Policy
Permanent staff	Political parties
Exemplary function	Accountability
Personal interests	Professional interests

BUILDING RELATED ISSUES

Urban level	Building level
Personal use	Representation function
Innovation	Functionality
History	Future
City heritage	Personal remembrance

PROJECT RELATED ISSUES

Quality	Time and money
Construction	Maintenance
Process partner	Design partner
Primary budgets	Secondary budgets

PROCESS RELATED ISSUES

Decision responsibility	Advisory rights
Supervisory function	Commissioning body
Political support	Content of decision
Control	Risks

Figure 6.5 Overview of four categories in pairs of potential conflictive interests found in the cases

Most of the decision makers belonged to different stakeholder groups. The members of the Provincial executive board for example, were employees of the province, but also the representatives of the client body. Therefore they were stakeholders and decision makers at the same time with both professional as well as personal interests. The director of the School was part of the jury panel, the project team and future responsible officer for the building project but the director also represented the children, the parents and the teachers during the process. The interests of the stakeholder groups related strongly to the client, project, process and building related issues that were displayed during the decision process. On a general level distinctions were found between shareholders (with mainly financial involvement) and stakeholders (with practical user involvement), between decision makers (involved in the decision process) and passive stakeholders (often represented in the decision process), and between personal interests (with private responsibilities) and professional interests (with official or work related responsibilities).

In order to make a final decision several of these potential conflicting issues needed to be addressed within the certain period of time. Delaying the tender is usually no option because of legal or financial implications and cancelling would mean loss of face for the board and council members. In both the City Hall and the School case the clients expected difficulties between the interests of the stakeholders; they therefore intentionally structured the decision process well and defined clear responsibilities and expectations of the stakeholders. The resulting successful process regained trust in the professionalism of the municipality in the City Hall case and created goodwill with the supervisory parties of the School. During the process also decision makers gained understanding in each others interests which seemed to be a promising start for the building project. In case of the Provincial Governmental Office the process revealed the lack of confidence in the starting point of the whole project and brought to the surface the large differences in the aims and approach between the project team and the steering committee. The cancellation of the tender after the selection phase is an unlucky consequence of the underlying uncertainties in the project. This confirms that a tender needs to be carefully designed in order to be successful. In order to prevent the candidates from wasting energy, it might be best to delay the announcement of the tender in situations with severe doubts about the project.

6.6.4 Project management

In two cases, the School and the City Hall, the commissioning client collaborated with other public organisations. The findings of the cases imply that the implementation of decision making benefits greatly from a transparent structure for this collaboration with clearly defined roles and responsibilities. The most important reasons for this are the changes of board members at the end or beginning of election periods and the high turnover of governmental staff in relation to the long time span of construction projects. For example, preparations for the City Hall project started in 2002; the delivery of the building was initially planned for 2013. None of the project team members for the tender in 2006 were still working with the municipality in 2008. Elections are held every four years. The principles of transparent governance make decisions politically sensitive, especially just before or after an election period. This could provide an additional stimulus for agility and quick decisions, or might cause serious lack of support of decision making. In the City Hall case the project champion, the Mayor, wanted to mark his term by selecting an architect that would design a building that would 'add value to the city'. In the case of the Provincial Government Office the Executive assigned to the project had to execute a decision about the type of contract made by his predecessor. This decision proved to be one of main reasons to cancel the tender. This shows that a decision cannot be taken without considering the context in which it was made because it only makes sense in that specific context.

The cases show different motives behind the planning of a tender project. An important aspect of the planning of the tender was the amount of experience of a decision maker in his or her position within the election or governance period. The data suggest that priorities could have differed if for example the Mayor of the City Hall had not been a streetwise politician at the end of his term but a young, recently elected alderman. The timing of the tender also related to the status of the building brief and the point of view of the client on the involvement of the architect. In the School case they wanted to involve the architect in the briefing phase of the building project and therefore organised a tender in a rather early stage of the construction project with only a concept building brief available. The Provincial Government Office wanted to award the Design Build contract to a consortium based on their design proposal including the building specifications. Because they needed a detailed brief with detailed cost estimation, the tender was announced after the building brief was finalised.

Because the tender is an essential part of a complete building project, the tender needs to be coordinated with the planning of the whole building project and with the governance policy. In the case of the City Hall and the Provincial Governmental Office these processes were not always in line. This meant that, for example, parts of the building brief needed to be drawn up in a relatively short period to be ready for use in the award phase. For the School it was no problem that the building brief was not totally finished because they needed it on a conceptual level only during the tender. In the City Hall project, the project manager recalls that sometimes he had to act very brusquely to get permission to pursue with the tender. On the other hand, because they wanted to finish the tender before summer, they had a clear deadline. In case of the Governmental Office it seemed as time passed, more and more reasons were found not to pursue the tender. When the planned deadline approached, the pressure on the Executive increased and he became more anxious about the legal implications of the tender.

6.7 Findings about the tender design

The findings about the design of a tender address the role of the tender brief, the development of the tender procedure, the involvement of stakeholder in the tender project and the evaluative decision process. Results illustrate that award decisions are based on a combination of proposal, person and process related issues. The client determines the balance between these aspects, which should be reflected in the tender procedure. Because decision making is mainly based on the information received from the tenderers and the perception of stakeholders, the format of the procedure is an important part of the tender design. However, during the preparations of the tender design the implications of certain decisions are not clear yet. Decision makers need time to realise the aim and importance of the decisions that they have to make during a tender process. The implementation of a tender procedure therefore includes numerous implicit decisions in iterative decision phases that determine the direction and format of the proposals that the architects submit. This section shows how panel members that represent the commissioning client body make sense of their aims and the intentions of the architects that showed interest in designing their future building. This process resulted for two clients in a match with an architect and for one client into cancellation of the tender procedure.

6.7.1 Tender brief

Analysis of the competition rules for the tender projects showed that in most cases the following aspects were addressed in the call for participation and call for tenders:

- Description of the client and the context.

- Budget and planning of the project.

- (Summarised) project brief including the future location.

- Expectations about the role of the architect in the building project.

- Description of the tender procedure including the planning, and the composition and rights of the jury panels for the selection and award phase.

- Description and relative importance (if applicable) of the minimum requirements and exclusion grounds, selection of the candidates, and allocation of the contract.

- Format of the documents to be provided and the legal or formal declarations if applicable.

Procurement law distinguishes three kinds of requirements in the selection phase of a tender: 1) grounds for exclusion, 2) minimum requirements for selection, and 3) selection criteria. Not every call for participation made a clear distinction between these kinds of requirements. The findings of the cases suggest that in practice the main difference between the minimum requirements and the grounds for exclusion is perceived as differences in the impact of the evaluation – not fulfilling the exclusion grounds leads to exclusion automatically while the minimum requirements could lead to lower chances of getting selected. This is accordingly to procurement law. However, the findings also show a potential mix up between the selection criteria and the minimum requirements for a selection. The strictness of assessment of the requirements was found to relate to the character of the information and to the division of tasks between the project team and the jury panel during the selection and award phases of the tender. Categorical information (nominal measurement level) was treated more strictly that non-categorical information (ordinal, interval and ratio level), which can be related to the measurement standards and clarity of categories: either you have an insurance policy or not. In most cases the assessment of the categorical information about financial, technical or managerial was assigned to the project team because it was seemingly easier to assess than non-categorical information.

The main challenge in the selection phase was found to be the evaluation of the level of professional and organisational abilities of the candidates. This assessment task of non-categorical information was usually the main task of the selection committee. Observations showed this requirement was perceived as a mixture of expertise and experience. Four indicators of ability were identified: 1) (recent) comparable projects, 2) recent (comparable) design experience, 3) specific professional competences, and 4) professionalism in general. These aspects were sometimes elaborated in relation to the firm, and sometimes in relation to the individual architect who would be assigned to the job. In all cases clients appeared to look for architects or firms that submitted projects of reference similar to the subject of tender.

After the candidates (six in the School case and five in the City Hall case) were selected to participate in the award phase, the call for proposals was sent to the tenderers. The tenderers prepared a proposal based on the tender brief and the other instructions in the call for proposals. The structure of the data of the City Hall case made it possible to systematically analyse the underlying importance of arguments in the award phase. Based on the structure of the DQI, fourteen design aspects were identified in the documents relating to the three basic Vitruvian areas of functionality, build quality and impact and its resource envelope of time, finance, and natural and human resources (Gann, et al., 2003; Whyte & Gann, 2003). Most phrases of a total 388 extracted from the documents about the assessment process in the City Hall case concerned the areas of impact (39%) and functionality (36%) of the design. The build quality was only mentioned in 4% of the phrases while the project constraints and the professional abilities and reputation of the architects were referred to in 13% and 8% of the comments. The analysis reveals that most of the arguments used to ground the award decision related to the official award criteria. However, about twenty percent of the arguments included managerial and personal aspects about the project and the entrant of the tender, criteria that were not officially included in the competition rules.

For every aspect of the tender brief in the City Hall case an overview of meanings could be derived. For example, within the area of functionality, use aspect phrases concerned the allocation of the activities and office concepts according to the brief, the quality of the workplaces, and the workspace climate (see Table 6.5). The potential quality of developing new activities and the flexibility of the floor areas were also mentioned. Phrases concerning the functionality issues addressed the positioning of the activities, especially the location of the library, the recognisability of the departments, the readability and orientation within the building, and flexibility in the use of spaces. Some attention was given to security after office hours, facility management, the functionality of the materials, and the recognisability of the entrance from outside.

Comparatively little attention was given to the build quality of the designs in the City Hall case (see Table 6.6). Although the brief asked for an environmentally friendly solution and a clear choice for the demolition or renovation of the existing buildings, only nine phrases were identified that related to these issues during the award process. The energy efficiency of the future building was mentioned only once and some concerns were expressed about lighting and heating. The

Fundamental element	Design aspect (N = number of phrases)	Description of design qualities
Functionality (N total = 139)	Use (N = 63)	Allocation of activities according to brief Flexibility of floors Application of office concept Quality of workplaces Potential activities Light and depth of floor spaces
	Access (N = 4)	Secure entrance after office hours Recognisable entrée Facility management and security between areas
	Space (N = 72)	Positioning different floors and activities Walking distances Recognisability of departments Flexibility and possibilities of space use

Table 6.5 Quality perception aspects in the City Hall case, concerning the element of Functionality

Fundamental element	Design aspect (N = number of phrases)	Description of design qualities
Build quality (N total = 16)	Performance (N = 1)	Energy sufficiency
	Engineering Systems (N = 6)	Inner climate system Natural and artificial lighting possibilities
	Construction (N = 9)	Application of environmental friendly solutions Demolition or renovation options

Table 6.6 Quality perception aspects of the City Hall, concerning the element of Build quality

Fundamental element	Design aspect (N = number of phrases)	Description of design qualities
Impact (N total = 152)	Form and Materials (N = 11)	Materialisation and shape in context Perception and associations (e.g. glass is cold)
	Internal Environment (N = 8)	Atmosphere in building Relation between light and space Working climate at workplaces
	Urban and Social Integration (N = 42)	Positioning of the building in location Recognisability of building Respect for historical context and use of materials Interaction with private and public areas Renovation choices Conflict of size and shape vs. (in)visibility and integration
	Character and Innovation (N = 91)	Idea (e.g. original, innovative, creative) Cleverness of solution Conspicuousness and perception (e.g. cold, organic, outdated, transparent, surprising, challenging, daring) Clear vision and quality of concept Hospitality, charm and representative Adding value and eye catcher Associations and nick names

Table 6.7 Quality perception aspects in City Hall case, considering the Impact element

limited number of aspects that related to the future build quality of the building can be explained by the early stage of the building project in which a tender takes place.

The Impact element of a design concerns the effect of the future building in terms of its (visual) impact on its surroundings and the effect on users, visitors and the general public (Table 6.7). For the City Hall the form and materials of the design had to fit the context and evoked reactions about the warmth of the building. Only seven phrases related to the internal environment of the building, considering the atmosphere in the building and the matter of transparency. Urban and social integration of the design was one of the most important issues. The positioning, recognisability and interaction with the private and public spaces were evaluated, as was the impact of the shape and size of the building on the surroundings. It was felt important that the design showed respect for the historical context but at the same time had an identity of its own. Most attention was given to the character and level of design innovation (91 phrases in total, 24% of the total). The judgements had a lot to do with the reactions evoked by the design. The aesthetics of the winning design were judged as being 'striking', 'surprising', 'original' and 'daring', while the other designs were described as 'massive', 'unnoticeable' or 'old fashioned'. Appreciation was shown for clever, strong, charming and original ideas. However, for the final and most abstract evaluation the design had to add value to the city and its image in such a way that people would talk about it, coin nicknames, and build dreams around the new City Hall building.

Beyond the basic Vitruvian categories of the DQI model, two other aspects of potential quality of the bids were identified in the data that related to its resource envelope. These concerned the consistence with the requirements set out in the brief and financial constraints on the one hand, and personal characteristics of the architect and the quality of drawings and presentations on the other hand. Among the project constraints aspects such as the existing zoning plan or budget conformity were discussed (see Table 6.8). Some proposals for the City Hall did not include enough information to make cost calculations. Therefore they had to be partly based on simulation and calculations were therefore considered to be less reliable. However, the information about the accuracy and reliability of the cost calculations did not seem to matter during final decision making.

Some documents congratulated the architects on the high quality of their designs and their inspiring presentations. In total 24 phrases concerned the consistency of the plans with the brief, the quality of the plans and presentations and the great cooperation during the whole tendering procedure (see Table 6.9). The personal experience and personality of the architect did not seem to be of much importance in the evaluation of the designs but still five phrases were dedicated to the personal characteristics of the designers.

The final preference of most stakeholder groups and the expert committee of the City Hall seemed to be based upon an overall judgement of all potential qualities of the design, integrated into the 'most appealing proposal'. Decision makers looked for the cleverness of the design, the impact of the design on the public, users and urban surroundings, and emotional reactions such as 'love at first sight' and 'surprising and exciting concept'. Interviews confirm that these

Fundamental element	Design aspect (N = number of phrases)	Description of design qualities
Project constraints (N total = 52)	Legislation and requirements brief (N = 35)	Consistent with brief Possible conflicts with zoning plan Number of square meters and parking places According to budget
	Financial means (N = 17)	Budget within limitations Check on budget

Table 6.8 Quality perception aspects in the City Hall case, concerning the element of Project constraints

Fundamental element	Design aspect (N = number of phrases)	Description of design qualities
Qualities of partner (N total = 29)	Professional abilities and reputation of designer (N = 5)	Experience with office and civic buildings Character in collaborative projects Commitment to the project
	Quality of plans and presentations (N = 24)	Allocation of activities consistent with brief Status of the design (sketch – concept) Demonstrated cooperative attitude Presentation quality

Table 6.9 Quality perception aspects in the City Hall case, concerning the element of Qualities of partner (professional and presentational abilities and reputation of the designer)

emotions were felt by the participating architects as well when they were present-ing and listening to the other tenderers. One of the members of the award com-mittee described the emotional impact of the design proposal in the public debate preceding the meeting of the awarding committee:

> "It was love at first sight. First I wondered what it was; a bee hive, a space ship, maybe a centipede. But then I saw it: It is an Ark. Wade-able. The heart was touched....".

The press release of July 14 2006 describes the official argumentation of the integral award decision (translation from Dutch by the author):

> "The winning design of Neutelings Riedijk is full of contrast and very func-tional. A design with charisma that seamlessly fits into the historic sur-roundings. The award committee praised the fact that parts of the old the-atre and school are given a second chance. The façade of the old city office is retained and integrated into the design. [...] All in all this leads to a very good integration of the new design in the exiting city fabric. The visibility of the different functional activities (library, front office, back office) is as-sured by the proposed concept. Situating the library on the minus 1 level is very surprising. Interweaving the current basement functions into a new concept is an invention."

6.7.2 Process procedure

To design a tender a procedure needs to be drawn up on the basis of procurement law. The call for proposals needs to address the actual assessment process including the people, systems and means that will be used in order to secure transparency and objectivity of the procedure for the tenderers. The number of phases of the tender project depends on the tender procedure that is chosen. As expected from the case selection, in all cases the client decided to apply the restricted tendering procedure with a separate selection and an award phase. The contracts were to be awarded on the basis of the economically most advantageous tender. The reasons for these procedural decisions that were mentioned during the interviews were that *"a restricted procedure suited the process and the brief"*, they *"previously had good experience with this procedure"*, and *"it seemed practical"*. These reasons imply a habit rather than a well considered choice. Only in case of the Provincial Government Office was the contract not traditional Design-Bid-Build but integrated Design-and-Build. In this case the client decided to apply the minimum financial requirements only to the contractors and let the reference projects of the architect be the deciding factor. Both the School and the City Hall project teams thought about a total engineering contract (one contract person for all design activities), but wanted to fill in the details of the contract after the tender, so no definite statements about the contract were made beforehand. The School obliged all submitting parties to sign a statement that they would be willing to take on the contract on total engineering basis.

The design of a tender procedure has to be in line with the EU principles of equal treatment, transparency, objectivity, and proportionality. Although everyone would agree on these basic principles, implementing them seems more difficult than acknowledging them. In all cases the management consultant designed most of the procedure, including the selection and award criteria, and acted as contact person to the entrants. All procedural aspects were discussed and finalised in the steering committee. Only in the case of the City Hall did the project team design the criteria themselves, and the city council determined the composition of the award committee. In most cases the call for proposals with the actual assignment for the bid was not yet final at the moment the tender was announced and the call for participation was made public. This meant that in all cases the candidates were not aware of the interpretation of the award criteria and the relative importance of the criteria when they submitted their requests for invitation. From a legal perspective this can be considered as not very transparent. From a psychological perspective this shows that a tender process is often not based on a standard format and that it is also a 'work in progress' that is adjusted based on increasing insights and changes that occur during the process. The aim of the selection phase is different than the aim of the award phase. Architectural firms that are interested in the tender should be able to understand what will be expected of them in terms of the contract and their chances on being selected as a tenderer, but do not have to be informed about the assignment in the award phase from the beginning of the process. There are limitations to transparency of the procedure but the findings indicate that clients have to explore these boundaries by themselves.

In all three cases the strategies for the design of the procedure appeared to be different. In the School case the procedure provided a pragmatic basis for the tender process. It offered room to include all stakeholders and supported the aim and stage of the building project. In their line of thinking the involvement of multiple stakeholders and the explicit inclusion of decision criteria and their weights during the decision process would increase objectivity. In case of the City Hall the consultant provided a standardised format which was adjusted to the characteristics of the project by the project manager. Within the legal obligations the project team of the City Hall tried to create as much room for evaluating the design proposals as possible. They acted from the idea that transparency would benefit from a democratic procedure. The project team of the Provincial Government Office tried to make the steering committee aware of the difficulty of judging architectural quality but the executives' main concern was the limitation of risks and finding away to control the design process after contracting. This client acted from the belief that they would benefit from a legally sound procedure.

In all cases the preparations for the tender project were made by the project team. Consultants played a large role in setting up the procedure. During the selection phase the project team checked if the exclusion and minimum requirements were fulfilled, and prepared the submission documents for ease of reading by the panel members. In both the School and the Provincial Government Office cases three entries were therefore excluded from further participation in the tender based on the exclusion grounds and/or not fulfilling the minimum requirements. In the City Hall one entry was excluded for selection. Most entries were excluded of further participation because of a lack of information or invalid project references (too old, too new, wrong sector).

The main documents required in the calls for participation were portfolios, specific project references and Curriculum Vitae of the architects that would be assigned to the project. In the School case the projects that were used as reference images had to be designed by the same architect that would design the building of the client. Only in the case of the Provincial Government Office were the references judged anonymously. But because the level of requirements for selection was so high (and therefore the projects and their architects were relatively well-known) almost all project designers could be identified by someone familiar with contemporary Dutch architecture. The School did state all (sub) criteria for selection in the call for participation but they were sometimes 'hidden' and not directly referred to in the headings of the competition rules. They mentioned for example the extent to which the reference projects would have to be comparable to the tender brief or show experience with sustainability, but did not list these issues under a heading of selection criteria or evaluation process. Procurement law requires clients to determine and publish the relative weight of the decision criteria beforehand if this is applicable for the procedure. The weight can be equally divided (as in the Provincial Government Office and City Hall) or vary with relative importance (as in the School case). In case of the Provincial Government Office four equally important aspects of architectural quality were specified: the level of public character, timelessness, human scale and professional character. These aspects would be assessed on a level from 0 (not sufficient) till 3 (excellent).

The School used nine criteria in total in the selection phase; three of these were assessed on a scale level from 6 to 9½ similar to school grading, others were more categorical (yes or no). The School did not publish the exact weights of the selection criteria beforehand, but did use them in the justification of their decision. To someone (such as the author) without legal training this might appear to be a violation of European Tendering law. However, the relative importance was also part of their sensemaking process in preparing the selection decisions.

The reference projects and organisational characteristics of the firms appeared to be the main sources of information used by the selection committee during assessment of the proposals. The School had the luxury but also the burden of choosing from 32 candidates. The supervisor of GEM Vleuterweide said during an interview that he pleaded for a low level of requirements for the selection phase in order to offer small and more innovative firms a chance at the job. The relatively low level of requirements, combined with the great potential and large size of the project, could explain the rather large number of interested parties during the selection phase of the tender. In retrospective interviews the project manager and School Board members stated that they were positively surprised by the amount of interest shown in their project. Although the large number of entrants did create a higher administrative burden on the project team and the selection committee, they did not regret their decision about the requirements. In the case of the Provincial Government Office there were only five parties left for the selection phase after exclusion and check on minimum requirements. Because procurement law requires a minimum of five tenderers, they had to invite all five for the award phase. This could be partly explained by the high requirements and to some extent by the low interest from the market in the Design-and-Build contract at that time (the end of 2007).

To comply with European procurement law a client is obliged to allow the candidates and tenderers to raise questions about the call for participation and call for proposals. Most of the questions raised in these cases related to the interpretation of the procedure and the criteria. Answers were dispersed via official minutes. In the School case and the Provincial Government Office case the instances of direct communication between the client and the entrants were restricted to a presentation during the award phase. This formality of communication was due to the project team's interpretation of the EU principles. These require that all candidates receive the same information. Yet this procedure was also dictated by habit and previous experiences. The tenderers in the City Hall case were invited to visit the future construction site at the start of the tender phase while being welcomed by the Mayor. During the retrospective interviews with the tenderers most of the architects said they appreciated the site visit and the opportunities that were offered to meet the client.

The format of the presentation of the proposals as included in the design of the tender was influenced by the status of the building brief (developed in concept or in detail), the aims of the tender project (design versus architect), and the phase of the building project during the period of tender (start of initial stage versus start of design stage). The presentation of the proposals in the award phase varied from a simple presentation in which the architect explains a strategy or vision for the project (the School case), to the presentation of a sketch design with a scale model (the City Hall) and to a detailed design (the Provincial Government Office).

All tenderers had to present their proposals to the award committee themselves. In the School case the advisory members of the award committee (former members of the selection committee) were present at all presentations. The tenderers did not submit any documents beforehand. In the case of the City Hall the participants needed to submit the sketch design beforehand on A0-posters and as a scale model, accompanied by a detailed cost estimation and several other documents describing the concept. In the City Hall case the tenderers had to present their proposals in front of four different audiences; the steering committee, the city council together with the project members, the employees, and the citizens. A summary of their documentation was placed on the internet page of the municipality.

In all cases the presentations were combined with a question-and-answer session with the tenderers among those present. This social and interactive aspect made it possible to sense a potential click between the representatives of the client and their potential partner in design. Yet, apart from their appreciation for the possiblities to interact with the client, the City Hall procedure was also perceived as over elaborate. The large number of presentations and the frequent discussions with stakeholders during the awarding phase turned the whole tender process into a spectacle, or rather a 'presentation orgy' as one of the participating architects articulated in an interview as a response to a question about the general opinion about the procedure:

> "A warm procedure with some signs of over-design. That presentation orgy, they might consider it very important but for us it could have been more condensed. It was a bit of a pseudo democracy."

Although the tender briefs and presentation formats for the City Hall and the School cases differed, two kinds of presentation strategies of the tenderers could be distinguished: 1) a focus on an introduction of the firm and the architect that would be assigned to the project by presenting a rather open idea about a potential solution for project, and 2) the presentation of a complete design solution for the project. Both strategies implied a risk to the architects. The presention of a complete solution the client makes it possible for a client to judge the potential competences of the firm in the context of the building project and make a statement that is supported by a visualisation of the future building. This enables the client to get a taste of what's in store for them once they sign a contract. The architect however runs the risk that the client does not like the design. One of the advisory members of the award committee of the School explained during an interview why this is important for clients:

> "Clients hire an architect to visualise their ideas and therefore they are looking for an architect in their line of thinking."

On the other hand some of the School decision makers indicated that the architects who proposed a complete design solution showed 'inflexibility'. One of the members of the School Board literally mentioned 'opportunities' as a reason to select the chosen architect. In the City Hall case some tenderers left some aspects of the design open or they presented two options to demonstrate their flexibility. Reactions to this strategy varied. Some people reacted rather short-sightedly to the level of detail of a proposal – "*this design does not have a bicycle shed*" - while others did not appreciate the sketched options that were sketched because "*the design was not finished*". The final decision in the City Hall case seemed to be mainly based on the exterior and concept of the design. More experienced decision makers stated in the interviews that one should not judge the details during the tendering procedure because each firm "*will throw the design as presented into the wastebasket*" if the actual design process starts because "*the character of the starting points are still of limited use in such an early phase in the project*". From their experience they know that a design process should be based on a constructive dialogue between the client and the architect. A tender procedure usually does not provide this interaction. The actual value of a solution presented during a tender process thus remains somewhat unclear; does the design provide the actual basis for the design of the building project or it is used as a means to assess the quality of the tenderer? If the latter is the case, is a (detailed) sketch design the right format of submission, considering the amount of time and money the tenderers have to invest? By law clients are not obliged to offer financial compensation for the activities they ask from the tenderers. In these cases the decision to compensate the activities of the participants was found to depend on custom within the sector (in the school sector no compensation is usually given), the perception of what it takes to be a professional client (proposed by the project team in the Governmental Provincial Office and City Hall), and/or the amount of work the entrants are being asked to deliver (sketch or detailed design versus a vision presentation).

EU law requires that in each phase of the tender process the decision (sub) criteria and their relative importance have to be published beforehand. In all cases the official award criterion was the 'economically most advantageous tender'. In the case of the City Hall the 'flexibility and synergy of the programme', the 'intelligence and innovation of the solution', and the 'added value for the context' were the sub criteria as announced in the call for proposals. These aspects were all related to the design proposal as presented by the tenderers. Further analysis based on the DQI system of Gann et al. (2003) shows that the first criterion corresponds to functionality, and the second and third criteria to impact of the design. Build quality and exogenous issues such as budget and delivery expectations appear to be subsumed into functionality. The prominence of impact among these criteria could be a sign that the conception of success was markedly different from what might normally be expected under the theme of 'economically most advantageous'. In this case the fee was fixed before the selection process as part of the contract conditions, which indicates that the estimated budget and fee proposal were mainly considered as minor importance compared to the quality of design.

Decision makers in the School case clearly focused on finding a partner. This was reflected in the kind of award criteria; four out of eight aspects of the award criteria as described in the competition rules refer to the professional abilities and reputation of the architect (communicative skills, presentation of the vision on the brief, the portfolio, the plan of action), of which the communicative skills seemed most important. The other aspects of the award criteria (vision on the brief, sustainability and urban plan) can be considered as aspects that relate to functionality, build quality and impact of the design proposal. According to the justification of the award decision the financial means had a relative weight similar to the communicative skills of the tenderer. This implies that in the School case the process of the project and professional abilities of the architect were just as important as the proposal itself. The proposed criteria for the Provincial Office were architectural quality, functionality, sustainability, maintenance, price and plan of action. So both in the School case and the Provincial Government Office the award criteria related to the basic aspects of the DQI (functionality, build quality and impact), completed with aspects from its resource envelop (human resources, natural resources, time and finance).

From a legal perspective the award decision should focus on the offer of the tenderer in relation to the requirements in the call for proposals. Further analysis from the call for proposals in combination with the observations during the award phase shows that in practice the most economically advantageous tender contains a mixture of proposal, person and process related issues (see Table 6.10). The proposal related issues have to do with the affective responses to the proposals and the match between the expectations and opportunities that are offered. The person related issues deal with selecting a partner. The other arguments depend on the course of the process and the preferences of the stakeholders. These findings indicate that the design of the tender procedure should depend on the reasoning behind the aim of the tender: a search for the 'most appealing design proposal' or looking for the 'most appealing design partner'. In this context a tender brief and procedure should be focused on selecting a 'design' or a 'designer'. The tradition of the design competition in which anonymous designs are used to evaluate the competences of an architect, causes a mixture of interpretations of the regulations. This mixture causes confusion for a client trying to get the best of both worlds but ending up in a jumble of expectations and disappointments.

The observations of the meetings in the selection phase of the School case and the Provincial Government Office case showed that decision makers use one of two strategies to perceive and judge the information provided by the candidates: 1) first view and evaluate the information individually, and then start a discussion, and 2) view and evaluate the information as part of a group discussion. In the first scenario the decision maker relies on his or her personal frame of references during the value judgement of the documents, whereas in the second scenario the frame of references of the decision makers are influenced by other group members during value judgement. In the selection phase of the School there were so many documents and so little time (one day) that it was not possible for the decision makers to individually study the documentation before the discussion with the group started. During the award phase for the City Hall the same documents that were handed in beforehand were distributed among the decision makers. The presentation and scale models made available to the decision makers as group. In the case of the Provincial Government Office an advisory expert guided the panel members in viewing the A0 posters with visualisations of the reference projects. He 'translated' the images into information that would be of interest for the mem-

Proposal related issues	
	Content of the vision for the brief or the design proposal
	Fit of the proposal with expectations and dreams of the client
	Surprise or delight offered by the proposal
	Fee of the architect and/or calculated budget
	The level of detail in the proposal
Person related issues	
	Personality of the architect
	Demonstrated competences as a designer or lead architect
Process related issues	
	The course of the decision process
	Support of the decision by other stakeholders

Table 6.10 Overview of issues that were employed during decision making in the award phase

bers of the selection committee. Based on that information and their own perceptions they individually judged the suitability of the architects. These judgements were input for a phase in which the decision makers deliberated about the possible implications of a certain candidate.

In the selection phase for the School the perception and judgement strategies differed per party. The School Board viewed the documents first individually and then discussed their findings with each other. One (more experienced) decision maker of the municipal Department of Education discussed the documentation with a less experienced person while educating her about design. Then they commonly judged the requests on the selection criteria. These different strategies of individual and collective perception suggest that the interpretation and judgement process is influenced by the amount of information (depending on the number of proposals), the format of information (visuals versus text), the ease of distribution (digital, physical, or virtual), the availability of time of the panel members (several hours to several days), and the level of expertise of the panel members (less versus more).

Apart from the fact that the level of detail of the information as provided in the tender brief, the format of the documents also influenced the level of detail of the proposals. In all cases strict requirements were set for the format of the proposals in the selection and award phase. Observations showed that a standardised format improved the comparability of information but decreased the possibilities for the architects to present their individual characteristics. During the perception of the information as provided by the architects, information is retrieved from different verbal and visual sources. In all cases the images and texts as produced by the architect were the main source of information during the selection and award phase. During the award phase this information were combined with explanations from the architects themselves and input from other stakeholders. Observations in the City Hall case showed that a presentation could distract people from the documentation, but could also confirm or counteract assumptions made based on the perception of the originals. Narratives of the architect were found to give the proposals more meaning in the eyes of the panel members. The presentations enabled the architects to communicate the main issues and values of the proposal to the decision makers which often led to apparently new information about the proposals. All together this means that in these cases the information sources for decision making during the tender process were:

- Images and texts produced by the architect.
- Discussion with other stakeholders based on the submitted proposals.
- Explanation of the architect about the original images and texts.
- Explanation of the experts, consultant or stakeholders about the original images and texts.
- (In)Formal contact between client and the tenderer (e.g. during location visit or after the presentation).

The architects participating in the City Hall case indicated in the retrospective interviews that they would have appreciated more input of the client in interpreting the tender brief because they were not sure for what the client was aiming for. Yet the project manager and the client emphasised in their interviews that they were definitely looking for an icon and that this ambition had been widely announced during the kick-off meeting (including the site visit) with all the tenderers. These differences in interpretation could have been caused by the fact that contrary to regular design processes very little interaction is possible between the client and the architect during a tender. A location visit can contribute to the understanding between the client and the tenderers, but it does not replace the natural process of starting the design process together.

The legal context necessitated a strict planning of the tender process; not complying to these regulations increases the chance of law cases. In all cases the panel members relied on the support staff (such as the project team and internal committees) to prepare the assessment phase in their decision making process because of a lack of time both in terms of the calendar and in their availability. Observations showed that sometimes there was simply no time for the decision makers to verify the information as presented by the project team or tenderers, even if the regulations allowed. Apart from the fact that it would have been almost impossible to get a jury panel together for more than a few days, the questions remains whether more information or time would improve the decision. As one of the more experienced panel members for the School, an architect himself, stated:

> "If an architect cannot sell his ideas to a small client panel in half an hour, how then can he sell an idea in a construction meeting of one and a half hour?"

The cases show that the more information assessed and the more decision criteria needed to be applied, the more structure was needed to make a good judgement. In the School case a matrix sheet with a list of the criteria on the one axis and a list of the entries on the other created several benefits during the assessment process. According to the interviewees this led to a structured discussion, efficient use of time, and a ready recording tool to collect the most important and striking aspects of the proposals. From their perspective the sheet contributed to the objectivity of the decision and the transparency of the process. During the selection phase the matrix sheet supported the reduction of the amount of information viewed on the individual and group levels, while during discussion in the award phase it structured the process for the decision makers. Observations show different ways to use the form in different phases of the decision process. Some decision makers judged each entry on each of the sub criteria and then summed the judgements of all topics to create an order of preference among the candidates. Others perceived all information per request and then created an overall judgement per candidate. Those people usually took notes on the form when they perceived the information to support them on making a judgement. Some people used the matrix sheet in two ways: first to survey plusses and minuses in the sub criteria and then to total the score. During the award phase of the School case a second matrix sheet was distributed, but it was only used by some decision makers individually. Aggregated scores were not discussed in the group. In all cases some form of list of the participants was used. The use of the matrix sheet also caused a side effect because it stimulated strategic scoring behaviour. Observations and interviews indicate that expert decision makers especially used this opportunity to increase the weight of their preferences.

The following methods were found on how the panel members reached their final decision:

- Voting for a number of entries.
- Ranking the entries.
- Discussion among the panel members.

Usually a discussion was combined with ranking (with or without the use of weighting factors on the separate criteria) or voting to reach a consensus. The final decision was taken either by voting or by consensus. The decision methods and decision steps were not mentioned in the calls for participation or proposals but decided by the panel members during the evaluation process. In two cases (the School and the Provincial Government Office) the criteria were used to structure the discussion about the proposals, especially during the selection phase. In the selection for the Provincial Office the proposals were discussed per criteria; therefore the four criteria created four discussion rounds. In the selection phase of the School the input for a big matrix sheet was based on the judgements per criterion per party. The sheet automatically incorporated the weight of the criteria and outcomes were input for the discussion. In the School case the members of the selection committee first reached a consensus about the first three or four firms on top of the list and excluded the firms at the end of the list. Then they discussed the most striking differences in judgements per party as shown in the sheet. Based on that discussion they decided to select six entries for the next phase of the tender. During the award phase the jury panel started with ranking the six entries holistically and then started a discussion among the parties. The final decision process of the jury panel in the City Hall took place behind closed doors. According to the interviewees the decision makers started with voting and then discussed the major differences in order to end with a voting round. The findings of the cases suggest that especially when negotiations have foundered or large reduction steps have to be made, a different decision method is chosen to get the process going. One cannot beforehand determine which method will be applied in what order because this choice is part of the process of making sense of the opportunities that are provided by the submitted proposals.

6.7.3 Stakeholder involvement

In the analysis of the data a distinction was made between the actual actors in decision making and the other stakeholders (see also Chapter 6.6.3). Several aims were found for involving stakeholders in the process of assessing the information of the entrants:

- Increasing the involvement of stakeholders (School, City Hall).
- More efficient information processing/ improving the readability (Provincial Government Office, City Hall, School).
- Verifying the information (City Hall).
- Gathering different perspectives and opinions (City Hall, School).

All actors in the cases belonged either to the steering committee, the project team and/or the jury panel. The other stakeholders are not making decisions but are more passively involved in the tender. These stakeholders included the parents or neighbours of the School, the Queen's commissioner and the Department of Interior and Kingdom Relations in the Provincial Government Office, and people from the (local) architectural community in the City Hall case. These stakeholders become involved automatically when the commissioning body initiates the project because they are part of the governance structure of the client and users of the built environment. Tender candidates and tenderers also have a rather passive role during the decision process of the client in a tender because they only provide input for the decision process and do not make decisions on the clients' behalf. They become involved in the tender by reacting on the call for participation and have to follow the procedure as set out in the competition rules. However, they do steer the direction of the tender and the building project by the designs they propose. Their submissions determine the number and character of the options that clients have to choose from in deciding about a winner. Therefore they determine the intrinsic basis of the decision task. The number of stakeholders that are involved in the tender process and the role they have in decision making was found also to be part of the tender design.

Table 6.11 shows an overview of the position and the role of the stakeholders per case. People who are part of the steering committee, selection committee, award committee of project team are actors. Other people are either not involved, have an advisory role for the actors, or are represented in the committees by the actors. In the case of the City Hall several stakeholders were involved on a consultative basis in the assessment process of the proposals during the award phase. Different formats of involvement were found that related to the format in which their advice was communicated to the jury members. The expert committee, the municipal user group and the library user group provided the members of the award committee with a memo including their opinions, preferences and argumentation. The employees and citizens were asked to fill out a survey on the internet or in the theatre where the proposals were presented. The project team provided the members of the award committee with a memo about the results of

* No award phase included in case	School with sports facility	City Hall with library	Provincial Government Office*
Commissioning body	Needed for approval of award decision but no actors during decision process	Represented in the award committee	Needed for approval of award decision but no actors during decision process
Representatives of the client body	Part of steering committee, selection committee and award committee	Part of steering committee, selection committee and project team, member of party represented in award committee	Part of steering committee and selection committee
Shareholders and super-visors of the project	Part of selection committee, advisor for award committee	Not directly involved	Not directly involved
Daily users	Represented in selection and award committee, part of project team	Advisors for award committee, part of project team	Part of project team
Non-daily users	Not involved	Advisors for award committee	Not involved
Representative groups of the users	Not involved	Indirectly represented via commissioning body and advisor for award committee	Indirectly represented via commissioning body

Table 6.11 Overview of stakeholder and actor involvement per case.

a requirement check and two budget checks by independent experts (as counter checks). Just before the award committee of the City Hall met a public debate was held in which the members of the award committee were able to justify their decision. In the case of the School the main stakeholders were represented in the selection committee, which meant that they had decisive rights in part of the decision making. During the award phase the members of the selection committee who were not part of the award committee had the possibility to advise the award committee. In the Provincial Government Office case the steering committee consisted of the same members as the selection committee. This meant that in most cases the decision makers were also the future users of the building. In the pragmatic and democratic cases the other stakeholders were somehow actively involved in decision making while in the very political case a limited group of people was involved. Only in the selection phase of the School case were stakeholders involved in decision making who would not use the future building themselves.

The format in which the preferences of the stakeholder groups were gathered differed per group and per case. During the award phase of the School everybody seemed to agree on the winner, so decision making mainly consisted of an exchange of preferred firms instead of a content based discussion. In case of the City Hall the survey created an overview of the preferences of the citizens and complete staff, but only on the aspects that were included in the survey. The preferences of the citizens and employees showed a scattered image. The representatives of the user group and the experts provided the award committee with a memo including their most important findings and an advice about the proposal that would best fulfil the requirements of most economically advantageous tender. Because the memo was referred to during decision making more in the City Hall case, a memo appeared to be more meaningful to the members of the award committee than the survey. The results of a survey always display average preferences on certain topics and an image of the overall preferences. In the City Hall this image was scattered and did therefore not steer the decision makers into a certain direction. The influence of the preferences could however also have related to a power difference between the stakeholders groups.

The issue of stakeholder participation raises the discussion about the position and role of the stakeholders or there representatives. The data suggest that clients find it hard to estimate the importance of experience, expertise, and knowledge in decision involvement. They realise that the decisions they make could have major implications for a large population because a lot of people will be daily users or visitors of the future building. However, this does not automatically mean everybody should be involved in the tender process. A difference was found between having expertise as a decision maker, having expertise about architectural design, and having expertise in tender regulations. Most actors in these cases (other than the consultants) had been involved in a tender or selection procedure before, but only a few actually selected the services of an architect in an EU context. Further analysis of the findings in the cases displayed different kinds of expertise in several levels and with multiple perspectives. Expertise tends to be domain specific. Even the legal advisor of the Provincial Government Office, a noted person in construction law, found it hard to deal with the subjectivity of the tender brief and the selection and award criteria. During the assessment of the reference projects during the selection phase of the School the panel members with relevant experience in construction and tenders but without a background in architecture did not acknowledge the same qualities as the decision makers with a background in architectural design. This created difficulties in the discussion. On the other hand the different backgrounds and perspectives of the decision makers reflected the complex nature of making decisions about buildings with a public character. The data give reason to belief that the more diverse a jury panel, the more complete the assessment will be.

The findings showed that a lack of expertise with a certain group was often counterbalanced by involving other experts in the jury panel or support on a consultative basis. In the case of the City Hall and the Provincial Governmental Office the politicians made use of special advisory committees with employees that focused on a specific item. Because the clients did not have staff members with the specific expertise about architect selections, they all hired management consultants to support them for the tender project. Yet, these consultants did not always have the most specific knowledge either. The background of the management consultants in these cases varied from civil engineering to real estate management and architecture. The legal advisors seemed to have limited experience with tenders in architecture too. The authority level of the consultants acting in the project teams appeared similar to the other members of the project team. Sometimes legal, architectural, or real estate professionals outside the project team were asked to provide the steering committee specific strategic advice on issues that related to the building project. These 'additional' consultants appeared to have more influence in the final decision process than the consultants who did most of the work on the procedural design level in the project team. The consultants that participated in the case did not have any decisive rights, and therefore no specific responsibility to the client organisation apart from their own integrity.

The perception of an individual's own level of expertise influenced the position he or she found him or herself in during the decision process. The data of the cases suggest that the personality of the decision maker is important for the degree in which one is aware of specific differences in expertise. For example, the School director indicated that he only attended the committee meetings passively:

"The whole day I sat there with my eyes and ears wide open. I am a layman; I did not find time to say anything useful about the entrants I saw. I just sat there for the show."

But the project manager who chaired the meetings indicated that the perception of the contribution of the School director to the discussion differed from his own:

"Novices are very often very capable of conveying a coherent image. He [the School Director] heard terms he did not know but he was able to make a judgement. These kind of people know what design quality is about. Then intuition counts."

The findings suggest that the level of expertise also differed per position or role that decision makers play. One of the School Board members – a real estate management consultant in daily life - explained the differences between being a consultant and being a board member:

"It is totally different. As a client we do not only have the authority but also the responsibility. That is why we are more aware of the choices we make. Now I understand why clients find it difficult to make decisions.... It is definitely harder. I feel responsible. As consultant I am responsible but I do not have direct influence on the decision. That freedom, you do not experience that as administrator because you have to act as you decided."

A similar phenomenon was found in the City Hall case. For most council members being a politician is not their primary profession – they usually have other daily jobs. However, this does not mean that people they therefore are lesser decision makers. Or as the process manager of the City Hall indicated about the decisions taken in the award committee:

"I think it is really clever of the award committee how they balanced all preferences. Although a pharmacist is not an urban planner, it did turn out well."

All jury panels in the cases consisted of people with different backgrounds and different levels of expertise that acted from different positions with diverse interests. The jury panels of the City Hall and the School consisted of a mixture of architects, administrators and managers with different decision rights. The data showed differences in the behaviour of decision makers based on their interests and level of expertise, for example about money: panel members who were not responsible for the eventual costs or had more experience in the field of architecture showed more risk taking behaviour. On the other hand, other administrators showed more risk aversive behaviour, presumably because they can be held personally accountable for their decisions. The analysis in the City Hall case clearly showed different interests between the stakeholder groups by the number and type of aspects they used in their argumentation. In total 14 aspects of design quality were identified during the analysis of the case on the elements of functionality, build quality, impact, project constraints, professional abilities, and reputation of the architect. The project team and the independent financial consultant reviewed only the aspects they were assigned, such as possible conflicts with the zoning plan or budget, and did not express a preference for an architect.

The survey of the citizens was developed beforehand and focused on the integration, materials and character of the design in the context of the city by using closed questions. Therefore the prospective building users with no professional background in architecture (the citizens) were able to present their preferences on four aspects of design quality. This was about half of the aspects that the expert committee and the selection committee used. The employees of the city completed the same survey with additional questions about the use and attractiveness of the offices (six aspects in total). The user group extended these aspects with an evaluation of the air conditioning systems and the interior climate (light and heating) because of the consequences of these aspects for the quality of the workplaces. The library employees focused on the position of the library in relation to the other parts of the building and the recognisability and image of the library from outside. They underlined the requirements from the brief as a way to evaluate the design qualities and used ten out of the fourteen aspects. It was the expert committee that considered the highest number of design aspects (eleven) of the advisors, focusing on feasibility and the contribution of the design vision of the quality of the city, but excluding finances, performance, and building services. The award committee seemed to have followed the expert committee in their judgement but also stressed the financial limitations. In the public debate they also stated that the current state of the design was to be developed further in dialogue between the architect and the client. In their press release, thirteen of the fourteen aspects of design quality were mentioned. Overall the aspects of urban and social integration, forms and materials, character and innovation were mentioned by all stakeholders. Performance, system engineering and financial means were only brought up by a few stakeholders.

Differences in expertise were also found in the perception and use of information. Observations showed that experts were better able to see through the images and needed less time to interpret information. The School Director, one of the less experienced members of the School panel, expressed to have been a bit more convertible if additional information would have been available and would have appreciated more time to think before having to take the final decision. The inexperienced members of the Provincial jury panel asked the expert for more information about other exemplary buildings of a specific architect, the flexibility of the architect, the capacity to listen to clients, and the innovative and stylistic expectations of the possible new building. This way they created an image of the tenderer and their firm in general and they used the expert to translate the images and implicit knowledge about the architects. During presentation of the City Hall proposals observations showed that for some members stakeholder groups it was difficult to evaluate the sketch design and understand the vision of the architect. Stakeholders sometimes focused on very human issues or detailed functional design characteristics, such as the location of the bicycle shed and staff entrance, while the design in the sketch stage usually does address these kinds of items. Besides, not every aspect of potential design quality can be examined from the information provided in the tender phase. Even the advisory expert involved in the Governmental Office case admitted having had difficulty in judging some of the criteria based on the images of the reference projects. He explained that every project needs to be evaluated in the context of its budget, brief, client and location. According to him the quality of a design can never be fully assessed without having any personal involvement in a project.

In all cases the discussion between the panel members influenced the frame of references of the group and the individual. Opinions of the more experienced panel members were often taken relatively seriously during a discussion but 'novices' could also make the other members think about certain arguments. The data indicate that less experienced panel members used more arguments provided by the other panel members than did the experts in building their frames of reference. The lack of an existing decision frame could explain the rather passive and reactive attitude of the less experienced panel members in the School case. In the Provincial Government case the Executive tried to solve every uncertainty with a legal measure, while the Director of the School *"trust[ed] his other group members on this"*. Experts already have their own frames of reference. Although they seemed to adjust their frame to the tender brief and the aims of the other stakeholders, they depended less on the other panel members and seemed certain of their competences. In both the School and the Provincial Government Office cases the experts seemed to take the lead in discussions and putting arguments on the table. They were also able to use the decision procedure more strategically by using extreme scores or their veto rights during the decision process. Sometimes they used emotions like becoming angry or disappointed to convince other panel members. The politicians in the City Hall case displayed their ability to use metaphors and personal emotional expressions to make a statement. At the same time, in all cases experts were more capable of distinguishing between their own taste and the needs of a client in a particular situation and between one example of excellent design and the complete work of a competent professional. This implies that experts are better at taking balanced decisions with a long term perspective, while novices usually focus on the immediate consequences of a decision.

Some of the interviewees of the School mentioned that they had sometimes adjusted their judgements due to information provided by or opinions of other team members. These changes were minor, e.g. from 7 to 8 on a scale from 1 to 9½. Discussions also seemed to have supported knowledge transfer among the panel members and between the panel members and the advisors. The panel members sometimes complemented each other in expertise but it was also found that in some cases a particular knowledge field was missing. In the Provincial Government Office case sustainability was a major theme of the project. During the selection meeting, the panel discovered that none of the available panel members or experts had enough expertise to make a proper judgement on this aspect.

The same phenomenon occurred in the School case with sustainability. In the final decision phase sustainability did not seem to be an issue anymore. It can be concluded that ongoing discussions between the decision makers increased the appreciation of other interests, raised the awareness of the importance of a valid judgement, and exposed more qualities of the submissions. However, discussion cannot compensate domain specific knowledge that is not available with the jury panel. Therefore the composition of the panel needs to fit with the criteria that will be assessed and the sources of information that are provided by the participants.

6.7.4 Process of decision making

For a client a restricted tender procedure consists of a selection phase and an award phase. In all cases each phase of the tender process consisted of four iterating decision steps of 'initialization', 'confrontation', 'communication', and 'contract' between the client and the architect. This process as illustrated in Figure 6.6 resembles that of any basic design process (see for example Roozenburg & Eekels, 1995): the design problem is defined, solutions are proposed and then evaluated in order to identify the most suitable one. They also resemble the process as described by rational decision models: problem definition, identification of decision criteria, allocation of weights and evaluation of alternatives (e.g. Harrison, 1999). Yet one of the things that the basic design model and rational decision models abstracts from, namely the role of different stakeholders and actors in the design process, is made explicit in this diagram. Both in the selection phase and the award phase the client initiates the tender by putting the demand into the market by presenting a problem definition and an invitation for architects to join the tender. The architectural firms that are interested in the contract then start to prepare their documentation. After the architects have submitted their proposals, the representatives of the commissioning body are confronted with the number and content of the proposals. This confrontation between the brief and the submitted proposals can be considered as the actual start of the decision making process about the quality of the design(ers), which ends in communication and contract allocation. During this process the 'demands' of the clients are linked to the alternatives the market parties offer them and start to make sense to the decision makers.

On the side of the architects (the supplying actors), the initial decision is whether or not to join the tender once the problem definition has been made public. In the tender phase when proposals need to be prepared, the phases the architect goes through can be described as 'interpretation of the problem definition', 'presentation of the proposal', 'perception of the argumentation', and 'acceptation of the decision' (A1 – A4 in Figure 6.6). The actual contract allocation can start only after the decision of the client is accepted by the other participants and the legal term for appealing has expired. The process of interaction between a client and an architect during a tender differs from the process of interaction in a regular initialization phase of a building project. Most interviewees in the City Hall case stated that a good trusting relationship between the client and architect is essential for a high-quality building. On the one hand a tender project is perceived as

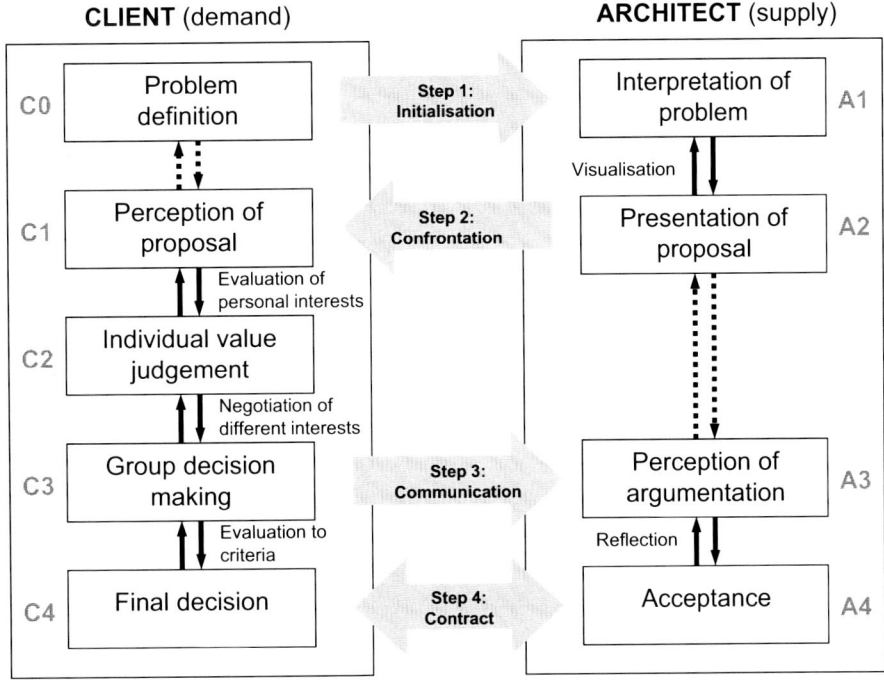

CLIENT (demand)   **ARCHITECT** (supply)

| C0 | Problem definition | Step 1: Initialisation | Interpretation of problem | A1 |

Visualisation

| C1 | Perception of proposal | Step 2: Confrontation | Presentation of proposal | A2 |

Evaluation of personal interests

| C2 | Individual value judgement | | |

Negotiation of different interests

| C3 | Group decision making | Step 3: Communication | Perception of argumentation | A3 |

Evaluation to criteria   Reflection

| C4 | Final decision | Step 4: Contract | Acceptance | A4 |

Figure 6.6 The decision processes of a tender as a result of the interaction between demand and supply in architecture.

a rather artificial way to build that relationship. On the other hand, according to one of the architects, winning a tender could accelerate the start of a project because *"a good emotion already exists between the client and the design proposal".*

The interviews and observations of the client decision makers indicated that psychologically an iterative process of perception, judgement and evaluation can be distinguished in both the selection and award phase of a tender process. The data imply that in the case of a design tender this process generally exists of four phases (C1 till C4 in Figure 6.6): 'perception of information', 'individual value judgement', 'group decision making', and 'final decision making'. Based on the perception of a proposal, the members of the jury panel evaluated their personal interests in respect to the proposal and frame their individual judgement within the context of the tender project. In the School and the City Hall case each party acted as a panel member in the group discussion - they first discussed their individual value judgements within their own party and then discussed their common judgement within the jury panel. The members of the committee of the Provincial Government Office acted as individual panel members without representing a group. In all cases negotiation between panel members led to a final decision that was communicated to the tenderers. In both the selection and award phase of the tender panel members needed several rounds of discussion and/or other means to come to a final decision. The process of client decision making can therefore be seen as an iterative process among the four decision phases consisting of constant evaluations between the problem definition, the proposals of the architects and the interests of the panel members, respectively as individuals, as representatives of a stakeholder group and as a jury panel. These evaluations contributed to the development of a common frame of reference between the panel members. This frame increased the understanding of the problem, the quality of the assessment of potential solutions, and the level of satisfaction over the final decision.

In all cases some members of the project team were attending the meetings of jury panels in the selection and award phase in a supporting role. Most of the time they presented the findings of a check of the proposals on the exclusion grounds, minimum requirements and format specifications on behalf of the project team, but did not participate in the actual selection process. The panel members of the School helped each other in finding information in order to judge certain criteria, in the Provincial Office case the advisory expert helped to read the data. No changes in the composition of the jury panels occurred during assessment, which is in line with procurement law. This however also meant that a potential lack of expertise within the panel members could not be solved by involving other people. Sometimes the information provided by the candidate or tenderer was incomplete, which decreased the possibility for a systematic comparison. In some situations this information was 'completed' by the jurors themselves, in others it was just left open. In the case of the City Hall and the School missing information did not prevent the winners from winning, but for some candidates and tenderers it did lead to exclusion or a low score on certain criteria. Observations indicate that the panel members (unconsciously) used a lack of information or conflicting information for their own benefit, especially during the group decision making. Sometimes conflicting opinions among the panel members led to a more thorough analysis of the proposals, while in other situations it led to disqualification or limited preferences for a proposal. This also depended on the power balance within a group and the character of the chair; in the Provincial Government Office the chair did not intend to have a lot of discussion with the other panel members, whereas the chair in the School case specifically stimulated a structured debate about the proposals among the different parties that were attending the meeting.

During the assessment of the proposals in all cases decision makers experienced difficulties on how to deal with the boundaries and interpretation of the requirements stated in the competition rules in order to compare the candidates and tenderers as equal as possible. Panel members needed to make sense of their own requirements because they had not realised the effects of the requirements beforehand. In the School case, for example, some of the tenderers made a scale model or a fairly detailed sketch to present their vision on the tender brief. Others presented just a short analysis of the building volumes. Both formats were in line with the competition rules but it made equal comparison the proposals difficult because of the different information levels. The same problem occurred with missing or indefinite information. In the City Hall case one of the proposals did not deal with the cost estimate in a proper manner. To compare the proposal with the others the project team still tried to counter balance these costs while officially they should have excluded this proposal from assessment. Another interesting phenomenon that occurred was the effect of deviations of the tender brief on the chance of winning. On the one hand, a surprising element additional to the tender brief could convince the client to prefer a certain proposal, but on the other hand a proposal could easily be rejected just because of a deviation from the brief. These findings indicate that decision makers use this room in decision making to steer the process in a direction that would benefit them. This makes it difficult for tenderers to estimate the chances of winning a tender.

Decision makers in the School and City Hall case stated during the interviews that both in the selection and the award phase the amount of information provided by the proposals was sufficient to make decisions. At the same time several panel members indicated that certain aspects, like the capability of an architect to make a sustainable building, were very hard to assess. In the selection phase it was even for experts difficult to judge the potential qualities of the candidate because the reference projects needed to be assessed in relation to the context in which they were made. The data suggest that in the selection phase decision makers mainly relied on their own interpretation of the proposals, while they gathered more input from others in the award phase. This conclusion might be distorted by the fact that contrary to the award phase the proposals in the selection phase were not supported by a narrative of the architects. On the other hand, in the selection phase panel members used information and personal experiences that they gained before the tender had started about certain architects or their buildings, while argumentations in the award phase relate more to the content of the offers of the tenderers. This suggests that perception and judgement in the selection phase is more related to the reputation and style of the architects whereas decisions in the award phase are related more directly to the proposal. In the award phase panel members sometimes seemed to forget that they were evaluating proposals instead of actual designs and therefore took the information more literally or strictly than the tenderers intended. The format of the sketch or concept design encourages this kind of misinterpretation more than a vision presentation without a specific design.

The findings of the cases imply that the longer the panel members were involved in the procedure, the more they seemed to become aware of the importance of the tender in relation to the complete building project. During the tender process decision makers increasingly got familiar with the characteristics of the project, with each other, and with their own ambitions, needs and expectations. Because in the School case and in Provincial Government Office case the jury panels in the selection and award phase were similar, the selection phase was a sort of preparation phase for the award phase. This created a learning effect which increased the awareness of the decision makers about the implications of the decisions. Despite the fact that the panel members of the Province were not involved very much in the preparations of the tender, the meeting in the selection phase triggered the enthusiasm of the panel members:

> *"I am terribly curious about the ideas these five candidates will present in the award phase."*

Potential conflicts between interests of the panel members came in particular to the surface when room and flexibility in decision making was explored during negotiations and when the moment of final decision making approached. At the same time panel members were prepared for the fact that a decision needed to be made and certain barriers between the decision makers and their interests needed to be overcome. Both in the the School case and the City Hall case the more experienced decision makers indicated during the retrospective interviews that in the selection phase it is rather easy to reach a decision about the top three and the bottom ten of the candidates but that the middle category creates the most discussion. But they also stated that a final decision will always be made. The findings imply that the main differences between the selection phase and the award phase are the character of the final decision and the amount of latitude within the decision. The aim of the selection phase is selecting several potential candidates while in the award phase one winner has to be chosen. Therefore there seems to be more room for decision making during the selection phase than during the award phase. On the other hand, there seems to be more room for interpretation in evaluating the proposals in the award phase than in assessing the requests for invitations in the selection phase because the requests for invitation include more formal statements and financial information. The decisions about design quality in the award phase give the panel members more latitude to follow their intuition.

Both during the tender and the award phase of the School case the panel members first discussed their individual judgements among the members of their own party. The plusses and minuses of each option were put in a bigger picture and compared to each other. Then their preferences and judgements were discussed with the other parties and decision makers became aware of the preferences and interests of the other parties. The findings suggest that knowing the reason behind a decision was beneficial for acceptance of the decision and increased the amount of support for the decision. The power relationships between the members of the jury panel of the Provincial Government Office were different than in the School case. Compared to the School case the project team did not have as much influence on the decision process. This could be due to the fact that the panel members of the Provincial Government Office were more acquainted with each other (they were all members of the same Executive board) than in the School case. The observations indicate that during the meeting of the jury panel of the Provincial Government Office certain behavioural patterns occurred in order to prevent further conflict within the group, which could be interpreted as group think. In all cases the decision makers indicated in retrospective interviews that the discussions during the panel meetings contributed to the strength of the tender project and offer an excellent starting point for further collaboration.

In case of the School and the Provincial Government Office the decision criteria were used very explicitly during the beginning and the end of the process, but very implicitly during the discussions in between. One of the School panel members explained about her experiences during the award phase:

> "You only use the criteria because you have to be transparent and because you have to explain later on. But I did not look at the criteria during my judgement process; I would lie if I would say so..."

The criteria occurred to have initiated two kinds of processes: 1) structuring the decision process, and 2) interpreting the aims for the decision process. In the selection phase the matrix sheet of the School case strengthened the effect that the criteria had on structuring the decision process in the selection phase. Once the proposals were ranked in the sheet original judgements did not seem important anymore. Everybody trusted the digital sheet and focused on the total scores. This was needed to take the discussion on to a higher level and find a balance between the selected candidates and the personal interests of the decision makers. The aggregation system of the separate judgements was taken for granted and not critically assessed by the decision makers. The ease in which the results of the matrix sheet with judgement of parties involved in the School case was accepted, indicate that the weighting factors as incorporated in the sheet were not explicitly applied but taken as a given during the decision process. The decision about the relative importance of the decision criteria therefore had important but implicit implications on the direction of the selection decisions.

The findings of the cases suggest that selection and award criteria as mentioned in the official competition rules only cover the basic requirements of the client. Priorities shifted during the process because the meaning of the decision task increased. This indicates that the interpretation of the criteria increases with building a frame of references within the group. This meant that sometimes criteria were interpreted more widely, and sometimes more narrowly during the process. Every now and then the previous experience of a panel member with a certain firm or architect was used to convince other panel members, especially by the more experienced decision makers. The panel members of the School used the argument of experience in the educational sector a lot during the selection process while the official selection criteria for the reference projects were not specific on (primary) education. In case of the City Hall the public debate in the award phase clearly showed a change in interpretation of the decision criteria. This was also confirmed in the interview. First functionality, financial aspects and social traits of the designers dominated the discussion among the participants of the debate. At the end of the debate the political parties discussed the overall judgement without referring to any of these possible constraints. Emotional responses to the designs came into play and arguments became more general, more subjective and more based on intuition. The decision makers referred to the criteria of 'most appealing design' as their main decision criteria instead of 'economically most advantageous tender'. The 'most appealing design' seems to include rather intangible criteria such as personal connection, faith in, and affinity with the architectural firm, a design and/or their designers. Also in the School case the main argument for decision making appeared to be the 'click with the architect', which shows the meaning they gave to the official award criteria.

The actual implementation of the criteria as stated in the competition rules could only be partly retrieved from the physical observations and documents that were available. The implicit argumentation was derived from interviews and informal conversations with decision makers. In case of the Governmental Office the pressure to take decisions, the lack of ambition, and the lack of specific expertise seemed to overrule the actual assessment of the proposals. For some of the selection criteria, such as 'timeless' and 'business like', resemblance was experienced in the interpretation of the criteria, this did not make applying the criteria easy. However, because there were only five candidates left, they did not have to phrase arguments which would justify their solution. Compared to the other jury panels, the panel of the Provincial Government Office experienced considerable conflicts with the perceived amount of freedom within the EU procedure. This was encouraged by the fact that a Design-and-Build contract would be awarded. The decision to award an integrated contract was made in a previous term of office with a different Executive board. Awareness about the limitations of this kind of contract in combination with a changing market situation eventually led to cancellation of the tender.

The experience and/or expertise of the firms, the experience and/or expertise of the proposed architect, the appeal of the design style, and previous personal experience with a firm, architect and/or their buildings were found to be the main decisive factors of selecting particular candidates for the award phase. The data from the School case suggest that the decision makers in the selection process focused on assessing the affinity of the firm and/or architect with the function, context and process of the future building. Analysis of the data showed that decision makers had difficulty with the distinction between expertise and experience, and between the architect and the firm that is represented. The panel members of School case dealt with the following issues in the selection phase:

- Amount of experience or having experience in general.

- Specific experience in a building sector (education, utility, offices etc.) or experience with similar jobs and projects (a primary school with similar size) or experience as a professional designer in general (how many and which projects have been realised).

- Experience as a design firm or personal experience of a designer.

- Assessment based on competences or assessment based on concrete examples.

Many decisions showed signs of strategic behaviour. In all cases the jury panels implicitly tried to enlarge the range of architectural types for the award phase by selecting at least a well reputed but less experienced firm and a smaller, international and/or more radical firm in the selection phase. At the same time they tried to make sure that at least a few of the selected candidates would be a reliable form to allocate the contract. An interviewee in the City Hall case indicated that they selected five high quality firms to assure the quality of the design. This meant that the award phase would be primarily a matter of taste. The decision makers in the School case were aware of the fact that they wanted to collaborate with the other parties further on in the project so they decided to select six firms for the award phase, even though they did not reach a consensus over the suitability of the sixth candidate and they had intended to select five. According to the members of the School board they gave this sixth relatively young and less experienced firm a 'wild card'. This strategy characterised the attitude of the client towards the tender procedure. The client of the School case was flexible, willing to take some risks if it would not cost too much and focused on the future. The wild card was a pragmatic decision to avoid further conflict.

Some of the decision makers in the Provincial Government case displayed a less daring attitude. One of the five submissions in the Provincial Government Office case, the one with the most controversial design, could have been rejected if a very strict interpretation of the competition would have been applied. The jury panel discussed the merits and demerits of allowing this submission. They balanced the weight of the diversity of choice and possibility of an unusual proposal against the additional compensatory design fee, the possibility of an unpopular proposal and conflict in discussions during the award phase. The chair of the panel was against allowing the fifth submission in whereas the other members were willing to take the risk. Finally the chair decided to have a second opinion from the legal advisors. This example was typical for the whole process of decision making in the Provincial Government Office case. In every conflict situation the situation was put on hold to search for legal or financial advice to support the decision outcome which increased doubt and the likelihood of cancellation. This led to a process which was shadowed by fear, a lck of decision support, and risk averse behaviour. One month after the selection committee had met the whole procedure was put on hold. The outcome of the selection phase was never published.

In the City Hall case the impact and functionality of the design proposals led to a definite 'click with the design' in the award phase. In the School case the personality and competences as shown by the architect were the most important in the award decision. Almost all decision makers in the School case reported in retrospective interviews that the final decision in the award phase was based on their intuitive judgement about the person and the potential competences as perceived by the client, which led to the 'most convincing party' and a 'click with the architect'. This 'click' was not an official award criterion but strongly related to the sub criterion of 'communicative skills'. Several advisory panel members of the School case indicated in the retrospective interviews that the current winner presented a weak vision in comparison to other firms. However they understood the choice of the School Board:

> *"The architect had a very charming personality and focused on the needs of children instead of architecture in general."*

The fact that in the award phase communication skills and sympathy of the architect were considered more important than the content of the proposal or previous personal experience with the architect confirms that the commissioning client was looking for a partner instead of a product. Because the fee of the most preferred architect in the School case turned out to be the lowest and the City Hall used a fixed fee, there is no evidence in these cases for statements about the influence of the fee on the final decision making.

All in all the data suggest that clients evaluate the potential of the full package rather than the actual measurements. Also in the City Hall case true excellence did not seem to be about answering to all functional, durable and aesthetical requirements but rather about the potential of the proposal, the designer and the firm all together. Therefore the winner of a tender reflects the design firm which the client believes will best fulfil their needs and provide them with the right architectural value. They judge the competences of the offer in relation to their needs, which can be considered as a first step in building a relationship. This indicates that building trust between the project partners is one of the main motives during a tender project. One of the participating architects in the City Hall case explained this kind of behaviour in tender situations during an interview:

> *"One always selects people who one trusts, who make one feel right. They will just see what would happen next."*

During analysis of the data several aspects were found that could have influenced the duration and character of the decision process:

- The structure of the tender process: the more clear and structured the tender design, the more efficiently the process went.

- The number of stakeholders: the more stakeholders that had to be involved and the more counter balances that needed to be performed, the more time was needed to collect all information.

- The number of entries: the more requests for participation and proposals, the greater the amount of information that needed to be assessed and the more time it usually took to view and assess all information

- The size of the tender brief: the more information provided per party, the more time was needed to assess the information in the proposals.

- The availability of the panel members: the less time the panel members had available to debate the decision, the sooner decisions needed to be taken.

- The differences in opinions: the less diverse and conflicting opinions and interests appeared, the smoother the process went and the less time it took.

- The amount and kind of emotions: positive or negative emotions sometimes made it easier to make decisions (e.g. 'falling in love with the proposals'); in other situations emotions caused more doubt (e.g. uncertainty about the implications of a decision).

- The characteristics of panel members: flexible attitudes, social consciousness, expertise, respect for the other panel members and understanding of client needs were found to increase the fluentness of the process and probably reduced the time needed to make decisions.

The findings indicate that decision making in tender decisions shows a significant resemblance to naturalistic decision making and the use of heuristics. Would this kind of naturalistic decision making be against the law? The findings of these cases show the difficulty of tracing the actual implication of the decision criteria and the relative importance of the arguments that are used during decision making. In both the School and the City Hall case the values and expressions used in the competition rules triggered the interpretation process of the panel members but decision makers did not always refer directly to the exact criteria. Mostly the criteria were applied intuitively but the criteria appeared to have created a structure to streamline the process of decision making, on individual level and at the level of the group. The analysis of the design aspects in the City Hall case showed that the expert committee in particular used criteria that corresponded to the official award criteria as described in the call for tenders. And although the three separate award criteria of flexibility, the intelligence of the solution, and contribution to the diversity were not used explicitly during the justification of the award decision, the analysis shows that the corresponding DQI aspects of Functionality and Impact were the aspects most mentioned by all stakeholders. Therefore the structure of the tender procedure acted as a steering method that enabled the client to stay within the context of the law. In the School case the findings indicate that the legal structure influenced the decision process more strongly than in the case of the City Hall. Because they used the matrix sheet to justify their selection and award decisions to the participants the arguments were reduced to official criteria that fitted the pre announced structure. In the context of transparency this devalued the message as communicated to the tenderers.

The data show that the selection phase has a different aim than the award phase. In the selection phase the clients aimed at a selection of candidates that showed interest in their project and would be qualified for the job. In the award phase the clients aimed at selecting the best architect to award the contract to. The findings suggest that awarding a contract concerns a conscious trade-off between quality and price, while selection focuses selecting candidates with a potential to meet the requirements of the client in general. So while in the selection phase questions like 'Do we know this firm?', 'Do they have similar experience?', and 'Would they fit our project in terms of architectural style and process approach?' appeared to be the rationale for decision making, the award phase was more focused on concrete expectations for the building and the building process, like 'Who is the person in front of me?', 'What to expect during the process?', and 'What do I get when the architect is gone?'. The School and City Hall decision makers also took into account a more strategic question of 'What do the other parties think?' During the selection of the Provincial Government Office the justification of the decision aimed at 'Can we justify this decision without unnecessary legal risks?'. In this case the possible implications of the selection and future award decision for the strategic position of the organisation were very important: 'How many of these firms will actually participate in the award phase and not retrieve in the tender?' and 'What will it cost?'. At the same time the question 'Is this what we want or not?' raised awareness of the actual aim of the project: 'What do we want anyway?'. Both in the selection and award phase increasing insight appeared to be an important phenomenon during the application of the decision criteria. The empirical results do not show a direct connection between the decisions and preferences of the first round selection and the winner of the second round, but in both the School and the City Hall case the winner belonged to the top three of the selection phase. In all cases the stakeholders agreed on the potentially high design quality of all five or six tenderers and therefore expected a serious offer in the award phase from each of the tenderers.

In all cases the steering committee or highest ranked executive of the commissioning body wanted to approve the outcomes of the award committee before the actual winner of the tender was announced. Then the final decision was communicated to the tenderers with a short motivation letter, in line with the legal obligations. The findings in these cases suggest that the degree in which a client goes into details about the decision depends on the political sensitivity of the tender process and the structure of the procedure. In the School case all criteria were judged quantitatively and the relative importance of the criteria was clearly stated. Therefore they were able to communicate a completed matrix sheet to the participants to underpin their decision in both the selection and the award phase. They felt that this matrix would justify their decisions, even though the matrix of the award phase was completed after the official meeting of the award committee. In the case of the City Hall and the Provincial Government Office they used a similar sheet in the selection phase, followed by a press release with a short statement.

It can be deduced from the observations and interviews that in both the School and the City Hall case the final scores of the judgements in the selection matrices were adjusted to fit the final decisions. They did this not because they intentionally wanted to elude the law but because they wanted to live up to the common expectations of the law conform their own interpreted. Only in the City Hall case the project team of the City Hall wrote a short report instead of a matrix sheet to motivate their decisions. Writing a memo about the decision outcomes is in line with the architectural tradition of the jury report. Tenderers in the City Hall case therefore had more information about the judgements of their proposals which might have benefit to the acceptance of the award decision.

6.8 Reflection

The three cases provided very valuable insights about the aspects that play a role in the process of decision making of public clients in the context of European tendering regulations. The cases clearly showed that decision making during a tender is a human process in a context of rational legislation. From a legal perspective selecting an architect appears to be far less complex than in real life. What proves to be an efficient and accountable tendering process from a legal perspective is not necessarily effective from a clients' perspective. This difference in interpretation probably originates in the subjective nature of the judging architectural design quality, the different traditions underlying the current regulations, and the governance structure of public commissioning clients. It appears that during a tender every client needs to discover the complex and sensitive nature of a selection process in architecture for themselves. An architect selection is therefore a process of sensemaking (see for example Balogun, et al., 2008; Weick, 1995).

Two kinds of decision processes can be distinguished during a tender: 1) decisions about the design of a tender, and 2) decisions about the quality of the (design) proposals that are submitted during the implementation of the procedure. Both these processes comprehend many rational and intuitive decisions that are often based on incomplete and conflictive information. In practice everyone's attention is mostly fixed on the assessment decisions about the quality of the architects, but for clients the preparations might be as important as the implementation. During the preparations of a tender many design decisions are made that influence the course of the decision process. Yet the course of the decision process also strongly relates to the sensemaking processes of the actors. A tender includes many assumptions about the future needs of a client. I found that the selection of an architect was often not the only goal of the tender; decision makers tried to fulfil strategic and personal aims as well. During the decision process numerous conflicts between the interests of the actors and stakeholders come to the surface. The legal structure of the tender procedure mainly determines the flow of the process and therefore needs to fit the client characteristics and the goals. The characteristics of public clients are closely related to the governance structure and depend on the competences and positions of actors that play a role in the project.

In the tender process the selection and award criteria, and proposals of the architects can be considered as input, argumentation during the process of decision making as throughput, and the final decision as output. The meaning of the criteria seems to increase during the decision process which sometimes caused a shift in the importance of certain design quality related aspects. Because preparations were usually made by others the panel members needed time to interpret the decision criteria and familiarise with the circumstances, including the promises that were made in the competition rules. The decision process in the selection phase was in essence similar to the process in the award phase. In order to make an award decision the panels had to go through an iterative process of several phases of perception, individual judgement and group decision making. During these phases voting, ranking and discussion were used. The observations indicate that the ongoing discussions between the decision makers increased the appreciation of each other's interests as well as the qualities of the proposals. Discussions also contributed to the level of satisfaction and acceptance among the decision makers about of the decision process and the decision itself. These effects increased the transparency of the decision process.

The selection and award criteria were used to build a frame of reference among the decision makers, but the final decision was mainly based on a holistic and often intuitive judgement that was supported by the other panel members. This suggests that the legal structure as compelled in procurement law provides a structure for the process of decision making but cannot replace the actual decision process of the client. The frame of reference of the decision makers was influenced by the characteristics of the panel members (e.g. the level of expertise and kind of interests), the content of the submissions (e.g. visuals, narratives and/or texts), the design of the tender procedure (e.g. a presentation, the format of the documentation, the level of requirements), and the involvement, role and preferences of the stakeholders (e.g. actors or advisors; and being represented or only informed afterwards). The results indicate that information sources, time constraints, and different kinds, and levels of expertise and interests need to be carefully considered while designing a tender.

Chapter 7

PROCESS DESIGN - A COMPETITION CASE AND VALIDATION

7.1 Introduction and research questions

The results of the three tender cases described in the previous chapter explained the most important elements in the design of a tender procedure. They also illustrated the difficulties and conflicts that can arise during implementation of a tender. When the building of the TU Delft Faculty of Architecture was destroyed by a fire on 13 May 2008, an opportunity arose to apply the experiences of the tender cases in a situation that was built on the competition tradition in architectural design. As an employee of this university I became a participant observer in the organisation of an open international ideas competition 'Building for Bouwkunde' for a new building for the Faculty of Architecture. This chapter describes the decisions that were made in organising this ideas competition for a Faculty Building and critically reflects their effects and the implications for selection procedure design.

The central research questions are:
1. How does a public commissioning client decide on the procedure for the selection of an architect?
2. What are the implications for the design of procedures for the selection of architects?

The structure of the chapter is based on the framework with which the three cases from Chapter 6 were analysed. In the first section the research methodology is described. Then the results of the analysis of the project characteristics, the context and the actors of this case are reported in one section. The following sections describe reflections on the results on tender design, in this case the competition programme, on the competition brief, on the tender procedure, on stakeholder involvement and on the decision procedure. Where applicable, the findings of this fourth case are used to reflect on the findings of the three tender cases in Chapter 6 and the theoretical insights from Chapters 5. The chapter ends with a discussion about this fourth case in relation to procurement law and the context of the research.

7.2 Research methodology

This chapter describes the results of a single case study. As described in Chapter 5, the method of studying cases makes it possible to study decision making in a real life context on different levels of individual, group and organisational decision making (Hackman, 2003; Yin, 2009). This case can be seen as part of the research strategy in which several cases are compared (Eisenhardt, 1989; Eisenhardt & Graebner, 2007). It differs from the cases described in Chapter 6 in terms of the method of data collection and the approach to data analysis. The competition had first been intended as a European tender, but was changed into an ideas competition to provide input for a future procurement process. The sudden cause and rushed character of the competition meant that the context differed from a regular tender. However, the situational characteristics and processes also show great similarities, which make it possible to compare the results with the other cases in this research. This case adds the competition tradition as one of the roots of architect selections identified in Chapter 4. The most important differences with the other cases are its international character instead of a restricted European tender, that proposals were submitted anonymously instead of a presentation by the tenderer, and that the client was very knowledgeable and widely known in the field of architecture. Therefore the impact and the aim of the decisions made to select an architect differed. These differences are used to explore underlying principles of the complexity of selecting an architect.

In this case the representative of the commissioning body, the Dean of the Faculty of Architecture, was aware of the PhD research and invited me as a researcher in architect selection procedures to take an active role in organising the competition. This created a revelatory case, "a situation in which an investigator has the opportunity to observe and analyse a phenomenon previously inaccessible to social science inquiry" (Yin, 2009, p. 48). I was already a member of the overall organisation in which the case took place, but since the project was conducted under the direction of another department, none of the project team members were direct colleagues and the organisational culture was different. This allowed for sufficient 'otherness' to conduct the participant observation (Sanger, 1996). Yet, the entry-exit problem was therefore relatively easy to overcome (Bechtel & Zeisel, 1987).

A large set of data was collected by using different methodologies (see Table 7.1). I was involved as one of the project coordinators and a full member of the project team for 32 weeks in order to organise an international ideas competition. During this period I kept a research diary. At least once a week I recorded notes about the activities of that week, noted the considerations and arguments that led to a certain decision, and filed all documents. Personal reflections on the events were noted in a special section of the log. After the ideas competition had ended, I conducted six semi-structured interviews with the jury members and the project leader. To create a certain distance to the data I conducted the analysis a few months after the project was finished and data collection had ended. All data were first analysed in Atlas.ti, a software application for supporting qualitative data analysis. The framework as established in Chapter 6.3 also proved to be a good structure for analysing these data. This means that actors, project characteristics and tender/competition design were taken as the main themes for the analysis. The data analysis focused on the design of a tender competition from a project team perspective, not on the content and submitted results of the competition. Therefore some parts of the data, such as the database with submissions, were not used for the purpose of this chapter.

Type of data	Specification of data sources
Observations	Participant observation for 32 weeks noted in a logbook Observation of jury meeting.
Documents	Competition programme Internal report analysis report of the submissions Database with submissions Press releases and news paper articles (e.g. de Volkrant, Algemeen Dagblad, B_nieuws) Debate on internet (e.g. www.archined.nl; www.architectenweb.nl; www.archicentral.com; www.architectenwerk.nl) Jury report TU Delft publication about competition 'Open to ideas'
Interviews	6 semi-structured retrospective interviews with jury members and project leader
Degree of involvement	Whole process from first project plan to award ceremony and publication

Table 7.1 Overview of research methods and available data

Most of the data were collected by participant observation. Participant observation studies are "in the tradition of 'verstehen' sociology and cannot by repeated in the experimental manner of the natural sciences" (Jackson, 1983, p. 44). My role as participant-researcher was known to the other actors and it was not prominently concealed during the project. As a project member I often had first-hand experience but could not always record 'private' information as it occurred (Creswell, 1994). I experienced several benefits and limitations of this research approach during data collection.

On the one hand, being directly involved in a project complicated the collection of data. I got personally involved, including the mixed feelings, responsibilities and emotions that this sometimes brought. These emotions became part of my process of learning to make accurate observations from a qualitative research perspective (Bechtel & Zeisel, 1987). A bias could have occurred due to manipulation of events (Yin, 2009) or personal involvement. Data collection was time consuming and there was a constant time pressure because of the tight deadline of the competition, and therefore limited time to reflect on actions.

On the other hand, the case illustrated the complexity of the phenomenon because I could personally experience decision making, including the conflicting issues. It allowed me to distinguish interpersonal behaviour and motives more carefully (Yin, 2009), and by being able to study the mundane world, the value of fine details were more appreciated (Silverman, 2007). Acting as a participant observer also created an opportunity for me to apply the findings of previous cases and get in contact with organisations and persons that play an important role in the context of architect selections. Compared to the non-participant observation in the cases of Chapter 6, being present during the jury meeting without being seen as intrusive was one of the most beneficial aspects of the participant approach in relation to data collection about expert decision making. In the end I felt that being involved at an early stage of tender development proved to be very beneficial in analysing the complexity and sensitivity of the context of architect selections.

7.3 Case characteristics

7.3.1 Description of the context

The Faculty of Architecture is the largest faculty of Delft University of Technology and one of the largest in its field in Europe. Delft University of Technology employs over 2,700 people of which over 600 belong to the Faculty of Architecture. The Faculty of Architecture is generally known as 'Bouwkunde'. Of the 15,000 TU Delft students about 3,500 are Bouwkunde students. The Faculty has four departments: Architecture, Urbanism, Building Technology and Real Estate & Housing. Both at the national and at the international level, the Faculty works together with universities, private companies and public bodies. Furthermore, there is an extensive exchange of faculty members and students with other architecture departments, both in the Netherlands and abroad. The number of students who choose to study at the Faculty of Architecture in Delft has steadily increased over the past years. Students are educated to become architectural designers who on the one hand contribute to the growth of scientific knowledge regarding architectural issues, and on the other are able to design practical solutions for tackling these issues.

The former building of the Faculty of Architecture was located at the Berlageweg in the middle of the campus in Delft. The building was designed in the tradition of Functionalism by van den Broek and Bakema, both former teachers at the faculty. The design was selected by a competition among all design professors at that time, who also acted as the jury member. This overlap of roles and potential conflict of interest apparently created no concern at the time. The building was realised in 1970 after substantial changes to the design concept due to increasing student numbers. Many students, alumni and staff members saw the building as a second home and a source of inspiration with consistent detailing and a sparkling atmosphere (Maandag, 2008).

A fire broke out in the morning of 13 May 2008 as the result of an electrical fault. Due to its age the building was not equipped with sprinklers and when the fire service arrived they were ultimately unable to extinguish the fire. Thus "this sturdy concrete building, which was full of valuable collections and always bustling with activity, was reduced to ashes" (Faculty of Architecture, 2008). In both a material and an emotional sense, the destruction of the building was undoubtedly a major loss for the architectural community. The fire was widely covered in (inter)national newspapers and on television. Memories of the old building were written (Maandag, 2008) and new dreams were built in a creative festival in June 2008 (Vergu, 2008).

Soon rumours started about the opening of a new building and the selection of an architect for this new faculty building. However, the sudden and unanticipated initiation of such a selection meant that there was no vision for the new building yet. The Faculty and the board of the university (the formal commissioning body) had to sharpen their ambitions first before they could start an official European tender. In July 2008 two trajectories were chosen to sharpen the ideas for a new faculty building: an Open International Ideas Competition called 'Building for Bouwkunde' and a national Think Tank with experts. Both projects were guided by the same steering committee and had to provide input for the brief and plan of action for a new faculty building in the future. This created a special situation, since people involved in a Faculty of Architecture typically have a special relationship to architecture and the built environment in general. Almost everybody using the faculty building is trained in one or more specific areas related to the field. A great part of employees at the faculty also run or participate in a design or consultancy firm on a part time basis. Because the faculty in Delft was established in 1904 and there are only two architectural faculties in the Netherlands, alumni represent a large part of the professional community. In the context of the Think Tank, the faculty building of Bouwkunde was considered as a case study for redevelopment of the campus strategy. The municipalities of surrounding cities (Delft, The Hague, and Rotterdam) and parties from the industry were actively involved in the Think Tank.

Immediately after the fire, the Dean of Bouwkunde started preparations for new accommodation for the several thousand students and over 600 employees of the faculty. In the meantime university staff worked hard to set up temporary accommodation, while staff and students of the Faculty of Architecture were housed in tents and at other faculties. The temporary housing of the faculty was established in the former head quarters of the university at Julianalaan in the northern part of the campus. The renovation was immense and required a great amount of workforce and energy in a short period of time. This created a positive chaos in which everything seemed possible. The design team mainly consisted of architects who were also educated at and employed by Delft University. The briefing and construction team also consisted of employees of the faculty. In September 2008 the renovated building was opened for the new first-year students, and by November 2008 almost all other students and employees had moved in. Because there was not enough room available to provide every employee with their own workplace, the concept of flexible workplaces was introduced. At the time of writing, the faculty of Architecture is still housed in this building.

Being part of Delft University of Technology, the Faculty of Architecture is subject to university governance. In the Netherlands, universities are mainly financed by the Dutch government, supplemented with money from industry and the European Union. Most of the buildings (including the Faculty of Architecture) are owned and insured by the University. The Ministry of Education, Culture and Science decides on the budget directly allocated to the universities and competitive funds for the Dutch research council. Because of the scale of the financial and intellectual loss, the Minister decided to allocate an additional € 25 million to create not just an ordinary new faculty building but an 'icon'. However, in September 2008 the consequences of a previous restructuring of governmental finances emerged, which meant that Delft University had to seriously revisit its spending and cut back on expenditure. At the same time the effects of the financial crisis were felt. So apart from the emotional shock of the fire, the financial situation created an atmosphere of uncertainty about the future of the Faculty. This changed the mindset of the Faculty. The ideas competition started just before the financial situation got worse. Therefore the structure of the competition was set up according to the standards of the former situation, in which money

Type of stakeholder	Specification for this case
Commissioning client body	TU Delft (Executive Board)
Representatives of client	Dean of the Faculty of Architecture
Shareholders & supervisors	Dutch Department of Education, Culture and Science, Municipalities of Rotterdam, Delft and The Hague
Daily users	Faculty employees, Faculty students
Non-daily users	Visitors, Students and employees of other (international) faculties, Alumni, Neighbours, Municipality, Architecture community
Representative groups	Student board, Works council, Dutch Royal Institute of Architects

Table 7.2 Overview of stakeholders of the Faculty Building case

seemed of secondary importance for creative and innovative activities. At the time the submissions had to be assessed and the winners were announced, the interim accommodation had been successfully brought into use and budget-cutting plans were to be implemented at the faculty. Until the insurance reimbursed the university, it was still uncertain if the clause that stated that the reimbursement had to be spent on a newly constructed building was to be enforced. Not applying that clause opened up the opportunity to think about the insurance money as a financial windfall in a more cash-strapped situation. In June 2009 the insurance company paid the university € 118 million without the obligation to reinvest the money in a new building. About half of this amount had been spent on temporary housing and interior refurbishment.

Table 7.2 provides an overview of the stakeholders that are involved in this case. Because of the unique and unprecedented character the fire can be seen as a rare event through which organisations can learn in order to revitalise their organisation (Christianson, et al., 2009).

7.3.2 Description of actors

In the Faculty Building case the main actors were a steering committee, a project team and a jury panel. The project team consisted of a project leader (externally hired), the Dean of the Faculty of Architecture (chair), an additional project co-ordinator (me), and the Head of Marketing and Communication. The website and submission system of the competition were developed by a consultancy firm, ICOP, and the project team was supported by a secretary of the Dean's office and several student assistants. The steering committee consisted of the President of the Executive board, the Head of the real estate department of the TU Delft (FMVG), a professor of Real Estate of the faculty of Architecture, and the Dean.

In line with tradition in architecture, an international jury was assigned comprising the Dutch chief government architect (chair), three architectural design deans/professors from abroad (USA, India and China), the Dean of the Architecture faculty, two Dutch professors of Architecture (also partners in architectural design firms), the director of the Netherlands Architecture Institute (NAi) in Rotterdam, and a Dutch MSc student of Bouwkunde. One of the international professors was unable to attend the jury meeting, he was not replaced. Because the Dean was a member of all actor groups, he acted as the connecting link. However, right after

Actor	Team members
Steering committee	President of the TU Delft Executive board, Head of the Real Estate department of the TU Delft (FMVG), Dean of the Faculty of Architecture, Professor of Real Estate of the Faculty of Architecture
Jury panel	Chief Government Architect (chair), two Dutch architect/professors from Bouwkunde, two international architects/professors (India/China), MSc student of Bouwkunde, Director of the NAi Unable to attend: Dean/architect of Bouwkunde, International Dean/Professor from USA
Project team	Dean of the Faculty (chair), external project leader, project coordinator (the researcher), the head of Marketing & Communication, administrative support, website designers and legal consultant

Table 7.3 Overview of actors of the Faculty Building case

the opening of the competition in September 2008 the Dean fell seriously ill, and remained absent during the rest of the competition period. He was not replaced in the jury committee. His responsibilities were assigned to another member of the steering committee, a professor in real estate from the Faculty of Architecture and chair of the Think Tank, who was not as deeply involved in the competition as the Dean. This meant that after September 2008 not everything about the competition was discussed on the same detailed level with the official representative of the client as before.

Main task of the project team was to write the competition programme, provide the information for the website, prepare the jury meeting and coordinate the whole competition. On several occasions the steering committee provided input on the competition programme and on some of the managerial aspects (e.g. finances). The jury panel acted as official 'award committee' and decided on the distribution of the prize money among the winning entrants. In preparation of the jury evaluation, two teams of employees were assigned to assess the submissions of two perspectives. Table 7.3 provides an overview of the actors in the Faculty Building case.

7.3.3 Case description

During the first few months the competition programme was set up mainly according to the ideas of the Dean, and on the basis of the model for competitions as described in the Dutch publication 'Kompas' (van Campen & Hendrikse, 1997). Kompas refers to a model of a competition programme as a result of a Covenant for Competitions in the field of architecture, urban planning and landscape architecture. The covenant was developed and signed in 1997 by almost all professional and governmental bodies in the Dutch field of relevance a few years before European tender rules and regulations were actively implemented. The model seemed appropriate for the aim of this competition and created structure for the project team. The project team first used the Dutch version of the model, and translated it to English before the launch of the competition.

The competition programme consisted of two main parts: Part A. Competition Brief and Part B. Competition Rules. Part A of the competition brief included:

- An introduction to the problem description.
- The assignment.
- The brief at building level.
- The conditions for the location.

Part B of the competition rules included a description of:

- The competition and objectives.
- The requirements of the entrants, method of registration and submission, the language, a time schedule with deadlines and the method of questions and answers.
- The names of the jury panel, the prizes, the evaluation criteria, evaluation procedure and publication of a jury report.
- The follow-up to the competition.
- An indication about the publicity, publications and exhibition.
- The rules concerning copyright, use, ownership, and disputes.

The official objectives of the competition were:

1. To stimulate research by design.
2. To encourage creativity among the important younger generation of designers.
3. To stimulate scientific development in the field by means of critical reflection and debate.

The Dean wanted to involve as many people as possible in the ideas competition. He saw the competition as a chance to enhance the reputation of the faculty. One of his strategic goals was for the faculty to become one of the most famous one in architecture in the world. The competition also provided a forum for all people in the community to voice their ideas. This diversity of goals is characteristic for the concept of a design competition, as stated in Chapter 4.3.1. Several means were used to reach these diverse goals, which will be addressed later in this chapter:

- Communication in English rather than Dutch.
- A competition website for communication, registration and submission.
- Low entrance requirements.
- An interesting jury panel.
- A substantial amount of price money.
- Launching the competition on the Architectural Biennale in Venice.

The opening at the Architectural Biennale in Venice resulted in a strict deadline to prepare on necessary documentation before that date, 13 September 2008. This created considerable time pressure for the project team and resulted in rushed decision making. Entry was open to registered architects and urban planners and to students following a degree programme in the area of architecture, urban planning, civil engineering or industrial design. The offices of the jury members were excluded from competition. The competition was set up as an open procedure with no pre-selection of the submissions. The whole competition was managed via a website containing the competition programme, some statements of renown

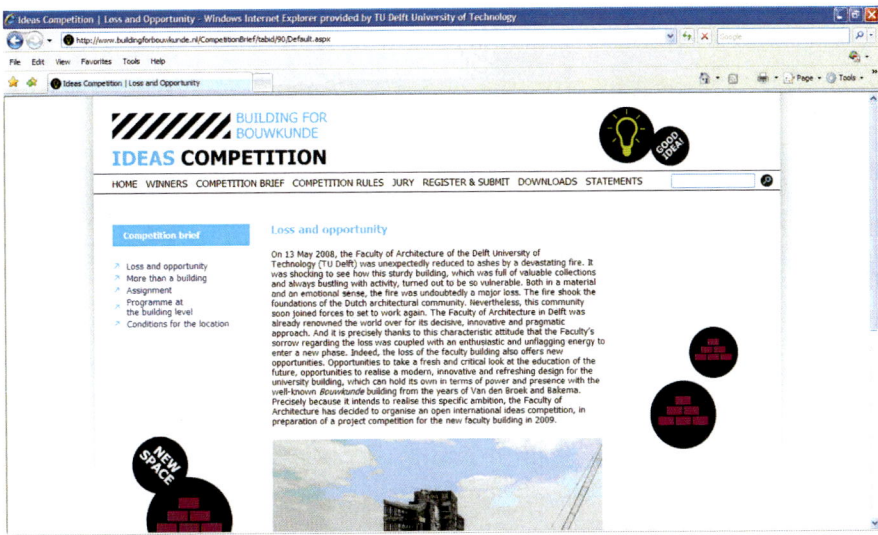

Figure 7.1 Home page of the competition website www.buildingforbouwkunde.nl

(international) professionals about the assignment, additional information about the campus and the faculty, registration and submission and the facilities to ask questions (see Figure 7.1).

The entrants were asked to:

> *"Formulate both in text and images (maximum two A1 posters and one A4 text) a vision on the two competition themes: 'new concepts' and 'dynamics of city and campus'. This vision should be presented in a sketch design for a new Bouwkunde building on the existing site, or on a well-argued, alternative site."*

Sustainability, connection with the urban and social context, and personality were important issues in the description of the assignment. An indication of the gross floor area (55 – 60,000 m²), the budget (€ 2,000/m2) and an overview of several functions to be included in the building were provided. General and historical information about the faculty, maps of the former location (scale 1:500) and the campus (scale 1:2,500), and a master plan for the campus renewal were available for download on the website.

The competition was officially launched by the Dutch Minster of Education, which created a lot of media attention (first picture in Figure 7.2). The deadline for submission was 13 November 2008. Potential participants were entitled to ask questions via the competition website until three weeks before submission. This resulted in 194 questions, which were all answered by the project team via the website. The website proved very popular and a total of 1,380 participants registered. Of these, 466 valid submissions were received originating from 50 countries. 5 submissions were technically invalid because they could not be opened or did not contain any drawings related to the brief. 65% were of the submitters were professionals and 77% were male.

Figure 7.2 Images from the opening of the competition in Venice (left, showing the minister of education), the jury meeting in Delft (middle, showing the submissions), and the exhibition in the NAi in Rotterdam (right, during the opening)

The evaluation procedure consisted of an assessment phase and an evaluation phase. During the assessment phase the submissions were analysed by two internal analysis teams on the content of the proposals and checked against the rules and assignment of the competition by the project team. Only 42% of all submissions fulfilled all requirements as stated in the competition programme. The chair of the jury, however, decided to include all technically valid submissions in the jury evaluation process. The results of the assessment, a typology of the submissions (see Figure 7.3) and a quantitative analysis, were made available to the jury for evaluation on 14 and 15 January 2009.

The competition prize money totalled € 60.000, to be distributed among the winning entrants by the jury at its own discretion. All submissions received a registration number and were evaluated anonymously by the jury in the former Techniek Museum of the University in Delft (see middle picture in Figure 7.2).

Three criteria were published in the competition programme on the basis of which the jury would evaluate the submissions:

- Visionary power, i.e. the originality and innovative character of the sketch design, including the issue of sustainability.

- Architectural quality, i.e. the spatial composition, the embeddedness in the urban environment, the expression and materialisation, and the consistency of the sketch design.

- Economic and ecological viability, i.e., the functionality and feasibility of the sketch design.

The jury selected six prize winning submissions and two honourable mentions in two rounds based on an integral judgement. During the first day 50 submissions were selected; on the second day these 50 were reduced to 8 nominees and finally six winners: three first prizes of € 15.000 and three second prizes of € 5.000. After the jury members had agreed on the outcome of the competition, the names of the winners were retrieved by the administrator of the website. Among the winning submissions were three Dutch, one Belgian, one Finnish, and one French submission. Almost all winners belonged to the younger generation; two of them were students. The jury members were positively surprised by the fact that so many of the winners were young professionals. The project team checked

ONE

one building volume – 46%

The **MONOLITH**: a type of building based on a solid volume. No clear structure can be recognised and the volume itself is undoubtedly single. (78x)

The **ATRIUM**: the type of volume that has been arranged around one open space. This space can also be covered. (64x)

The **SPONGE**: singular volume as an amount of enclosed spaces. It absorbs open spaces like a sponge. (44x)

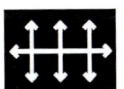

The **LINE ORGANISATION**: the organisation inside the building is primarily based on a linear element. This can be placed in a horizontal or in a vertical direction. (16x)

The **SPIRAL ORGANISATION**: the organisation inside the building is based on a spiral. This can be placed in a horizontal or in a vertical way. It is even possible that the building as a whole has the shape of a spiral. (8x)

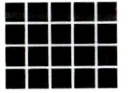

The **GRID ORGANISATION**: based on the principle of the grid organisation. This can be placed in a horizontal or in a vertical direction or a combination. The lines can be placed diagonal as well. (5x)

SOME

two or three volumes that are related to each other – 25%

The **BASEMENT & VOLUME(S)**: a basement with a volume placed above. They are physically related to each other. (62x)

The **TWO OR THREE VOLUMES**: the arrangement of two or three (easy to count) volumes that are physically related to each other. (34x)

The **LINEAR STRUCTURED**: some volumes that are arranged in a linear way. This structure could have for example a reference to an urban structure or could be a volume itself, but it is always a physically relation. (19x)

MORE

several volumes that are related to each other – 21%

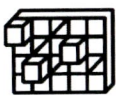

The **WINE RACK**: an arrangement of volumes that are placed upon each other in a structure that is three-dimensional. It is about how the volumes have been positioned, not about the organisation inside. (3x)

The **COLLECTION & BORDER**: this shows a collection of building volumes that are arranged within a border. This border could be for instance a roof, a glass box or an urban line. (34x)

The **COLLECTION AD RANDOM**: a collection of building volumes that are arranged without any structure. There is no physical connection and no border can be defined. (43x)

The **COLLECTION AS A CITY**: a collection of building volumes that are arranged in a pattern that could be interpreted as a city. A city structure is more or less recognisable, but the volumes are not physically related. (18x)

JUST IDEAS [6%]

The type **JUST IDEAS** includes plans that could not initially be classified as 'buildings'. The idea is more or less elaborated than the building. (31x)

RE-USE [1%]

A few examples of **RE-USE OF BUILDING(S)** are addressed: Van Nelle Rotterdam, Julianalaan TU Delft (3x), EEMCS TU Delft, Post offices Rotterdam + Amsterdam + railway station in Delft. These plans have been classified as a special type, because the existing building is dictating the typology. To express their ideas, a separate typology was introduced. (5x)

Typology types of the competition entries (pictograms Koehorst in 't Veld)

Figure 7.3 Overview of categories as identified in the assessment phase (designed by Koehorst in 't Veld, in Faculty of Architecture – TU Delft, 2008)

Characteristics	Specification for this case
Contract	Prize money € 60.000
Building costs	Max € 120 million
Gross floor area	55,000- 60,000 m2
Submission requirements	Registered architects and urban planners or design students in relevant areas
Basis for awarding	Design sketch on poster
Competition phases analysed	Open procedure: one phase
Compensation for participation	€ 0
Characteristics of case	Sudden cause due to fire, ambitious and rushed process, looked at by external and internal community

Table 7.4 Overview of Faculty Building case

if all winners fulfilled the requirement of an official registration in their country's architect register or at a recognised educational institution. In line with architectural tradition and to provide for enough time for publicity and preparation, the names of all nominees were announced first. The winners and their ideas were announced during an award ceremony two months later. At the same time a jury report was published.

An exhibition in the Netherlands Architecture Institute in Rotterdam was opened on the day as the award ceremony, 14 March 2009. The exhibition included all submissions of the competition as well as the results of the Think Tank (see third picture in Figure 7.2). It attracted a lot of visitors. The best 50 ideas, sixteen ideas for discussion and the results of the Think Tank were included in a publication called 'Building for Bouwkunde – Open to Ideas' (Faculty of Architecture - TU Delft, 2009). The whole process of idea generation (Think Tank and Ideas Competition) was closed out with a debate and the presentation of the publication on 13 May 2009 to coincide with the anniversary of the fire. In that meeting the results of the Think Tank and ideas competition were offered to the Executive Board on of the TU Delft, which will decide about the next phase of the project. Table 7.4 provides an overview of the characteristics of the Faculty Building case, which started the end of May 2008 and officially ended 13 May 2009.

7.4 Competition design

7.4.1 Competition brief

The competition programme of the Faculty case consisted of the competition brief and the competition rules. The competition brief plays an important role in the direction of the submissions of the competition. Reflecting on the findings of the competition in relation to the development of the competition brief, a few themes become visible that will be addressed in this section:

1. Balance between tradition and innovation.
2. Influence of time and changes in the context on the direction of the brief.
3. Availability and awareness of information during preparation.

4. Frame differences between jurors and participants.
5. Cultural and political differences in the decision frame.

Ad 1) Balance between tradition and innovation

In order to reach the objectives of stimulation of creativity and debate, the assignment for the competition brief was kept relatively open to any direction possible. During the development of the competition programme the project team was constantly faced with conflicting issues on client and building related issues, which were also found in the tender cases (Chapter 6.6.3) and by Kreiner (2006) and Rönn (2008). In defining the competition brief the project team experienced a need for innovation as well as a need to maintain the positive qualities of the old situation. For example, the old building structure provided a 'street' on the ground floor for meeting and group gatherings. At the same time this structure created vertical traffic in the building that did not encourage interaction between the departments and lacked flexibility on the floors. In the assignment social interaction and flexibility therefore became central themes.

Another example with potential for innovation was the location of the faculty building. The old faculty building was situated on the southern part of the campus. On the campus map this location seemed relatively central, but in practice most activities and facilities are situated in the northern part of the campus. The project team did not know at the time of publication which location would be best and therefore this issue was left open. The assignment created the possibility to choose 'a well-argued, alternative site' but only 13% of the submissions proposed another location than the Berlageweg.

The participants instead seemed to focus on the old situation in an attempt to analyse the future needs. Remarkably none of the questions asked by the participants in the period before registration and submission were about ambitions or strategies other than mentioned in the brief. A few participants explored specific location options but in general the requirement was not questioned. However, the submissions did raise new issues and not all submissions fulfilled all requirements. About half of the 194 questions raised by the participants related to the competition rules (registration requirements, categories of registration, registration as a team, the format of the A4, statement of intellectual property, the requirements of the entrants). Another large category of questions related to the content or interpretation of the assignment and the characteristics of the faculty of architecture and its users, including the structure of the old building (the use and interpretation of the maps, numbers of students divided over groups, specific aspects of the assignment, amount of themes, opening hours, floor plans of the old building). A few questions related to the follow up on the competition (e.g. relation with the European tender selection) or the evaluation procedure of the jury (presentation of the posters). This shows that indeed a competition brief is read as half stimulation and half instruction (Kreiner, 2006) and therefore acts as a communication filter (Heynen, 2001) that steers the direction of the decision process. This direction is based on an interplay between the client and the participants and therefore dynamic and relatively unpredictable. In this case the brief was used to explore the options, which is in line with the aims of an ideas competition.

Ad 2) Influence of time and changes in the context on the direction of the brief

At the start of the competition, the competition brief was communicated via a website. For several weeks potential participants of the competition could inform themselves about uncertain or missing information. The project coordinators were responsible for providing answers via the website. Most of the questions were easy to respond to because the answers were obvious or already included in the competition programme. Sometimes questions needed to be redirected to other people in the organisation because they required specific expertise (e.g. the division of the MSc students per department). As project team we were aware of the fact that the content of the answers would steer the direction of the submitted ideas. Too much deviation from the original assignment could place early starters at a disadvantage against those starting late. As the Dean was on sick leave, he could not take a stand in this matter or answer questions.

However, during the period that the participants were working on their submission, the context at Delft University had changed. The temporary accommodation of the Faculty was put into use during the course of the competition. Local participants were aware of the positive experiences of this accommodation and therefore able to use this information in their submissions. Because the project team wanted to focus on a new faculty of the future, we chose to answer the questions in an as neutral manner as possible by staying close to the original text. This meant that no additional information about the old faculty building or the building at Julianalaan was provided.

Additional information to improve or complete the assignment was offered, such as student numbers per space function or technical details about the location. Reflecting on the communication to the participants, one of the Dutch jury members mentioned in an interview that the participants could have been informed about the positive experiences and new insights about the temporary location. This would have caused a disadvantage to participants who started the preparations for the competition early, but would have decreased the discrepancies between local and international participants. This dilemma illustrates the dynamics of the matching process: the context can change over time and this affects the availability and importance of information during the process of the competition design. A client has to decide while the competition is ongoing in which direction to steer information and therefore also on the variation of information among the participants.

Ad 3) Availability and awareness of information during preparation

As a project team we experienced how hard it was to define a clear ambition for the tender brief about the future accommodation based on the existing information. Gathering factual information about the client's organisation was also not easy on such short notice (e.g. number of staff members and students, mission, strategy). On campus level, developmental plans existed within the real estate department of the TU Delft (FMVG), but these were obviously not adjusted to a future without the former faculty building at the Berlageweg. The brief of the temporary accommodation was used as input for the functional requirements of the competition. The Dean provided the project team with most of the input for the vision and ambition of the competition assignment and criteria in three sessions. His input seemed to be a combination of the needs of a faculty as a client, the expectations of the professional community about the role of the faculty as a client, and his own experience and expectations as an architect about a faculty building for Architecture. Reflecting on the process of ambition formulation, it appeared questionable that only the Dean, probably influenced by some of his close contacts and network, provided input for the ambition and brief. The project team decided to ask an employee with a background in journalism to turn the notes about the ambition from the project team meetings into an appropriate text for the competition programme. Although this person did a marvellous job at turning wild ideas into a consistent story, the brief was not supported by any other individual or group than the Dean and the members of the project team.

In the tender cases in Chapter 6 a similar situation was found. At the beginning of a process decision makers were usually not aware of the importance of the brief while the brief is one of the main steering instruments of the competition. Participants rely on the information in the brief being valid and reliable because theoretically it should provide the basis for their ideas. Especially for foreign participants the information on the competition website was the primary source of information. For them it was almost impossible to gather inside information about the Faculty. This is illustrated by the fact that of the winning participants, five out of six (had) studied or worked at the Delft Faculty of Architecture as student, exchange student, or employee. In line with Kreiner (2006) these findings show that specific information and a feel for the relevance of information can improve the chances of winning because the ideas provide a better match with the assignment.

Ad 4) Frame differences between jurors and participants

Participants typically expect that the information in the brief would also serve as a starting point for the jury members in their assessment process. This would imply that all jury members are fully informed about the competition and use a comparable frame to assess the participants. However, as the literature on decision making showed jury members, especially experts, take along their existing frame of references and apply this on a particular situation such as this competition (Hutton & Klein, 1999; Kazemian & Rönn, 2009). The results of the tender cases in Chapter 6 showed that if jury members would be involved in the preparations of the competition they could align their frame of reference, or shape that of the competition programme. Such an exchange would increase the chance that the jury is aware and mindful of the client's ambition rather than their own preconceptions.

During the design of this case, the project coordinators and the Dean therefore decided to inform the chair of the jury about the competition programme and rules. For the other jury members this was not possible due to time constraints and practical reasons. Due to the absence of the Dean during the jury meeting, there was no longer a personal connection between those who provided input for the ambition and the members of the jury as it had been intended. Those who were present, such as the project coordinators, the author of the jury report and the author of an essay for the publication, did not have a vote in decision making nor a chance of providing additional information during the assessment process.

The results of this case imply that complete equality of frames among the participants and jury members is not a realistic aim. This effect was reinforced by the fact that the launch of the competition was at least two months before the deadline for submission and the submission a few weeks before the moment of assessment. At the time of the competition launch, only a few people could have predicted that the winners of the competition would include two submissions that proposed to renovate and transform the building at Julianalaan. Although the sudden cause and rushed character of the competition might not be typical for all cases, these findings suggest that there will always be a difference between the frames with which submissions are designed and with which they are assessed. Within these different frames jurors' preferences for alternatives could also differ. It will therefore be very hard for participants and outsiders to predict preferences of a jury panel beforehand.

Ad 5) Cultural and political differences in the decision frame

The amount of information about the current situation of the Faculty also differed per jury member. Retrospective interviews with the jury members showed that only a few of the members used the competition brief explicitly to prepare themselves for the meeting. The two foreign jury members arrived just one evening before the jury process started. They visited Delft once or twice before but in different context. All the other members had studied at the faculty and/or were still involved or familiar with the situation at the Faculty. Only a few of them had visited the new temporary housing at Julianalaan. In the beginning of the jury meeting the jury members attended a presentation about the campus and the current housing situation of a staff member. This was done to align the frame of the jury members.

Because of the absence of the Dean, the frame of reference about the assignment was defined by jury members who had not been directly involved with the phrasing of the assignment nor the current university governance. This meant that within the jury panel the political context of decision making was simplified and inside information about the current state of the finances and insurance issues did not influence the argumentation as much. The jury could therefore act as an independent committee with the interest they thought would benefit the Faculty. However, the observations indicated that some of the Dutch jury members took the changing financial, political and social context into account and included this into their frame of decision making during evaluation of the submissions. The foreign jury members were less aware of these changes but were informed by the other jury members about the positive experiences with the temporary building at Julianalaan. Therefore they were also able to take some of these contextual changes into account in their decisions.

Cultural differences may, however, have also affected the chances of winning, even in the situation of anonymous reviewing by an international jury panel. The outcome of the competition (winners from Netherlands, France, Belgium and Finland) and the distribution of the best 50 submissions (mainly originating from Netherlands, France, United States and other Western countries) imply that the frame with which the jury members assessed the submissions had an orientation similar to the context in which the assignment was set up. This meant that for participants who were not familiar with this frame, the chances of winning the competition could have been reduced. On the other hand, in the current digital era a lot of local information is easily available all around the world. Certain aspects of architecture culture might even be uniform around the world.

Cultural differences in framing could have implications for European procurement law and the principle of equal treatment. Based on the findings from this case it remains unclear if the differences of framing only concern issues of information about local characteristics, or display cultural differences. Related to the issue about fair competition is the language of communication. In this case all communication was conducted in English to involve as many people as possible. This is in line with the principle of equal treatment of procurement law. However, many Dutch clients prefer to communicate in Dutch, and therefore include communication in Dutch as a requirement during selection, which excludes a significant part of the market. Further research on this topic is needed.

7.4.2 Process procedure

The competition rules determined the kind of entrants, amount of information per entry, the process of assessment, the communication of the justification of the decisions, and the utilisation of the submissions after the closure of the competition. Reflection on the findings of designing the procedure of the competition raised several issues to be addressed in this section:

1. Project management in a dynamic context.
2. Communication in a digital era.
3. Composing a jury panel.
4. Balance between professionalism and ambition.
5. Relation between assessment process and the submission format.
6. Determining and carrying out the rules of the game.
7. Basic principles of procurement law in practice.

Ad 1) Project management in a dynamic context

The launch of the competition at the opening of the Biennale provided a strict deadline and meant that the competition needed to be set up within two months. There was no clear deadline for the end of the competition, apart from the dates that were announced in the competition programme. The context of the fire meant that no real overview of the budget and costs of the competition project were provided. The planning was based on the intention to start the new building project within two years after the fire because of the conditions of the insurance policy. As the competition progressed, the time frame for assessment was extended and the financial, managerial and psychical context had changed. This did not lead to an extension of the submission deadline but changed the date that the nominees and winners were announced. Additionally it seemed impossible to set a date for the jury meeting within the original time phrase due to their busy diaries, and the evaluation was scheduled for January instead of November.

The data of the competition show that the dynamic context of a competition could cause changes in the ambitions of the clients. Without a clear goal and a reliable network of possible actors these changes could cause serious deficiencies with the competition rules. In the School case and the City Hall the responsible officers had a clear ambition and goal, in contrary to the officer in the Provincial Government Office case. In these cases contextual factors were found - just as in the competition case - that reinforced or weakened the willingness to reach a final conclusion. The legal context offers only limited possibilities to adjust the competition rules in case this becomes necessary to fulfil the actual aim of the selection process. A thorough preparation of a competition could therefore contribute to the success of a tender.

Ad 2) Communication in a digital era

The project team decided to manage the competition via a website. A website would lower the costs for the entrants because they would not be required to print and send their submissions by mail and it would decrease the barriers to join the competition. For the client, choosing a website reduced administrative costs and it created more flexibility to analyse and prepare the submissions for the jury evaluation. At the same time the digital format created an interesting database for future research. After the launch of the competition the website turned out to be an excellent medium to improve the profile of the competition. The website was easily accessible for people from all over the world and several newspapers and internet forums provided a link to it. The website had 91,000 visitors in total until July 2009, with an average of 307 per day (Figure 7.4). Statistics of the website show about 18,000-20,000 visitors in September and October 2008 (the launch and submission period) with 750,000-800,000 hits. Then hits dropped to increase in January (the month of the jury meeting) unto 11,000 visitors in March (the month of the award ceremony). In July 2009 the website still had

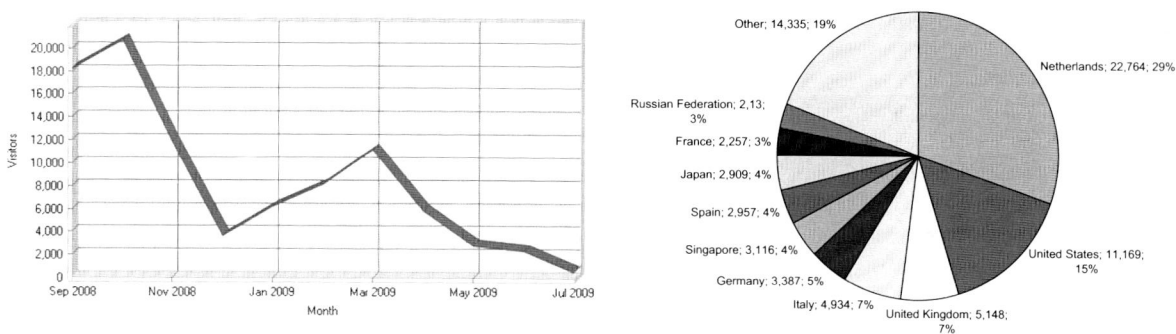

Figure 7.4 Visitors on competition website per month and per country

about 2,500 visitors a month. Most visitors originated from the Netherlands, the United States, the United Kingdom and Italy, with a similar distribution found for the submissions. It can be concluded that the website was definitely worth the investment and fitted the aims of the competition.

Ad 3) Composing a jury panel

The project team aimed for an interesting, internationally renowned jury panel. The Dean knew from experience that an interesting jury panel would increase the attractiveness of the competition and therefore the amount of entrants. The project team also knew from the cases in Chapter 6 and previous experiences that the jury panel should consist of a balanced and diverse group of people who could address all aspects of the assignment. Finally the project team decided to invite nine jury members. An uneven number of members would make voting easier, and this size also gave the faculty the opportunity to get a significant amount of experts involved.

During the meetings of the project team several options for the composition of the jury panel were discussed. In the beginning the idea was to invite the chair of the TU Board and the minister of Education, Culture and Science to be part of the jury, next to several renowned (inter)national architects. However, the minister declined his invitation for political reasons. To prevent the jury for an unbalance in the direction of client representative, the project team decided that the Dean would represent the client, together with a student and a professor (and practicing architect) of the university. In the end the network of the faculty enabled the project team to attract some famous Dutch architects, the Dutch Chief Government Architect, and rising international professionals as jury members for the competition. The simultaneous process of a Think Tank made it possible to invite other key players to play a different role in the process of rebuilding the Faculty. This also suited the aims of the competition. Most of the Dutch jury members who accepted the invitation as a jury member were personal acquaintances of the Dean. Some of them had already actively approached the Dean about their ideas for a new faculty building and the concept of the competition. The project team also approached several international professionals but they not very interested in participation. In the end most of the international jury members originates from the network of the Director of the NAi, who also was invited to be a member of the jury.

Until the last week before the launch of the competition the composition of the jury composition was not final. Especially the international members were not easily confirmed because the project team did not have direct connections with them and they had busy schedules and travelled all around the world. The data show that composing a jury panel relies as much on the design of the panel as on the availability of experts. The project team was aware in shaping the jury panel, the distribution of fields of expertise, group representation and the amount of experience of the individual members should be balanced against each other and related to the aim of the competition. However the project team experienced that composing a team that would meet the theoretical requirements of an expert team was almost impossible without a proper personality assessment and knowledge of how each person would carry out their function. Strategic and political aspects also have to be taken into account. Yet in my opinion the composition of the jury panel for the competition can be considered as the best possible result in the light of the different aims of the competition.

The Dean contributed greatly to both the design of the panel and the process of securing jury members. The data suggest that he applied his tacit knowledge and prior experience as a chair or a jury member in this matter. The composition of the project team developed in a similar manner in the aftermath of the fire: the Dean asked several persons to be involved, and depending on their availability and willingness to participate the project team was composed. The project-based and rushed character of this competition probably contributed to this pragmatic strategy to compose a project team. The findings from the tender cases in Chapter 6 also suggest that in practice it is hard to strategically design jury panels and project teams in the field of architecture, because clients either feel too confident or too insecure about their knowledge in the field. When clients feel confident they think they do not need support, and clients who are insecure do not know whom to invite.

The professional level of the client could have contributed to the success of the team composition in the Faculty Building case. Strategic team design opens up possibilities for building expertise and professionalism within an organisation (Salas, et al., 2006). In the situation of competitions and tenders the panel member could learn from the feedback they receive from other team members, participants and the media. If juries and project teams are not able to perform repeated tasks and receive feedback on their actions, their experiences cannot contribute to the idea of organisational learning (see for example George & Chattopadhyay, 2008). Although the selection of an architect selection is a rare event for most clients, I think not creating a learning environment around this kind of event is a missed opportunity for further development of the client organisation and the professional field.

Ad 4) Balance between professionalism and ambition

In the competition rules a balance needs to be struck between the ambition and the actual potential. The more the project team realised the potential of the competition, the more enthusiasm and ambition they showed. At the same time the project team was aware that they could not to ask too much of the participants and the jury members in relation to the professional image of the faculty. We aimed for a significant amount of submissions and a satisfied jury panel that would benefit the image of the faculty. Two examples about the submission and presentation format illustrate the considerations of the project team in order to balance their ambitions with the professional image of the client: the level of detail of the submissions and the presentation of the submissions to the jury panel.

A first example of a well considered decision about the competition rules is the balance between the detail of a design proposal and the conceptual power and aims of an ideas competition. On the one hand the project team wanted to collect concrete ideas as input for a project brief in a future tender. This would require detailed solutions and clear designs, but therefore also a significant amount of work from the participants and the jury members. On the other hand, the competition acted as a start of a new building project and aimed at conceptual ideas on a general campus level. Therefore it would be a waste of energy to ask for detailed proposals if the basic assumptions (such as the location, budget and amount of floor area) were not yet determined. These kinds of decisions influenced the format of the submission. The first intention of the project team was to ask for ideas on the level of a sketch design because this would create a typical architectural design competition. In line with the competition tradition the Dean was in favour of also requiring a scale model from the participants. However, at the start of the project it was decided that the whole competition would be done digitally via the website. A scale model would not fit a digital format. Next to that it would oblige the entrants to mail the scale model which would results in additional costs for participation. After consulting other organisations that organise international design contests more regularly and further discussion within the project team it was decided that a scale model would not be asked. The example shows that the design of the competition rules should be aligned with the aims of the competition.

The decision about the presentation of the submissions to the jury panel also relates to the balance between modesty, professionalism and ambition. The digital submission format offered several more innovative options for the project team to shape the assessment process. The jury members could, for example, have assessed the submissions online, an individual voting system could have been developed and a virtual jury meeting would have been possible. This proved to be a bridge too far for some members of the jury and of the project team. Therefore we decided to select a secured location near the university to present the posters. This would ease the preparations without the risk that submissions would become public.

The location of the former Techniek museum of the university was coincidently available due to scheduled renovations, and turned out to be a very inspiring ambiance during the jury meeting. Our first intention was to lay all posters on the floor. The idea behind this was the flexibility of rearrangement, an informal atmosphere during assessment, and lower costs for preparation. The chair of the jury panel indicated than an expert would need only a glance at the posters, which could easily be done from above. Finally the project team chose to hire rather expensive panels to present the submissions because this would be more professional and would not create possible back problems of the jury members. Still jury members could walk along the posters (see Figure 7.5) and personal contact among the jury members was ensured. The concept that jurors physically meet during the assessment process is part of the competition tradition. Reflecting on the jury process the panels and physical jury meeting appeared to have been beneficial for the cohesiveness of the jury panel. It also increased the awareness among the members of the jury panel and the project team about the importance of the competition and the large about of work of the participants. This example shows that clients sometimes have to acknowledge the tradition of the professional field of architecture, even if it requires an additional financial investment. At the same time the digital format showed that new developments can also improve practice (see also point 2 of this section).

Ad 5) Relation between assessment process and the submission format

The literature about decision making shows that the format of the information presented should be adjusted to certain traits of the decision making, such as the level of expertise (see Chapter 3.4.3). The findings of the tender cases in Chapter 6 suggested that clients asked for more information about the candidates to decrease the uncertainty about the potential level of quality of the tenderers, which is in line with the findings as reported in literature (e.g. Betsch, 2005; Mosier & Fischer, 2009). In this Faculty case it was clear from the beginning on that a jury would assess the submissions. However, the format and presentation of the entry documents was never directly connected to the composition of the jury. For example, if the Minister or the President of the executive board - people who are less experienced in reading designs - would have participated in the jury the format could have been adjusted, for example by requiring less sketch design and more text or more abstract visions. Compared to the findings in Chapter 6.7.3, experts or more experienced jury members could have had a more prominent role in 'explaining' the submission. The large amount of information was also created by participant behaviour: 90% of the entrants submitted two posters, while the format provided freedom to submit one poster only. This indicates that participants often choose to submit as much information as possible within the competition format while this does not automatically increases the chances to win: Further analysis showed that the amount of submissions using two posters instead of one remains equal in the final 50 (92%). This suggests that the chances of winning were not influenced by amount of information as submitted by the participants and the submissions were judged by the contents. The observations in this case and the experiences of the other cases in Chapter 6 indicate that in making judgements in design the level of expertise of the assessor is more important than the amount of information, but only if the level of expertise of the panel members fits the information means.

One of the main differences between the tender cases in Chapter 6 and the Faculty Building case was the anonymous judgement of the submissions. Anonymity requires a very structured format to link submissions to participants at the end of the competition. The registration numbers were required as part of the submission format, which were used to structure and monitor the large amount of submissions during preparation and assessment. During the assessment phase a typology was developed by the members of the analysis team to structure the large number of submissions. The jury members received lists of the submissions ordered by type and increasing registration numbers. They also received a handout of the results of the quantitative study with the most important topics as addressed in the submissions. In the retrospective interviews, almost all jury members expressed that they appreciated the structure provided by the typology but that they did not agree on the content and structure of the typology. During the jury meetings the panel decided that the typology should not be part of the jury report because it was only an assessment means and not an outcome of their evaluation process. Neither the winners nor the best fifty represent the distribution of the submission over the types. These findings suggest that a structure can help to interpret and process submissions but that this structure does not automatically determine the direction or typologies of the outcome of the jury assessment.

In line with the competition tradition the submissions were assessed anonymously. The findings show that because there was no information about the designer provided, the jury members focused on the content of the idea instead of the background of its creator. Although the jury members were sometimes very curious about the background of the participants, they explicitly wanted to wait with opening the database that linked the registration numbers to the participants until after they decided about the winners. In this respect a competition is not the same as a tender. In a tender a client searches for a partner to further develop ideas for a building project, while an ideas competition aims at collecting design concepts independently of the person behind the design. In an ideas competition a relatively small amount of money is awarded, and compared to signing a contract the implications of a decision about the winner are limited. This reflects the need to assure the qualifications of the designer. In a tender the award phase should be focused on the offer instead of the qualifications of the tenderer. This implies that anonymity would suit the award phase but not the selection phase. At the same time, the theory and results from the tender cases show that an offer always consists of a 'package' of the person, the product and the firm (see Chapter 6.7). Next to that it was found that the personal explanation of the designer increases the understanding of the proposal. This means that the role of anonymity during a tender in architecture requires further research.

Ad 6) Determining and carrying out the rules of the game

The competition rules determine the rules of the game for a competition. In this case the use of the Kompas model created the structure and overview of the issues for the development of a competition programme. However, the project team also experienced that the use of a model could weaken the specific characteristics of the competition and hide possible weaknesses or imperfections in the model. One of these issues concerned the payment of the prize money. The project team had decided to rely on the model for this issue but regulations about VAT had changed. The steering team had just approved the documents of the project team when it became clear that the new VAT regulations would lead to an increase of the budget. These findings indicate that the use of a pre-developed procedure is only appropriate if the project team is able to detect necessary adjustments and the steering committee has the expertise to judge the risks for the organisation. This requires specific skills and up-to-date knowledge about regulations and conventions in a professional field.

The 'playing field' of a competition is determined by the design decisions. The project team experienced during implementation of the competition rules that some of the violations of the rules, such as missing registration number, did not influence the playing field of the competition but only caused administrative issues. Other rules actually caused deviations in the playing field, for example not able to compare the submissions because of a difference in the scale levels of the sketches. This raised two issues that required sensemaking during the implementation process: 1) How strictly should the rules be applied? and 2) Do the rules negatively influence the character of the competition?

For the project team it was difficult to estimate the effects of a certain rule on the number and kind of participants and submissions. For example, the obligation to be registered in an architect register could have shut a lot of (international) professionals out, but it is remains uncertain how many potential participants will be affected by this requirement until after the deadline of registration. Even with the available expertise within the project team a lot of time was spend on how to design the competition in order to reach the aims of stimulation creativity, research and debate. Therefore it can be concluded that a call for participation is always a guess, and it is only after the submission deadline that a client becomes aware of the response of the market. Then officially the moment has passed to make adjustments to the requirements and submission format. Participants have to decide for themselves if they feel the balance between costs and benefits is worth taking the risk and investment of participation. We hoped for a response of around 350 submissions but did not expect the competition to have such a large impact in the field. In line with our expectations not many of the major players in the field joined the competition, but both the project team and the jury considered a competition with 471 submissions and a lot of media attention a very successful outcome. Therefore the jury decided to judge all submissions that could be printed and presented an idea inline with the aims of the competition.

The competition rules of the Faculty Building case provided room for increasing insights that would respect the positive response on the call for participation. A strict legal procurement context does not allow for the same kind of flexibility. In order to assure competition the tender regulations do not allow less than five tenderers in the award phase but a client is free to invite as many tenderers as they want as long as they fulfil the requirements. During a tender the chair of the jury panel would not have been able to accept the submissions that did not meet the requirements into the assessment procedure (about 58%). The need to react on the response of the market on the call for participation was also found in the School case. Because of the large amount of candidates that responded to their call, the School Board invited sixth candidates for the award phase instead of five. The need for adjustments is part of the process of sensemaking (Weick, 1995).

Trust in the organisation also appeared to be an important issue in implementation of the competition rules. In this case there were only two official complaints of participants who could not trace their submission. They assumed that their submission had not been assessed by the jury because of a mistake of the project team, which can be related to trust in the commissioner. Because jury meetings happen behind closed doors, participants never know for sure if the jury personally looked at their submission. Only those participants whose submissions were included in the publication and addressed in the jury report have proof that their submission was evaluated by the jury. The project team relied on the database of the website when they created the lists of participants and the jury panels trusted the project team not to have made mistakes. All entrants received an automatic confirmation message after submission. Such automatic messages are more likely to get lost in spam filters, yet the project team felt that with this large number of submissions non-automatic processes would have caused more administrative problems.

The communication could have improved if the party who developed the website had been part of the project team. On the other hand the communication among the actors in this case seemed to have largely depended on (a lack of) clear leadership and structure within the project team and not explicitly on being a member of the team or not. The issue of trust between a commissioning client and the tender candidates is most likely more important in tender than in an ideas competition because of the legal consequences of possible mistakes. It is therefore plausible that a jury report contributes to a larger extent to building trust than a quantified score because it displays that the jury panel actually looked at the proposals.

Ad 7) Procurement law in practice

Because the Faculty Building case was based on the Kompas model and stated as an open ideas competition, the legal context seemed quite clear. However, the relation of the competition to the European tender of the future new building still remains a topic of discussion. During the preparation phase the project team suspected that the link between the competition and tender would be important to attract (more established) professionals from the field. Some members of the jury and the Dean tried their very best to officially connect the outcome of this competition to the future tender, but the externally hired legal advisor successfully prevented this. The competition rules therefore only included a general message that "the winners are emphatically invited to take part in the (future) project competition" (Faculty of Architecture, 2008). At the time of writing it is still unsure if there ever will be a new building project. Yet the jury members and participants still expect the winners of the ideas competition to benefit in the next phase of the project. It is now up to the steering committee to decide on this matter as the project team was dismantled after the award ceremony had taken place. So officially no legal commitments were made but a client should be aware that by organising a competition a client does create professional obligations to the participants and the professional field.

During the composition of the jury panel and the Think Tank, the legal principles of equal treatment and objectivity proved to be a barrier to collaboration. Some of the experts that were invited for the jury panel and the Think Tank even declined because this would mean that they would not be able to participate in future activities of the building project. In this sense a conflict between the legal context and the ambition and wishes of the client seemed to have occurred. In practice a tender is preceded by numerous other activities to analyse the needs, ambitions and possibilities of a client concerning the future building, such as this ideas competition. Drawing up a business case or developing a brief are just some examples of activities that often require contributions of external parties that might want to be involved in the rest of the project too. Of course being involved in these kinds of activities could benefit the potential tender candidates and therefore threaten the legal principles of equal treatment and objectivity.

In the end the project team had to decide that people that were involved in the Think Tank or jury panel would be excluded from any future tender for the faculty building. For the competition no promises were made on further collaboration with the winners. It remains to be seen if indeed fair competition would not have been possible without this kind of measures. Awarding contracts takes place in the context of an organisation that often has multiple aims and stakeholders to serve. In all cases of this research several goals were identified underlying the tender or competition. The clients often also had more strategic goals in mind and used the tenders to improve the relationships with important stakeholders. Strategic behaviour is inherent to doing business decisions but procurement law does not seem to take these kinds of aims into account.

7.4.3 Stakeholder involvement

The involvement of stakeholders was also in this case an important element. Findings on the issue of stakeholder involvement in the competition raised the following issues:

1. Relation between participatory means and competition aims.
2. Team composition and participation in decision making.
3. Expertise and client obligations.

Ad 1) Relation between participatory means and competition aims

The primary objectives of the Faculty case were to collect inspiration for a new building brief, to encourage creativity among the younger generation, and to stimulate research and general debate. Because of the tremendous effect of the fire on the local architectural community, the competition was also organised to support employees, (former) students, and the professionals in the field of architecture in the process of coping with the loss of the old building. The Dean therefore wanted to focus on the future and involve as many people as possible. Ideas about the involvement of the press were strategically formulated from the beginning of the competition preparation period. Professional architects, employees of the faculty and architectural students were the main stakeholders. Historically competitions have proven to be a breakthrough for several architects. Therefore every architect silently hopes that their talents will be discovered by a renowned jury. The project team did not consciously design a participation strategy for the stakeholders of the case, but did develop several participation options with different levels of influence on the final decision about a new Faculty building (see Table 7.5).

In the table a distinction is displayed between decisive (1), advisory (2), and informative rights (3) on the direction of the future building (F), the competition rules (CR), and the competition outcome (CO). For most stakeholders, decisive or advisory options were only possible by personal invitation (e.g. part of the project team or jury panel). Other professionals in the field could only influence decision makers by submitting their ideas for the direction of the future faculty to the competition. Jury decisions were binding. The steering committee did not have a say in the competition as such, but did have to consider the results of the competition and Think Tank to advise the board of the University. Originally the

Stakeholders and options for participation: 1 = decisive rights 2 = advisory rights 3 = being informed and provide support CR = Competition rules CO = Competition outcome F = Future direction of Faculty accommodation	Steering committee (by invitation)	Think Tank (by invitation)	Project team (by invitation)	Jury panel (by invitation)	Competition participant	Analysis teams (by invitation and voluntarily)	Invited open activities (e.g. final symposium)	Open activities (award ceremony, exhibition, publication)
Commissioning body (Executive Board)	2-F	2-F						
Representatives of client (e.g. Faculty staff)			1-CR	1-CO; 2-F				
Shareholders and supervisors (e.g. Ministry, municipalities)		2-F						3
Daily users (e.g. employees, students)			1-CR	1-CO; 2-F	2-F	3	3	3
Non-daily users (e.g. alumni, exchange students & professionals)				1-CO; 2-F	2-F			3
Representative groups (e.g. BNA, student board)							3	3

Table 7.5 Overview of participation options and influence on the future building

Dean of the faculty was a member of the jury panel, but because of his illness the faculty was only represented by a student, a professor and an emeritus professor. Apart from the Dean, neither the steering committee nor the board of the university were part of the jury panel. I have the feeling this was mainly because of strategic and political reasons. In the end, the primary competition aims (inspiration, stimulation and debate) and hidden organisational project aims (providing a podium for stakeholders, increasing the Faculty's international reputation) seemed to have strengthened each other. This resulted in an overwhelming amount of high quality ideas for the future faculty as well as tremendous international exposure of the faculty to students and professionals.

The fire in the building at the Berlageweg and further financial development within the organisation changed the point of departure of the real estate strategy of Delft University. At the time of writing, Delft University is still reconsidering its real estate strategy based on the results of the Think Tank and the ideas competition. Most of the ideas of the competition concentrated on the design of the new faculty building and did not consider further implications on campus level or about educational strategies. The results of the Think Tank did include a lot of ideas and strategies on campus level (Arkesteijn, den Heijer, Vande Putte, & Volker, 2009). Therefore competition participants could have the impression that the results of the competition might have had a lesser impact on current decision making about the real estate strategy. Nevertheless, the fact that the jury panel decided to select two winners who proposed to renovate the Julianalaan building stimulated the discussion about the possibility to stay at the temporary location.

In January 2010 the Executive Board decided that the Faculty of Architecture would remain in the building at Julianalaan until further notice, and bought the building back from the real estate developer they originally had sold it to. A current lack of financial means and the positive experiences with the building appeared to have been the main reasons for this decision. However, the competition contributed to the sensemaking process of the Executive board and the Dean of the faculty.

Ad 2) Team composition and participation in decision making

Also in this case the project team, the steering committee and the jury panel were the most important actors. However, as displayed in Table 7.5, a lot of other stakeholders were involved in the organisation and implementation of the competition. Theoretically a lot of connections and roles are possible to let people participate in decision making. Traditionally competition jury panels mainly consist of architect experts. The jury panels that acted as selection and award committees in the three tender cases consisted of a mixture of experts, administrators and representatives of certain stakeholder groups (see Chapter 6.5). In current practice some people assume that architect experts are best able to judge the quality of design, while others are convinced that the responsible decision makers from the client organisation should be personally involved, even if they do not have domain specific expertise about architecture. In the Faculty Building case the tender and the competition traditions were joined because the client representatives were in the main also experts in the professional field. In the tender cases the balance between number of experts and other stakeholders appeared to reflect the attitude of the commissioning body towards the phenomenon of judging design quality.

The data also suggest that the interest of stakeholders to be involved in a jury panel also related to the potential impact of the decision (see Chapter 6.7.3). The impact of awarding a contract is larger than the impact of dividing prize money, which could explain why the members of the steering committee of the Faculty case were less interested in participating as jurors than in tender situations. For this same reason the composition of the jury panel could have also caused more discussion between the project team and the steering committee. Because jury members and their firms are usually excluded from participation, it is usually hard to attract architect experts for a jury panel. In this case the jury members were willing to give up their chances at winning the competition in order to be part of the jury panel. If this competition would have been a tender it probably would have been more difficult to find experts for the jury panel. More leading professional firms might have participated. This shows that the aims and implications of a tender influence the actors that take part and the position of the stakeholders.

The findings of the Faculty Building case indicate that the jury panel did not experience the same tensions between the legal obligations and the client expectations as found in the tender cases in Chapter 6.7.4. From the beginning of the competition the project team assumed that the jury panel would have decisive rights about the prize money. This meant that responsibilities were delegated to actors that were not part of the commissioning body. This was not perceived as a threat because the level of expertise of the jury members was considered as a quality guarantee. It can also be seen as part of the competition tradition that the criteria stated in the competition rules were not used as input for a decision matrix based on individual or collective judgements of the jury members. It was almost taken for granted that a jury report would be drawn up based on the discussions in the jury meeting. This shows the benefits of a proven concept of decision making within a certain profession, a tradition which still needs to be developed for tender procedures.

Ad 3) Expertise and client obligations

The case of Faculty Building is unique because of the direct connection between the nature of the competition and the core activities of the client. The level of expertise of the daily users of the future building was consequently relatively high and they were able to join the competition. In the design of the participation options no specific attention was paid to the different levels of expertise of the stakeholders in relation to their involvement in the process but during the preparation of the jury assessment support could easily be found within the organisation. Offering these opportunities for employees and students increased the participation options. Also in composing the project team the available resources within the (network of the) organisation expertise were used. As a former Chief Government Architect the Dean was aware of the potential of a competition, the regulations and traditions, and he felt enthusiastic about creating this opportunity for the field of architecture. The project team could build on this experience in considering the various participation options, which was very helpful.

In the tender cases of Chapter 6 the clients were not very familiar with (tender) competitions and architecture in general. It was found that while most public clients do have a special purchase, procurement, or real estate department they still decided to hire a consultant for the specific competences needed for architect selections. The results of the cases indicate that the perception of the opportunities that a tender can offer and the available support within the organisation could change the approach of the project. In the Faculty Building case, the School and the City Hall case the client all saw a lot of opportunities and tried to realise their ambitions in the (tender) competition. In the Faculty Building case, working with experienced and deeply involved stakeholders also created a lot of expectations about the quality of the competition. The competition therefore also had an exemplary function, which created additional pressure on the work of the project team and the jury panel. All jury members stated in the retrospective interviews that they were satisfied about the level of participation by the stakeholder groups. At the same time they did not seem to be very interested in this issue, or in the level of the requirements set for participation in the competition regulations. They seemed to consider participation to be a given in architectural culture: competing in a competition offers a lot of participation options already. The client sets the rules and the jury members comply with these rules, on the condition that they are able to do their job properly. At the time of writing, most resistance about the competition among alumni, students and staff related to the lack of communication from the university management about the accommodation strategy for the faculty and the follow-up of the competition. Based on these findings it can be concluded that at least stakeholders should be well informed about the developments and follow-up of the competition.

7.4.4 Decision procedure

The previous sections addressed the decisions that were made in the preparation and design phase of the competition. This section focuses on the assessment process of the competition. The observations and interviews about the processes of decision making gave reason to discuss the following issues:

1. Development of an assessment frame.
2. Stages of incremental decision making.
3. Subjective character of objective criteria.
4. Methods to express and justify preferences.
5. Differences between experts.
6. Connection between the jury panel and the aims, criteria, final decision of a competition.

Ad 1) Development of an assessment frame

The literature on decision making showed that decision makers have to develop a frame in order to make decisions (e.g. Beach & Connolly, 2005). The evaluation procedure as explained in the competition rules contained a strict distinction between the assessment phase (a check against requirements and analysis of information) and an evaluation phase (a value judgement about design quality based on different kinds of assessment). Only the jury was entitled to make judgements about the quality of the submissions and their decisions were binding. The jury panel did not have an active role in preparing and assessing the competition rules nor the submissions. During the check against the requirements the project coordinators developed a sense of direction about the submissions by assessing the first 10 or 20 submissions together. Decisions had to be made about when to approve the submission on a certain topic or not. This was conducted in three categories: 'yes', 'doubt', and 'no'. Due to the large number of submissions, after the first 50 submissions, the tasks were divided among the two project coordinators.

A team of 25 volunteer employees individually analysed all submissions with respect to how they dealt with the themes and requirements in the competition brief. They used an inventory list developed by the project team in collaboration with the chair of the jury. The inventory list was explained during an introductory meeting of the project team with the member of the voluntary team that could attend. The results of the analysis were collected in a database and analysed by two student assistants. The level of detail of analysis and amount of time spent on each entry differed greatly per employee and the reliability of the analysis appeared limited. However, it provided a useful overview of the submissions and a starting point for further analysis. The results of the analysis were discussed in a small core team of four employees as representatives of the issues addressed in the criteria: 'architecture', 'construction', 'sustainability' and 'economic viability'.

This team performed its own analysis in three meetings. Their first meeting was almost entirely spent on determining the aim of their team and their decision task. They mainly discussed how to best support the jury: by developing typologies, by distinguishing excellence on sustainability, economic viability or architectural quality, by making proposals or otherwise. In the end the analysis team presented the jury with 15 building types (see Figure 7.3), suitability scores per entry, and an overall overview of trends addressed in education, research and realisation.

The above findings show that any kind of assessment starts with the development of a frame of reference, which is based on the aim of the assessment in combination with the documents to be assessed. This means that preparations on an assessment frame can be carried out beforehand, but the actual assessment frame remains part of a sensemaking process. It therefore cannot be developed without the proposals that are created in response to the competition brief. The project team had a different view on the submissions than the analysis team, and the typology of the analysis team was in the end not used as a typology of the jury process. In every phase of the decision process a group of decision makers went through their own process of sensemaking. This structure and outcome of these processes are not predictable beforehand. In this sense assessment could be compared to the process of qualitative data analysis; the structure of analysis arises from the data. This is in contrast to most research on a quantitative basis, which typically consists of testing assumptions that already guided the process of collecting data. Procurement law assumes a quantitative process in this matter, while the findings of the cases suggest a qualitative process. The competition regulations also compel decision makers to make predictions about the decision process but these are interpreted within the competition tradition and not the legal context of tendering. These findings suggest that it is mainly the interpretation of the regulations that cause the potential conflict between the possibilities and the aims of the client in making decisions about design quality.

Jury members need a shared anchor point, which could be provided by the competition brief and regulations. Theoretically the chance that decision makers apply the criteria announced in the competition rules would improve if these criteria were already part of the frame of the jury members. In the current Dutch tradition jury members tend to spend one or two days maximum on discussing the submissions and deciding about the winner(s). Contrary to the Scandinavian tradition (see for example Kazemian & Rönn, 2009), preparations are usually not done in consultation with the full jury panel. In this case the project team discussed the most important competition rules with the chair of the jury panel but did not involve the other jury members either. Several mainly pragmatic reasons were contribute to the lack of preparation and assessment time of jury members.

First of all, the final composition of the jury remained uncertain until the actual moment of the meeting, in other words until after the submissions had been received. Secondly members of the jury were highly regarded professionals and therefore busy. It was very difficult to schedule a date for the two-day jury meeting. It would have been even harder to plan a meeting to discuss the criteria or discuss the criteria with all members individually before the announcement of the competition in September. This partly caused by the strict deadlines for announcing the competition at the Biennale, but findings from the tender cases in Chapter 6 showed similar planning issues. The project team anticipated the limited preparation time and organised a joint presentation about the campus and the faculty at the start of the jury meeting. This created a kind of shared anchor point and natural starting point for the jury members to initiate their assessment process.

Less involvement of the jury panel meant considerably more influence of the project team in setting up the competition rules and preparing the jury evaluation. It is generally assumed that the decision made in the preparation phase of the jury assessment are kind of objective which implies that any assessor would arrive at the same decision. In the Faculty Building case the assessment of the submission requirements and the first analysis of the issues addressed by the participants were done by the project team and the support staff in two different analysis teams. Because of the professional level of the client there was more support staff available than in the tender cases of Chapter 6. In the tender cases they had to compensate this with externally hired consultants who often also advised the panel members in assessing the proposals. In this case the influence of the project team on the evaluation process was limited because there was no selection phase and most jury members were experienced jurors. Therefore the panel members were able to make a distinction between the structure that was needed to create an overview of the submissions and the actual evaluation process. Yet the results of the cases indicate that the role and qualifications of the project team members should be critically monitored by the steering committee and the jury panel, because they implicitly influence the decision process of the client.

Ad 2) Stages of incremental decision making

The results of the Faculty case show comparable iterative phases of decision making as found in the tender cases (see Chapter 6.7.4). In this case there were seven jury members including the chair present during the two-day jury meeting. The project leader acted as secretary of the jury. Apart from the author of the jury report (also author of the competition brief and staff member), a staff member of the Faculty (author of an essay for the publication), the other project coordinator (me), and two student assistants nobody else attended the jury meeting. The chair of the jury panel clearly stated in her introduction that only the jury members would be involved in the evaluation of the submissions, but as colleagues it appeared hard for the authors of the jury report and the essay to keep silent throughout the whole evaluation process. Also during lunch and breaks the jury members spoke with the other attendants. During these discussions the jury members exchanged experiences about tenders and the potential impact of competitions like this one on the current procurement situation, which illustrates the strong connection between both types of selection processes.

The results of Kazemain and Rönn (2009) describe a judgement process of six stages: submission check, determination of order of work, choice and preliminary judgements, presentation of interesting contributions, ranking and decision making with architectural criticism. The findings of this case and the tender cases in Chapter 6 confirm these stages more or less on a general level, but also show a more incremental process of decision making from a psychological perspective. In this competition the submission check was done by the project team before the jury meeting started, but the chair decided to allow all submission for assessment by the jury. The determination of order of work was done by the project team in consultation with the chair of the jury panel. The programme of the jury meeting was based on the findings of the previous cases (see Figure 6.6) about the process of decision making. Therefore the members of the jury went through a process of: 1) initialization, 2) perception, 3) individual judgement, and 4) group decision making.

The pictures in Figure 7.5 display the start of the jury meeting with a general presentation (left), individual perception and judgement (middle), and the group discussions (right). During the two days that the jury met these four phases were repeated in several incremental rounds of decision making in order to make sense of the submissions and the aim of the competition. Choice and preliminary judgement was done in two phases: first a reduction to make a presentation of interesting contributions possible (comparable to the selection phase of a tender), and then a comparison to make a ranking among the best submissions and final decision about the winner (comparable to the award phase of a tender). The process of the first jury day can be described as 'selection by positive impression' agreed by some but not always the majority of the jury members. The second day can be described as a process of 'selection by comparison' based discussion among the panel members in which the jury members gave input based on their own perceived qualities of the submission.

The initialization phase can be seen as goal setting (Beach, 1990). In this first phase the jury members introduced themselves, and the chair introduced the aim and context of the competition and discussed the official evaluation criteria. A member of the temporary housing team presented a short description of the temporarily housing process and the current situation at the campus. The initialization phase mostly aimed at goal clarification at the beginning at a new stage but was sometimes also part of the evaluation of the decisions at the end of a previous decision cycle. The chair used the criteria and aims as stated in the competition programme to support these activities. During this phase the jury members often addressed their opinion about the competition, indicating a connection between the aims of the competition, and the overall impression of the submissions. They

Figure 7.5 Common presentation at the start of the jury meeting (left), individual perception and judgement of the posters (middle) and group discussion among the jury members (right)

tried to read the decision task (see Balogun, et al., 2008) and to translate the aims of the client into the competition context.

The perception phase (see for example Bell, et al., 1996; Robbins & Judge, 2008) on the first day was planned as individually walking along the posters. None of the jury members had seen the submissions in advance. The chair had planned to select around 50 submissions at the end of the first day for further discussion, and the jury succeeded to do so. Therefore the first round of assessment aimed at 'go/no-go' decisions. The submissions were structured by the typologies of the preparatory analysis (see Figure 7.3) and the jury members used lists to keep track of the submissions they had seen. Observations show that the jury members mostly looked at the A1 posters to gather the information about the submissions. The A4 with text was only consulted when something was unclear or for daring concepts. In total the jury members walked around individually for about three hours. During lunch the first impressions were discussed. Then a member of the core analysis team presented the results of their analysis. The jury members asked how and on which basis the typologies had been developed. All confirmed the supporting character of the typology. These findings confirm that decision makers use the competition brief and the submissions to develop their individual frame of references.

In consideration with the other jury members the chair decided to start day two again with an individual perception round. The project team had prepared this day by collecting all 50 selected submissions in the area of the discussion table and making these 50 also available on A3 format, including the A4 of text. Three of the Dutch professionals were absent in the morning; partly because of obligations elsewhere, partly because they felt that they did not need that much time to study the final 50. To prepare oneself for another voting round some of the jury members decided to study the submissions on A3 while sitting at the table, others walked around the submissions on A1 format. At the end of the first day the chair had proposed a secret round in which the jury member could propose some additional submissions to add to the final 50 after the second individual perception round. None of the jury members used this opportunity. This implies a high level of confidence about the decisions that were made during the first day. A comparison of the decision process as found in the tender cases and the perception process of the jury panel in this case indicates that the strategy of information perception as described in Chapter 6.7 ('first individually, than a discussion' versus 'perception as part of a discussion') depends on the level of expertise of the jury members. Experts already have a frame of reference with which they perceive the designs. Therefore the strategy of 'individual perception first' seems appropriate for expert juries. Less experienced jury members seem to benefit of a 'shared perception during discussion' strategy to build their frame of references.

Apart from the general programme as proposed by the project team the actual decision methods and process of decision making was designed by the jury members themselves during the two-day jury meeting. Group decision making on the first day was conducted by walking along each submission collectively and exchanging the preferences of the jury members about that particular submission. If jury members were in favour of a particular entry, the other members had to agree on giving it a 'go' post-it. After the first few minutes of collectively walking along the posters, it became clear that some submissions lacked quality but tackled an interesting topic or proposed a good idea. The jury panel decided not to select these submissions for the next round but to pay extra attention to them in the jury report. The jury members assigned these submissions an 'idea' post-it. The secretary noted all post-its. At the end of the first day, about a hundred submissions still needed to be discussed and the members of the jury became tired. Because the goals for the day were clearly set, the jury members stimulated each other to finish this first round of selection. Miraculously the jury selected exactly 50 submissions at the end of the day. The additional 'idea' category demonstrates the increasing insights that occurred during the judgement process that changed the structure of the assessment process. Therefore it can be concluded the assessment frame is built during the assessment process but the preparation provides the structure.

Just before lunch the jury members had to vote for their five favourite submissions. After lunch the jury panel started a discussion around the table. During this discussion the A3 documents of the 50 submissions were put in the middle on the table and used to focus the discussion on the design. In total two rounds of non-anonymous and one round of anonymous voting were used on the second day to steer the discussion among the jury members. After each voting round the results were discussed. The chair started to discuss the submissions that received just a few votes to reduce the number of submissions suitable for a prize. Then the submissions with a lot of votes were discussed on their ideas and impact for implementation. Remarkably some of the jury members voted for almost the same five submissions out of the final 50. Further analysis of these similarities did not show any pattern between characteristics of these jury members (nationality, amount of experience, personal interest), which suggests no shared frame of expertise among subgroups of the jury panel.

Observations showed that the jury members based their judgements on the information provided in the group discussion and information as perceived from the proposals. The discussions always aimed at acceptance of a decision by the other jury members. Before the voting rounds the jury members limited their discussions to statements like *"I think this is one is interesting"* and *"I agree"* as a quick exchange of opinions. After the first voting round the panel members started to negotiate by putting more explicit arguments on the table in order to convince the other jury members and to justify their preferences. Finally the jury decided to award two submissions an honourable mention because they could not convince all jury members, but they did display a high level of design quality and provided useful input for debate. At the end of the negotiation phase six submissions were left for a final anonymous voting round to determine the winners of the competition. These findings indicate that negotiation is needed to reach a decision about a winner. In this case the final decision was a consensus based on a voting round. The methods that were used to reach the decision are addressed under point 4 of this section.

The ideas competition was based on an open procedure. This means there were officially no separated phases for the selection of suitable candidates and determining a winner. However, the results show that even without an official selection phase a jury panel needs to reduce the number of potential candidates. Therefore similar processes were found as for the selection and award phase in the tender cases. The first day of the jury meeting aimed for a selection of 50 probable winners, while the second day focused on a comparison of these in order to decide on the best. In the competition both phases were conducted by the same group of people in the same time slot. In case of a restricted tender there are two strategies: separation of the selection and awarding phases by appointing two different committees, or connecting the phases by using the same committee for awarding as well as selection. The findings of this research suggest that, although seemingly more time consuming for the panel members, decision making would benefit from using the same committee members for both phases. This would enable decision makers to apply the same frame in both phases and connect the argumentations in the selection phase to the argumentations in the award phase. The sensemaking process among the decision makers could then be continued more easily.

Ad 3) Subjective character of objective criteria

In Chapter 2.2 a distinction was made between tangible and intangible aspects of design quality and in Chapter 3.3.2 the structure of the decision task was related to the use of intuition and level of expertise that is required to perform well on the decision task. The cases in Chapter 6 indicated that assessments of well-structured information with predefined categories are often assigned to a project team because of their presumably objective character. In this case the project team also performed the check of the formal submission requirements. The data however showed that even with these kinds of assessments, a lot of judgements require expertise and sensemaking among the decision makers. The format of the entry documents and the time of submission are just a few examples of presumably objective judgemental tasks that required interpretation of the decision makers and raised ethical questions: If the website fails, it is allowed to submit an entry on the day after? How many millimetres can the size of the submission deviate to still fulfil the requirements? What can be considered as a plan, an exterior view, and a location sketch? Because of international differences the requirements for registration in a professional or student register also raised questions. This shows that even with relatively simple assessment tasks a certain amount of expertise is needed.

The results of the assessment phase also showed differences between the analysis team and experts from the staff and the jury panel. Based on their general impressions the members of the analysis teams suspected that most of the submissions were created by students because a rumour had been spread that the overall quality level of the submission was not very high. The jury on the other hand was very impressed by the quality of the submissions and expected a lot of submissions to come from professionals. The chair of the jury even stated in reference to her experiences as Chief Government Architect that

> "The quality of most submissions exceeds the quality of an average European tender project."

A similar discrepancy was found on the theme of sustainability. The analysis teams scored the submissions on the level of sustainability to create an overview. The jury was also informed about these scores. One of the jury members remarked that

> "This score was almost opposite to the submissions I thought would be very sustainable."

The differences that occurred between the different groups of decision makers can be explained in terms of their role, interests, amount of experience, level of expertise, and background. At the same time they show the subjective and relative character of judgements about design quality. Who decides on the appropriate level of expertise and who is able to make a 'right' judgement? One would expect the staff members who were invited to join the analysis team because of their expertise on a certain area to be able to judge design quality, but apparently they approached the submissions from a different perspective than the jury members. The jury of this competition did not seem to be bothered by deviations from the requirements nor differences of opinions with the supporting teams; they judged the idea behind the submission from the perspective they felt was right. The competition regulations made this possible and they felt confident enough to follow their own intuition.

In the final debate the submissions were considered as options for the future faculty building and weighed holistically against each other on their unique qualities. During this final selection process the text documents that accompanied the submissions were often used as a means to convey the message about a design idea from the perspective of the participant. In his column in the competition publication (Faculty of Architecture - TU Delft, 2009, p. 63) Bouman (one of the jury members) refers to:

> "Certainly more than enough entries to allow for extensive comparison, from which one was able to discern patterns in the international development of the architecture profession [... which] validates the subsequent pronouncements about architecture's vitality as a cultural medium."

The kind of comparison Bouman refers to was also found on a lower level in several remarks made during argumentation on the second day of the jury meeting about to select the winners from the final fifty. Regularly statements were used by the jury members such as:

> "The other one has a similar idea but better quality", or

> "This was on my list too, but that one is better compared to this one."

One of the Dutch professionals stated that:

> "I almost always compare the solution to the old situation in order to find something better."

Other jury members used their personal experiences with the old building to convince the other jury members of the appropriateness of a certain submission with statements like:

"It [the accommodation situation] was really like that after the fire."

Associations with other environments or comparable buildings were also made:

"It [one of the proposed ideas] feels like a caravan exhibition hall."

At the same time discussion was raised about the need to rethink education in architecture and repositioning the faculty on a more general level. The jury realised that discussions on a higher level about shape and space were definitively needed to reach a consensus about the outcome of the competition. While discussing the submissions preferences were phrased by short statements such as *"really interesting scheme"*, *"interesting statement"*, *"attractive"*, or *"not very realistic"*. Common themes in the discussion were the separation of functions (education, research, supporting functions), a central meeting place, flexibility of the building in time, the concept of sustainability, connection to the 'outer' world, and the position of the faculty within the campus. Discussions about architecture in general were limited, although several remarks were made about the large number of iconic architectural objects among the submissions. In the end the jury members were almost surprised about the outcome themselves:

"It is probable, but surprising that these ideas came out [as winners]; none are Super Dutch architecture."

This implies that the jury members were able to connect the outcome of this competition with the historical context, something that can be considered as the highest level of expertise according to Parsons (1987). Findings of the tender cases in Chapter 6 indicate that the experiences of most of the panel members did not reach these high levels of expertise; the reactions of most decision makers in the tender cases did not exceed the positive experience as such.

The decision making process of the jury members displayed many of the characteristics described in the literature on Naturalistic Decision Making (e.g. Gore, et al., 2006; Lipshitz, et al., 2006). The observations of the jury meetings showed that jurors used patterns from their own frame of reference to make judgements about design quality in the context of the competition. They jointly developed a common frame in which the decision about the winners was placed. This means that judgements within this frame are inter-subjective and specific to the case - the frame cannot easily be transferred to another context. The frame is part of a sense-making process among the members of the jury. Both the results of this case and the cases in Chapter 6 raise the question if repeating the same assessment process several months later or performing the same task by two different jury panels would result in the same outcomes. The issue underlying this question is the issue of how objective a judgement in the context of architect selection can be in general. Clients need to take the relativity of decisions about design quality certainly into account when they design a (tender) competition.

Ad 4) Methods to express and justify preferences

There are different ways to express and justify the decisions that are made during a selection process. The decision to assign several winners fitted the aim of the competition: to gather ideas and stimulate debate and creativity among the younger generation. The competition rules provided this freedom. It also suited the statement the jury wanted to make: to show the wealth of the ideas and to make clients aware of the potential creativity of design as displayed in this kind of competition format. The jury panel was able to fill in the decision strategies by themselves. The chair of the jury was most concerned with the official evaluation criteria and the other jury members were mainly relying on her for that matter. The official decision criteria as stated in the competition rules were not explicitly evaluated during the final debate but merely interwoven into the frame of reference as developed during decision making. The jury used the criteria explicitly to initiate the decision process and for further sharpening the decision task in reaction to their individual perception of the submissions and collective discussions. When the jury needed direction in order to reduce the amount of submissions or initialise of a new phase in decision making, they also applied the criteria in between the different decision phases. This indicates that the decision criteria steered the decision process in making sense of the decision task. Because the criteria were not explicitly and orderly addressed during the decision process, it is very hard to trace the exact reasons why a certain submission was chosen as a winner.

During the different phases of the jury process the chair often proposed a decision procedure and discussed it with the other panel members. The proposals led to the use of several decision methods:

- Discussion with assigning post-its based on the lists with individual notations (day 1).

- Expression of a number of votes during a discussion round the table (twice at day two).

- Anonymous vote (end of day two).

Voting made it possible to define an individual judgement before the start of a group discussion and therefore stimulated the independent individual input of the jurors. A few categories of solutions evolved during the decision process: the location Berlageweg, the location middle of campus, and the temporary accommodation at Julianalaan. The final voting reflected these three categories, which made the jury choose for three first prizes and three second prizes. During the discussion the preferences among the selected submissions did not seem to be very obvious, but the voting expressed a clear difference between the first and second prizes. This implies that a lot of the arguments were not literally expressed during the discussions. It also suggests the use of intuitive and rather holistic judgements. These kinds of decisions are difficult to justify to people who were not part of the inter-subjective decision frame in which the decision was made.

Voting appeared to be a good way of expressing intuitive preferences during an assessment process. Although current interpretation of procurement law might imply otherwise, only stating the (number of) votes would not have justified the final decision of the jury panel. This case shows that a jury report actually communicates the story of the reasons why a certain submission was considered as a winner. But as Kazemian and Rönn (2009, p. 177) state "it is typical for practitioners to evaluate different submissions and distinguish the favourite one without needing to articulate too much about how and why". Because of the large number of submissions, only a small portion of submissions were addressed in the jury report. A reflection on the observations in relation to the jury report indicates that some aspects of the jury's argumentation and categorical structure were more explicitly stated in the jury report than during the discussions. This was for example shown in the structure of the five themes in the jury report: 'sustainable and technological innovations', 'dynamic landscapes of education', 'generators of social interaction', 'out of site: the faculty as a mobile community', and 'visions on future educational concepts'.

The project team decided to appoint a staff member to write the jury report because none of the jury members would have found the time to write a decent jury report. This is rather common in Dutch design competitions. A report is always different than the meeting itself and therefore the jury report including its structure and phrases will reflect the meeting through the eyes of the reporter. In line with the competition tradition, neither the project team nor the jury panel considered a quantitative score form the criteria and their relative importance to justify their decision. This issue raises a question about the dilemmas decision makers must face during the justification of their decisions. The jury report was written by the same author of the competition brief and approved by the chair after corrections from all jury members. Yet I believe that still a lot of the reasoning behind the final decision was said between the lines. This makes it very difficult to justify a decision to 'outsiders'. The jury felt they had done justice to the impressive and extensive amount of work done by the esteemed participants of the competition by providing an extensive jury report.

In relation to the tender cases in Chapter 6 these findings suggest that being part of the jury panel would increase the ease to explain and act upon a decision as made during the decision process. However, neither a jury report nor a matrix form reflect the full decision, especially when a lot of intuition was used within the relatively closed context of a group. This means that justifying a decision remains to be a matter of trust between the decision makers and the participants. At the same time it also raises a question about the amount of which a final decision reflects upon the whole process of incremental decision making of a jury panel. For justification and legitimacy would it be enough to be transparent about the development of the frame and the process of sensemaking, or should the full frame be justified? In which way and to what extent can a matrix sheet actually support the justification?

Ad 5) Differences between experts

The project team decided that the submissions would be anonymously judged by an expert jury to ensure equal chances for all participants, thus for young as well as established talent. This excluded face-to-face interaction with participants and meant that all judgements were based on the perception of the submitted materials only. In making value judgements a lot of assumptions are made about the potential quality of the submissions. This is inherent to the conceptual stage of an ideas competition or tender. The student member of the jury several times expressed her difficulty with evaluating all plans in such a short period of time, both during the jury meeting as well as in the retrospective interviews. She felt especially insecure about the time spent of on each of the submissions on the first day. The other members of the jury panel seemed quite sure about their decisions and did not make use of the chance to re-assign votes or change their initial preferences. Almost all the jury members stated in the retrospective interviews that they had enough time to make an accurate judgement. This difference can be explained by the fact that the student member had lesser patterns available to apply, which is in line with the theories on expert judgement (e.g. Klein, 1993; Tversky & Kahneman, 1974). Most of the jury members also expressed to trust the other members on not having missed out on interesting submissions. Based on a comparison of the findings in this case with the findings in the tender cases in Chapter 6 it can be concluded that experts need less time to make a value judgement about the quality of the design. By recognizing patterns (shown by reference to similar experiences) they seem better able to put the design idea into perspective and refer to certain 'quality standards' they built up during years of experience. They also trusted their peers in this matter.

Similar to the findings in Chapter 6.7.4 the data of this case indicate that having experiences as a jury member in other juries made decision makers more aware of group processes and strategic voting opportunities. From this angle a learning effect occurs among jury members. This learning should be taken more advantage from in tender processes because it decreases the amount of fear among decision makers. Several of the jury members stated in the retrospective interviews that they know that:

"Almost every jury process eventually leads to a satisfying result."

Most members of the jury were able to defend their preferences very well in the discussion with the other panel members. The chair of the jury stimulated the student member to explicitly express her opinion and in the retrospective interview the student stated that she felt strengthened by this support. Sometimes the jury members explained the criteria that they personally used to evaluate a submission to strengthen their statements and convince others of their opinion. Expertise of the jury members was for example shown by the ability to visualise the proposed idea, as articulated by one of the jury members during the discussion:

"I finish this project in my mind; I know you do that too, and it has so much potential. It will create a wonderful environment."

Some jury members also displayed that they were able to put their personal interests aside and explicitly mentioned that:

"It is not my cup of tea but it is a very clever idea."

Yet they also tried to get their own preference included in the final selection of potential winners. One of the jury members showed a great deal of interest in the category of 'just ideas'. The winners and nominations of the competition do not include any submissions of this category because the jury felt that a good submission had to answer all three criteria as mentioned in the competition programme, architectural quality included. During the retrospective interview he explained his interests for the idea category from his background in architectural and cultural history. His critique about the final outcome of the competition related to the difference of his interests in comparison with the other jury members. He expressed to have tried to steer the discussion in the direction of discussing the concept of the schemes rather than the implications of the physical shape connected to the ideas. The observations confirm this and also showed that this strategy was stopped by another member of the jury by saying:

> *"I notice what you are doing, and I went along in your direction for a long time, but this is the limit."*

The strategic character of jury discussions was confirmed in the interviews, mainly with the Dutch professors also working in practice. The observations showed that strategic behaviour was sometimes also made rather explicit, for example when one juror explained to the other jury members that:

> *"I liked that entry a lot but decided to choose this one because it would have better chances of getting selected."*

The jury members in this case were well aware of the voting behaviour and preferences of other members. In the interviews they talked about the balance between clients and professionals in a jury and the right moment to express a preference in the group to convince others. Two of the jury members explained during the retrospective interviews that panels with mixed disciplines and different expertise levels could lead to subgroups within the panel. This can be compared to the findings as for example described by George and Chattopadhyay (2008) and Robbins and Judge (2008) about the effects of group decision making. The expertise level of the jury panel of the Faculty Building case exceeded the expertise level of all tender cases. The observations indicate that the jury panel of the Faculty case acted more like an expert team in giving each other feedback and trying to consider the submission in a historical context (see also section 7.4.4). On the other hand their relative independent character could have limited the fit between the aims and ambitions of the client and the results of the competition.

Emotions and personal affection were sometimes used in argumentation:

> *"We will still be friends if you leave this out, but I still think it is very strong project."*

Yet emotional conflicts between the jury members did not occur in this case. The findings suggest that can be attributed to the fact that the character of the competition was not legally binding and the jury had the opportunity to divide the prize money among several winners. In the final phase of decision making the long term strategy appeared to be more important than personal differences. So in the end of the second day discussions proceeded on a more strategic level about the outcome of the competition:

"We have to think about the message", and

"If we take them both, we really make a statement."

Submissions with interesting topics, such as centralization of functions, changes in the educational system, and different philosophical statements about the profession that were not be part of the winning submissions, were noted to be included in the jury report. This gave the jury panels the opportunity to address the issues without further decision implications, which contributed to the flow of the decision process.

Ad 6) Connection between the jury panel and the aims, criteria, final decision of a competition

In the context of assessing quality during competitions, Heynen (2001) mentions the importance of aligning the project definition, the physical, legal, economical, social and cultural context, the shape, construction and materialization of the planned space, and the potential of use, meaning and performances. The literature in Chapter 3.4 suggests that therefore a decision frame should include a connection between the aims of the decision and the most important aspects of the decision task. In this case the project team intended to create a connection between the ambitions and aims of the client and the jury panel by making the Dean part of the jury, inviting several jury members with close connections to the faculty and involving the chair in the preparations. This did not guarantee a connection between the aims, the decision criteria, and the decision outcome, but suggests at least a well-considered competition design. If a client delegates the final decision to a fully independent jury, they have to make sure this panel should be very well aware of the ambition and aims of the client. An introductory presentation or more involvement in the preparations of the competition could ensure a close connection. However, the results of this case indicate that being part of the jury might be the best option to ensure a match between the aims, ambitions, needs and opportunities that are offered during a competition.

Reflecting on the composition of the competition jury, a relationship between the type of winning submissions and the composition of the jury seems visible. The expressions in the media about the results of the competition suggest that especially the lack of iconic architecture and the preference for low-rise buildings was specific for the kind of experts invited for the jury panel (Bockma, 2009; Vollaard, 2009b). The jury hoped to have revealed a trend in architecture, which hypothetically should be independent of the persons who signal it. A trend can however only be evaluated after acknowledgement in the next (few) years. Heynen (2001) suggests to select a diverse group of jury members who are competent at all relevant domains and have a vision about architecture that they can communicate during a debate. The findings of this case suggest that in line with the findings of the tender cases, only including a panel member with specific expertise such as sustainability ensures that considerations on these specific themes become part of the final decision. In this case further differentiation on the fields of expertise and stakeholder perspective could have increased the amount of information and perspectives that were brought to the table during the jury meeting. However, a balance should be kept between the members of a jury as a group because too much diversity could scatter the interests too much (e.g. Robbins & Judge, 2008). In the retrospective interviews most of the jury members expressed to have been satisfied with the composition of this jury.

The observations and the interviews confirm that both personal and professional competences are important for decision makers to function as a professional jury member (Kazemian & Rönn, 2009). They suggest that being a jury member appears to require a certain kind of willingness for critical reflection, openness to learn from others, to spend time on educating people, and to invest in the future generation of designers and society. These traits seemed related to the passion, enthusiasm, and other emotional responses by which a jury process seems to benefit from. The results indicate including several members who experienced a jury process before would surely benefit the process of decision making. The main reason for this is the faith that experienced decision makers have in their selves, in the jury as a group, and in the client organisation. This reflects on the positive atmosphere in the room when decisions have to be made.

Then the issue of the content of the decision criteria in relation to design quality remains open: What should be the balance between functionality and aesthetics, between costs and quality, and between viability, innovation, sustainability and durability? In Chapter 2 several models and lists of quality aspects are discussed that are confirmed by the findings in the tender cases. Literature on jury processes in design competitions shows that not only the separate qualities are important in assessing quality but also the totality of a design solution and its impact on the context (Heynen, 2001; Kazemian & Rönn, 2009; Spreiregen, 2008). The weight of the qualities should strongly relate to the ambitions and expectations of the client about a future building. The aspects of 'development potential' however refer to the potential value of the proposal for future use and the value during realization as well. During the jury meeting of this case some references were made to future potential of the building, but further arguments that related to the resource envelop (time, finance, natural resources and human resources) of the project were almost absent. This can be attributed to the character and aims of the competition that did not include the actual realisation of the designs as proposed by the participants. These findings indicate that because the aim of a tender procedure is to award a service contract, decision criteria should also refer to the future role of the architect. If the architect should play a specific or dominant role in further development of the brief or during construction, the tender brief and award criteria should somehow reflect these expectations.

7.5 Discussion

7.5.1 Case reflection

This fourth case study set out to apply the findings of the three empirical cases in a competition design. Apart from the aim and content of this case, I can conclude that the organisation and decision processes of a competition are similar to what was found for the tenders in Chapter 6. In competitions the stakes and implications are usually not as high as in tender projects, but in each of the four cases public money was spent and expectations were aimed at 'building dreams'. In this case, the client by virtue of being a Faculty of Architecture was very closely connected to the field and had an exemplary function to practice. This close connection to the field made it possible to use the existing network in selecting members of the project team and jury panel, to consult experts in the field easily, to involve the press in announcing the competition and the winners, and to aim for an international character. The exhibition in the NAi provides a good example of collaboration with the professional field.

During the competition two intertwining perspectives on the role of the client came to the fore: 1) playing a role in society as a stakeholder in the field of architecture and being interested in general debate as well as doing business, and 2) acting as a professional client, trying to satisfy stakeholders in a context of financial and organisational constraints. These perspectives do not necessarily harmonise well. In the three tender cases these dilemmas were also experienced by the decision makers but merely shifted towards legal responsibilities instead of responsibilities towards the field of architecture. This required that sometimes decisions needed to be made of which the adverse consequences were (implicitly) known. I believe that in the case of the competition the interests of being a professional client usually prevailed, just as in the other cases. Yet, in all cases the difficulties with the implications of the decisions are currently still shown, which indicates that deciding about design quality is almost too complex to do it perfectly right.

The composition of the project team was quite similar to the project teams found in the three tender cases. However, the research method of participant observation made it possible to gain knowledge and experiences from the insights perspective instead of reflecting from the outside only. During the organisation of the competition I experienced tensions that related to common project management issues, like the input of personnel, inter-organisational politics, and available financial means. A lot of these conflicts originated from dissimilarity between the temporary project-based aims of the competition and the long-term aims of the organisation. Due to a certain leadership style the tasks and responsibilities were not clearly defined among the members of the project team. To some degree trust some members of the team did not trust each other and appeared to have different styles of working. The rushed character and sudden impetus of the case surely also contributed to these tensions. Especially after the Dean had fallen sick, the number of differences in interests about patterns of spending and authority within the organisation between the internally and externally hired members of the team increased. The conflicts were, for example, revealed in relation to the accuracy of information and feelings of responsibility in communication to the professional practice. This resulted among other things in a mistake in a press release and an adjustment in the timing of the announcement of the winners. Overall I believe the competition can be called a success with so many submissions from all over the world. This implies that the project team did do a good job. Most of the aims of the competition have been reached, although one could argue if the full potential of the results is already taken advantage of in relation to research-by-design and stimulation of debate. An interesting database with submissions, observations, and other kind of documents awaits further analysis, also in relation to the results of the Think Tank.

All cases show that being a jury member provides an excellent learning opportunity from both the other jury members and the participants. The composition of a jury panel is therefore not only important for the success of a (tender) competition but also for the jury members themselves. This research created the opportunity to evaluate the jury process and close the learning cycle for the jury to act as an expert team. At least half of the jury members expressed during the retrospective interviews to have reviewed the competition processes and outcomes more carefully in preparation of their interview. Tenders and competition could also offer participants a learning experience, on the condition that explicit feedback is provided. A number in a matrix sheet or grade does not qualify for that. Writing a jury report could also contribute to the scientific and social relevance of competitions and tenders. The learning effect could be further exploited by international systematic research but clients and participants would have to open up their attitude towards architect selections and research in this interesting field.

7.5.2 Findings in the context of procurement law

Based on these four empirical cases I found that the main difference between organising a tender and organising a competition is the potential impact of the decision about a winner. This strongly relates to the aim of the decision and affects the decision process. This difference created a certain amount of flexibility that would not be available in a procurement case. In comparison to a tender, the competition offers the client more options to control the course of the process and therefore also the potential value of the decisions. Consequently the dynamics and social aspects of the governance context of the client could be taken into account and prevent information from being obsolete at the time of publication. Tender activities often cross organisational activities which create uncertainties among decision makers and a need to reframe the aims or decision methods. In my opinion the competition tradition made it possible to create room for interpretation of the brief and adjustments to the decision criteria; it created therefore room for sensemaking processes during the decision process. In both tendering and competition regulations the principle of fair play and an equal level playing field are important values which need to be retained. Yet, the current interpretation of procurement law seems to cause a feeling of restrictions among the representatives of public clients that does not benefit the process of decision making. I sometimes question if the current tender regulations are realistic because public administrators have to fulfil their social, legal, personal, *and* political duties in an area that is usually not their field of expertise.

The results of the four empirical cases show that decision making during an architect selection process is a process of sensemaking in which a number of implicit aims and ambitions, a potentially explicit brief, and several abstract proposals for future solutions are linked. From an economical perspective a tender connects the demands of a client with the potential supply from the market. Decision criteria should be a translation of the aspects that the client thinks will be important, but merely provide the structure for the decision process instead of steering the design of the decision process. Together with the competition brief the decision criteria should provide an anchor point for the jury. Jury members all bring in their own interests, opinions and expertise and try to act as representatives for the client. They aim for a consensus on their judgements but can sometimes hardly find the time and opportunity to gain more in-depth knowledge about the topic of discussion. The emotions that originate from the interaction with the proposals and the other jury members influence the process of decision making in a sense that cannot be predicted beforehand. In all cases satisfaction was felt among the decision makers when a consensus was reached about the winners, even tough they were beforehand aware that this consensus would never fulfil all the requirements and obligations. The results show that clients experience a conflict of two rationalities because they have to comply with procurement law by acting rational, while decision makers representing a contracting authority need a psychological process of sensemaking in order to reach a decision. This means that the characteristics of the process of architect selection tend to cause a conflict with the interpretation of the commonly accepted legal principles of transparency, objectivity, equal treatment, and proportionality.

7.6 Validation of findings

As stated in the previous section clients often experience conflicts in the interpretation of the legal principles of transparency, objectivity, equal treatment and proportionality. Based on the results of the four empirical cases and the theoretical framework I distinguished several sensemaking processes and situational characteristics that were found to be of influence in the decision process of public clients in the selection of architects (see Chapter 5). In order to validate the findings of this research a workshop was organised in January 2010 with ten experts from the field of architect selections. Some experts were selected from the network of the supervisors and the researcher, others were asked to join because of their reputation. We aimed for a mixed expert panel of Dutch public clients, legal professionals and well respected architects. In the end ten experts were present during the workshop: two representatives of a professional public client; one representative of the real estate developers; two legal professionals who both work in practice but also are professors (in building law and procurement law respectively). The other five experts were practicing architects who also had substantial experience as a member of jury panels because of their (inter)national reputation. I asked the director and senior researcher of Architectuur Lokaal - an independent national centre of expertise and information devoted to commissioning building development in the Netherlands - to observe the workshop and reflect on the findings at the end of the workshop. They did not actively engage in the workshop and mainly took notes. The supervisory team of the research project supported me during the different items of the workshop.

Sensemaking processes:	Main situational characteristics:
Reading the decision task	Complexity, Uncertainty, Time
Searching for a match	Control, Time, Affect
Writing a decision process	Time, Intuition, Expertise
Aggregating value judgements	Structure, System, Expertise
Justifying against different rationalities	Support, Trust, Control

Table 7.6 Overview of sensemaking processes and situational characteristics

The workshop was planned for two and a half hours but took almost three. The programme of the workshop was as follows:

- Welcome and short introduction of attendees.

- Presentation of the five sensemaking processes as a result of the research by the researcher, supported by eleven PowerPoint slides.

- First reaction and reflection of the experts on the findings (validation step 1).

- Modelling exercise with the five sensemaking processes, eleven situational characteristics and four legal principles (see table 7.6).

- Presentation of the findings of the modelling exercise by the participants (validation step 2).

- Reflection on the workshop findings by the external observers.

- Closure of the workshop.

Validation step 1 consisted of a reflection of the experts on the presentation of the findings of the research. The presentation of the findings was supported with a power point presentation in which the research was shortly introduced and the five sensemaking processes were explained from theories and empirical findings. All experts expressed to recognise every sensemaking process that was identified in the conclusions. During this round of reflection several experts shared their experiences about the current problems in Dutch practice. These anecdotic stories were to a large extent similar to the problems as described in Chapter 4.6. Several issues were raised as suggestions for further research about the differences between private and public clients, and between project development and architect selections, the proportionality about judgements of the competences of an architect, and the level of professionalism among public clients in general.

Validation step 2 consisted of a modelling exercise in which the five sensemaking processes, eleven situational characteristics and four main legal principles as distinguished in this research were used as ingredients for implications for tender design. The ten experts were divided in four groups. Each group was told to select one of the legal principles (equal treatment, transparency, objectivity, proportionality) and discuss the possible implications for the design of a tender, while taking the sensemaking processes and factors into account. Each group received one yellow post-it with the legal principle, five pink post-its with the sensemaking proc-

Figure 7.6 Validation step 2 in which each group of experts used post-its with the processes and situational characteristics to discuss possible implications for a tender design

esses, eleven green post-its with the situational characteristics, a sheet of brown paper, a short handout about the results and a marker. In about thirty minutes the groups thought about possible implications for a tender design. Then each group presented their findings to the others.

Figure 7.6 shows the post-its (on the left), the discussion of one of the groups (in the middle), and the presentation of their findings to the other groups (on the right). The task appeared to be difficult in the beginning though successful in the end. During the presentation of the findings the experts actively reacted on the relationships as proposed by the other groups and discussed possible implications for a tender design.

The positive results of the validation workshop showed that the distinction of the five sensemaking processes and the situational characteristics helped the experts to reflect on a more abstract level on the origin of the current conflicts in architect selection processes. At the same time the results suggested that the implications of the legal principles from the perspective of a client often conflict with each other (see Table 7.7). The experts indicated that proportionality for example could be positively influenced by the use of intuition, while intuition was indicated to have a negative effect on equal treatment. For structure this appeared to be the other way around. Time was not found very relevant, apart for the fact that everyone usually is busy. Therefore only relevant questions should be raised. The motivation of the decision was found to be very important, from a legal as well as from a professional perspective.

Legal principle:	Positively influenced by:	Negatively influenced by:
Equal treatment	Structure	Intuition, Uncertainty, Affect
Transparency	Respect, Openness, Clear goals and aims, Risk estimation	Uncertainty, Lack of Control
Objectivity	Control, Expertise, Affect, Intuition, Support, Trust	Lack of recognition, Support of decision
Proportionality	Expertise, Experience, Intuition, Control, Support	Structure, System

Table 7.7 Summary of the results of the modelling exercise in validation step 2

The main conclusion of the external observers was that the experts mainly reasoned from their own perspective, which is an inevitable part of the problem in architect selections. "*What sounds logical from one perspective does not make sense from another perspective*" they stated. They indicated that the professional level of clients needs to increase to overcome a part of the current problem. Maybe architects could play a role in this. Not all situational characteristics and sensemaking processes were considered relevant for each legal principle, which is in line with the findings of the research. Based on the results of the validation workshop I conclude that the framework as suggested in the workshop can be successfully used in proposing possible implications for future tenders.

In Chapter 8 the results of the empirical cases and the validation workshop are compared to the success factors proposed in the theoretical framework of Chapter 5 to draw conclusions about the process of value judgements and decision making in the selection of architects by public clients under European tendering regulations.

Chapter 8

CONCLUSION, RECOMMENDATIONS AND REFLECTION

8.1 Introduction

This research focused on the process of decision making of public clients during architect selections in order to address the challenges they currently face in the context of EU procurement law. A tender for a public building is more than a legal obligation and should be considered as a project embedded in the process of fulfilling the accommodation needs of a public client. In the selection of architects several actors play roles that represent different groups of stakeholders. Existing guidelines and models to support the tender process seem to neglect the fact that a decision to select a winner of a tender is preceded by numerous other decisions by a public commissioning body that relate to the ambition and execution of the tender as well as the building project as a whole. For a public client the justification of the award decision might be as important as the decision itself. A process of sensemaking is therefore essential.

For the theoretical framework for this research I drew on literature from the fields of architecture, industrial design, environmental psychology, construction management, management science, and organisational psychology to explore the origin of making value judgements and decisions about the quality of design. I gathered empirical data by using different qualitative research methods in four case studies and validated the results in a workshop with professionals. The main focus during data collection was on the cognitive and social processes of decision makers in the context of architectural design. In this chapter I compare the five sensemaking processes and the fifteen possible success factors proposed in Chapter 5 to the processes and situational characteristics identified from the empirical data in Chapter 6 and 7. Based on this comparison, I give recommendations for the design of future tenders in architecture. These recommendations take the form of a series of success factors. The chapter concludes with a reflection of the limitations and possible implications of the study and directions for further research.

8.2 Conclusion

8.2.1 Five sensemaking processes contributing to an interplay of rationalities

The evidence from the four case studies confirmed that the process of public clients selecting an architect is indeed best described as a process of sensemaking in a presumably rational world. For public clients this often led to a clash of the different rationalities that play a role in decision making. In all empirical cases in this research, decision making started when people became aware of a problem and experienced a need for a solution. A call for tenders is an important component in the process of realising a building project to resolve a client's housing problem. The exact results of such a project are not definite, nor is the meaning of the results for the organisation. The results of the cases confirm that in architect selection decisions can be seen from two different rationalities: a *legal* perspective and a *psychological* perspective. The underlying logic of the legal perspective is that an open procurement market and free movement of goods and services in the European Union would ultimately benefit all citizens. Equal treatment, transparency, objectivity, and proportionality are the most important principles for the procurement of services, works, and deliveries. Prior announcement of decision criteria and decision methods could help to instantiate these principles and inform participating companies what to expect. Current European procurement law is based on assumptions similar to the first generation decision theories (Beach & Connolly, 2005; Harrison, 1999). These models perceive the process of decision making as a sequence of problem definition, identification of decision criteria, allocation of weights to the criteria, development of alternatives and evaluation of alternative with the use of the decision criteria as set out in the beginning.

The psychological perspective adopted in this study, however, addresses how people actually make decisions and which situational characteristics influence these processes. The dynamics of the context implied that the information that project definitions were based on often had become obsolete by the time a judgement is made. This made the identification of decision criteria and allocation of weights to the criteria more complex than presumed in the legal model. The decision alternatives are developed by architects who submit a project proposal, with no possibilities for the client to influence or control the development of alternatives. The only opportunities a client has for controlling the quality of the service lie before and during the evaluation of the alternatives. The process of client decision making was found to be dynamic and incremental rather than chronological and static. The empirical cases showed that the decision process of selecting an architect was a result of the decision makers' interaction with the alternatives once they were confronted with them and began to make sense of the proposed designs. It is therefore almost impossible for clients to design a selection procedure and announce the criteria and weighting factors up front, as required by procurement law. In this respect the rationality of the legal requirements clashes with the psychological rationality of decision making. On the other hand the legal rationality provides a public client with the structure and room needed for successful decision making. The data showed that this interplay of rationalities results in sensemaking being an essential part of decision making for public clients in the selection of architects.

The findings of this research confirmed five sensemaking processes that contribute to a potential clash between expected and actual decision making behaviour, namely:

1.   Reading the decision task.
2.   Searching for a match between aim, ambitions, needs and opportunities.
3.   Writing the decision process.

4. Aggregating different kinds of value judgements.
5. Justifying a decision against different rationalities.

Clients act in the context of the tendering regulations of EU procurement law. In every case the steering committee, project team, and jury panel were identified as the main actors. In analysing the project context, the governance, the social environment, and project management frame influenced the decision process somehow. The tender brief, the procedure of the process, the involvement of the stakeholder, and the decision procedure were found to be the most important elements of the tender design. I found that clients endorse the principles of EU procurement law - equal treatment, transparency, objectivity, and proportionality - from a psychological perspective. However, the interpretation and execution of these principles could cause clashes. Based on the empirical data and theoretical framework eleven situational characteristics were identified to influence the sensemaking processes of the public clients: affect, complexity, control, expertise, intuition, structure, system, support, time, trust, uncertainty. These characteristics were successfully validated in a workshop with professionals.

The theoretical concept of sensemaking explains why actors often only experience the conflicts with the legal principles when the tender design is implemented during a procurement process. The complexity and uncertainty of selecting the services of an architect means that decision makers need time to understand the actual aims and opportunities of a tender process. Because of the dynamics of the organisational context in which a decision is made, changes could have occurred in the basic assumptions are originally framed in the call for participation. This potentially conflicts with the basic principle of equal treatment because the procedure and the participants cannot be adjusted to these changes while the decision makers do. Therefore clients are not able to play by the rules of procurement law. Because the rules of the game make it possible for tender candidates and tenderers to know what to expect, the sensemaking process is inconsistent with the legal interpretation of the transparency principle. Transparency, however, has different meanings when looked upon from different perspectives.

In sensemaking processes decision makers need time to go through several iterative and incremental stages of decision making. These stages are often accompanied by numerous negative and positive emotions as a result of interaction with the proposals and with the other decision makers and the use of intuitive decisions. In the search for a match every decision phase increases the possibility to control the process. The impenetrable characteristics of sensemaking processes could make architects feel that they were not treated equally. The positive emotions that are felt when a match is found between a client and an architect could also lead to feelings of unequal treatment with the other tenderers. Clients and winning architects tend to be very satisfied about the outcome of the selection process because the decision 'grows' upon them. The use of a predefined and structured aggregation system could prevent the participants from accusing the client of unequal treatment. However, these kinds of systems often leave no room for the added value of the totality of a service package and the dynamic course of the decision process. Contrary to expectations they may therefore decrease the

Sensemaking processes as identified in this study:	Main situational characteristics:	Potential conflict with legal principle of:
Reading the decision task	Complexity, Uncertainty, Time	Equal treatment, Transparency
Searching for a match	Control, Time, Affect	Equal treatment, Proportionality
Writing a decision process	Time, Intuition, Expertise	Transparency, Equal treatment
Aggregating value judgements	Structure, System, Expertise	Transparency, Objectivity, Equal treatment
Justifying against different rationalities	Support, Trust, Control	Transparency, Objectivity

Table 8.1 Overview of sensemaking processes, situational characteristics and procurement principles

validity and therefore the objectivity of the value judgements used in decision making.

Expert judgements are usually more easily accepted by stakeholders as objective, even if these kinds of judgements are not accompanied by a completely reasonable explanation. Organisations try to compensate for the fact that individual intuitive judgements might not be objective by using a group of decision makers. Group decision making often results in an inter-subjectivity consensus decision, which strictly speaking conflicts with the idea of objectivity. Making decisions in a team of experts supports the process of participants trusting the decision makers, and somewhat alleviates the possible lack of rational explanation. The involvement of experts also contributes to quality of the decisions and the control of the decision process. At the same time, the characteristics of an expert judgement prevent decision makers from explaining the exact reasoning and course of the process beforehand, or tracing a decision afterwards. This could conflict with the current legal perception of the principle of transparency.

Table 8.1 provides an overview of the main characteristics that were found to influence the sensemaking processes and displays the potential conflicts with the procurement principles. This overview illustrates the complex and dynamic character of selection processes in architectural design. Due to the explorative character of this study this overview of processes, situational characteristics and legal principles should be considered as an indication rather than formal proof.

In the next sections each sensemaking process is explained based on the theoretical concepts introduced in Chapter 1 to 5 and the empirical findings reported in Chapter 6 and 7. In the following sections the success factors proposed in Chapter 5 are evaluated and converted to recommendations for the design of a future tender in architectural design. The recommendations are numbered in the same way as the proposed success factors were.

8.2.2 Reading the decision task

Because a public commissioning body acts as a client rather than a customer, distinctive dimensions of architectural and legal language have to be analysed by the decision makers during the process of decision making in order to know what to expect. The sensemaking process of reading the decision task is based on the concepts of sensereading and framing as described by Balogun, Pye and Hodgkinson (2008) and Beach and Connolly (2005) and deals with the translation of the aims of the client into a tender procedure (Jones & Livne-Tarandach, 2008) (see Chapter 5.2.1). The development of a tender brief and the analysis of the project environment are important parts of this sensemaking process. The results of this study suggest that complexity, uncertainty and time are the main situational characteristics that influence the process of reading a decision task.

In all cases the tender documents that had to be prepared in the beginning of the tender processes were typically somewhat vague at the time when a client decided to start an architect selection process. One of the main questions the decision makers asked themselves during the cases was 'How should we select an architect and what should the focus of the decision process be?' The results indicate that a distinction has to be made between the brief for the tender process and the brief for the building process. In order to support the decision process, the level of detail of the tender brief needs to be aligned with the aim of the tender and the proposed procedure. This means that a sketch design requires a different kind of tender brief than a vision for a potential design project. The results of the participatory observation in the Faculty Building case showed that the dynamics and uncertainties of the situation made it hard to explicitly design a tender procedure and develop a brief because both the requirements and their interpretation changed during the course of the process. This was due both to the internal evolution of the understanding of the project and 'external' events that case a change to the scope or brief. Reading the decision task is consequently an ongoing activity that requires attention throughout the tender process.

The ambiguous characteristics of current tender procedures and underlying assessment systems still show the mixture of their roots in the competition tradition, the tendering of works, and the search for a partner in design (Strong, 1996). The public clients in the three tender cases all combined the procedures of these traditions without being aware of elements that would ensure fair play and equal treatment. The most important dilemma that clients faced was a distinction between the search for the right solution for their design problem as suggested by the tradition of design competitions, and the search for the right partner in designing a solution for their accommodation needs, as suggested by the tender principles (see Figure 5.3). In the cases of the City Hall and the Faculty Building, the frame in which decisions were taken focused on the search for an appropriate design solution. This search connected with the architectural tradition in which architects anonymously show their competences to the public and the client by means of a design or vision about the future building. In the School case, the most dominant frame was the search for the right partner that would be capable of designing the future building and realising their dream. Their search focused on the maximum value from a client's perspective with several parties acquiring a contract. The perception of circumstances and development of the decision frame can therefore be considered as essential part of the sensereading process.

Factors 1 & 2 as proposed in Chapter 5.2.1 for a successful reading process of the decision task relate to the idea that a decision about design quality will be based on a holistic judgement about the characteristics of the person, the future product that they will deliver and the firm they represent. In all cases existing values, structures, ambitions and needs of the decision makers played a role in relation to design quality and the potential value for the future (see for example Figure 6.5). Specific examples are the observation that the international jury members of the Faculty case sometimes showed more interest in the impact of the competition on the international community than in the consequences for the future users of the university building. Likewise the Executive of the Provincial Government Office appeared to be more worried about his position than about the successful conclusion of the tender. The analysis of the City Hall data clearly showed differences between the different kinds of stakeholders in the amount and kind of design qualities that were taken into account during the decision process (see Chapter 6.7.3). This confirms that design quality is essentially based on individual perception as a result of the interaction between an individual and a design. The data of the Faculty case illustrate that even seemingly objective requirements or criteria, such as the submission time and format of the entries, contained subjective and intangible aspects (see Chapter 7.4.4). The jury members used their expertise to overcome these issues of objectivity and potentially conflicting characteristics of assessing design quality. The observations displayed that all decision makers used several decision rounds in which (partial) judgements were combined in order to make a decision. Their final judgements were holistic by nature and motivated by decision criteria that related to the particular proposal.

In all cases the selection process of an architect enabled clients to reach multiple aims that were often more strategic in nature than the selection of an architect would imply. This is in line with the diverse competition aims and applications described by Spreiregen (1979) and Svensson (2008). This multitude of aims was usually reflected in the arguments that were used to justify the decision for a winning architect but not always made explicit in the tender design. Table 6.10 in Chapter 6.7.2 displays the proposal, person and process related issues that were found in the argumentation about the award decision during the tender cases. Even in cases with a clear design product to be evaluated, such as the City Hall, or in situations of anonymous evaluation, such as with the Faculty Building case, argumentation always related to the person and the firm behind the proposal as well as the offer that was presented. Clients were not able to make a clear distinction between these issues. The selection process of an architect thus appears to be aimed at selecting a 'package' of the person, the firm and the proposal they offer the client. The relative importance of the product in comparison to the person and the firm should direct the tender procedure and the tradition in which the tender is designed.

Based on the results I conclude that success factors for reading the decision task are:

1. Allowing for a holistic judgement in the tender design that incorporates potentially conflicting judgements.

2. Addressing the characteristics of the architect as a person, the proposed design, as well as the firm that they represent in the decision criteria of the tender design.

8.2.3 Searching for a match between aims, ambitions, needs and opportunities

Tendering is a way of granting contracts for projects based on the principle of an open market. This implies a process of matching supply and demand. The second sensemaking process of searching for a match (see Chapter 5.2.2) relates to the fact that during the selection process a client's architectural values are connected to the opportunities that are offered by the architects to be matched with specific goals and operational plans. This matching process of goals and plan is also part of the image theory (Beach, 1990). Although European procurement law aims at opening the market across the EU member states, experiences confirm that Dutch clients prefer to work with Dutch architects (Geertse, et al., 2009). The results show that the decision makers apply existing knowledge about the architects in order to create a sense of control over the situation. This implies that the match between demand and supply is influenced by the use of culturally dependent heuristics (Gigerenzer & Selten, 2001; Tversky & Kahneman, 1974). The results of this study suggest that control, affect and time are the main situational characteristics that underlie the matching process of aims, ambitions, needs and opportunities.

In the tender cases the architects roughly adopted two kinds of strategies to present their solution for the tender brief: 1) a focus on the presentation of the firm and the proposed architect by presenting a rather open idea about the assignment, and 2) a focus on the presentation of a complete design solution for the assignment. The first strategy fits the procurement principles with a collaboration perspective, the second strategy matches the competition tradition with a product perspective as displayed in Figure 5.3 in Chapter 5.2.1. For architects participating in a tender, it is difficult to estimate whether a client holds a collaborative or a product perspective. Presenting a complete design solution provides the client with a good opportunity to judge the potential competences of the firm in the context of the assignment and make a statement on the governance level. Focusing on the qualities and competences of a firm gives a client the opportunity to control the design process from the beginning on together with a partner. Clients did not consciously grasp the strategies that the architects displayed. The tender cases showed that clients either consciously or unconsciously applied their knowledge about the work and reputation of the most important players in the Dutch field of architecture during their tender process. Especially during the short-listing of parties in the selection phase, this implicit knowledge played an important role in the framing process of decision makers. Sometimes decision makers used their implicit knowledge to regain the loss of control they experienced during the call for participation, in other situations they used it for strategic reasons.

For municipalities, building relationships with entrepreneurs is a form of quality assurance in construction (Doree, 1996). In the current interpretation of European procurement law, this need for quality control is not fulfilled sufficiently. In every case I found multiple aims that were not always directly related to the tender or competition procedure. It was found that for clients the business goals that were subordinate to procurement often prevail over the legal obligations. From a project perspective the awarded contract is meant to play a certain role for the architect in designing a building project, but a client has to take responsibility for the product resulting from the project. The results showed that the more the tender preceded, the more decision makers felt that a decision about the selection of an architect could have major implications for the community they represented. From a rational perspective one would say that a commissioning client should be well aware of the characteristics that come with the nature of awarding a service contract in architectural design. Then the aims could be easily expressed in the constitution of the award criteria. However in psychological terms this differentiation appears to be extremely difficult, especially before the process of sensemaking has started. The observations indicated that decision makers started to realise the effect of their request for invitation only after they received the requests for participation and proposals that were submitted by the architects. Additionally decision makers had to deal with a lack of consistent and reliable information. All together this led to a searching process in order to find a match between the aims, ambitions and needs of a client and the opportunities that were offered by the market.

Factors 3 & 4 were proposed in Chapter 5.2.2. They concern the searching process as it relates to the fit between the aims, participation strategy, and the design of the tender. In the case of the Faculty Building the open procedure and the high level of expertise among the stakeholders made it possible for all kinds of potential users of the building (employees, students and professionals) to participate as participants in the discussion about a new faculty building. In this sense participation made it possible to reach the aims of stimulating young talent, debate and research. However, this also created high expectations about the quality of the process and critical views of the outcome. In this case the commissioning body delegated the final decision to a jury panel. In the three tender cases the public commissioning body consulted a jury panel but retained final control. The three tender cases were characterised as pragmatic, democratic, and political in terms of their potential fit of the aims with the participation strategy and design of the tender procedure. In the pragmatic School case and the democratic City Hall case, the project team consciously designed a participation strategy and assigned specific roles and responsibilities to different stakeholder groups. In both cases this strategy appeared to have been successful, although it remains difficult to explicitly address the success of a decision process of the design of the procedure. Other examples of Dutch cases (e.g. Houben, 2007; van Geels & Kriens, 2009) indicate that the success of a participation strategy depends on how much the preferences of the other stakeholder groups are in line with the jury panel.

The legal obligations and existing guidelines provided clients with a structure within which they determined the playing field and rules of the game. However, I found that even for the experienced project team members of the Faculty Building case it was difficult to make an estimation of the effects of a certain rule. This means that clients will find out only during the tender process if the rules successfully prevented deviations from the actual aims of the project. In a sense a call for participation is therefore always a guess, and it is only after the submission deadline that a client becomes aware of the response of the market. Then officially the moment has passed to make adjustments. I found that the proportionality of the financial, technical and organisational requirements in the selection phase strongly related to the client's need for control over the potential quality (see Chapter 6.7.4). In the School case and the Faculty Building the client felt they had to do justice to the large amount of reactions to their call for participation. The procedure of an ideas competition provided room for increasing insights in order to respect the response on the call for participation. The strict legal context of tenders does not allow for the same flexibility. The perceived simplicity and static character of the legal context appears to exclude the need of increasing insights and changes of the environment in time during the tender. The responses of the market and interaction with the participants affected both the course and the outcome of the tender process, as is shown in Figure 6.6 in Chapter 6.7.4. The actual tender project can thus indeed be seen as a design that naturally incorporates all changes that occur during the process in time. Decision makers may feel that they should obey the law, but what they see and learn during the tender process may change their intentions and place them in conflict with the legal principles of procurement law. The process of sensemaking can increase the amount of fear and cognitive dissonance among the decision makers. The results suggest that the nature of sensemaking in finding a match prevents clients from being transparent about their decisions. It also makes it difficult to design procedures that fulfil the proportionality principle of procurement law.

Based on these results I conclude that the following success factors contribute to the process of searching for a match between aims, ambitions, needs, and opportunities:

3. Ensuring a fit between the aims of the selection process and the design of the tender.

4. Ensuring a fit between the position and type of the stakeholders and their role in the decision process.

8.2.4 Writing the decision process

The third sensemaking process that can be identified in the selection process is the process of writing the decision process (see Chapter 5.2.3 This process is based on the concepts of 'sensewrighting', 'sensegiving' and 'framing' as described by Balogun et al. (2008) and Beach and Connolly (2005). This process entails the writing of the selection process of an architect by the client during a project. In general a distinction was made between the preparation of the tender in which the brief, procedure, stakeholders involvement and decision process was designed, and the execution of the tender in which the design was applied. The main situational characteristics that I distinguished as influencing the process of writing a decision process are time, intuition and expertise.

Observations showed that decision makers have to deal with a lot of uncertainty due to incomplete understanding, lack of information, and conflicting alternatives during a tender process, which makes it necessary to allow for flexibility in the decision process. In all cases the procedure of the selection process (restricted or open) determined the number of phases in decision making, but not the interpretation of these phases. The competition was an open procedure, which meant that there was no pre-selection of the submissions. This was in contrast to the restricted tender procedure of the other cases, which consisted of a selection and an award phase. Yet both procedures showed similar decision processes.

In the preparation phase most of the work was done by the project teams. As shown in Chapter 6.5.2 in all cases external advisors were hired for their specific expertise. In the School case and the City Hall the management consultants became a kind of mediator between the steering committee, the stakeholders, the project team and the jury panel. In the Provincial Government Office case, the project management was carried out by staff members. For them it was difficult to raise the awareness of the responsible officers for the tender. During the selection and tender phase the members of the project teams supported the decision makers in checking the requirements and structuring the information. In doing so they implicitly influenced the decision process. The data of the Faculty Building case displayed a discussion about the level and kind of expertise that would be required for certain assessment tasks (see Chapter 7.4.4). The judgement of the specialised staff members about sustainability and concepts of the ideas differed for unknown reasons from the judgements of the jury members about these issues. The work of the project team and supporting committee did lead to a structure that was used during the assessment process. On the issue of sustainability the School and Provincial Government Office cases indicated that without a panel member with this kind of expertise, this issue did not receive a lot of attention during the process. This notion of involving specific expertise in the jury panel could be extended to other specific areas of attention addressed in the tender brief.

European procurement law is currently interpreted as prescribing a rational decision approach. The findings of the cases do not fully confirm this rational and sequential process but do show that the tender procedure provided a structure for decision makers to write their own decision process. On a general level the findings of the execution phase resemble the six stages of the selection process as described by Kazemian and Rönn (2009): submission check, determination of order of work, choice and preliminary judgements, presentation of interesting contributions, ranking, and decision making with architectural criticism. In all cases perception and judgement was needed to make sense of the proposals and to rank the alternatives in order to make a choice. It was usually during the judgement phase that decision makers started with the development of a frame of reference based on the aim of the assessment and the documents to be assessed. This implies that preparations for an assessment frame can be made, but the definite assessment frame cannot be developed in the absence of the actual submissions. In this sense assessment could be compared to the process of qualitative data analysis: the structure of analysis arises from the data. This is contrary to most quantitative research that typically consists of testing assumptions that already guided the process of collecting data. Procurement law assumes a quantitative process, while my findings suggest a qualitative process.

In all cases the decision criteria were somehow used to build a frame of reference between the stakeholders. The findings about the decision process of the tender cases were used to design the program of the jury meeting of the Faculty Building case. This guided the jury members successfully through an iterative process of initialization, perception, individual judgement, group decision making, and consensus building on an individual and group level in order to make their final choices as shown in Figure 6.7. During the different phases of the decision process judgements were adjusted due to external influences such as opinions of other members of a group, changes in the context such as time pressure, or personal factors such as moods and emotions. Consequently, decision making was a dynamic, incremental and cyclic process based on several kinds of value judgements. In every phase of the decision process a group of decision makers went through their own writing process of sensemaking.

The frame of references of the decision makers appeared to be influenced by the characteristics of the group members, the content of the submissions and the design of the tendering procedure. A conflict arose if the decision criteria did not match the increasing insights of the decision process. This meant for example in the City Hall case that arguments were used during the award phase to motivate the decisions that were not part of the official award criteria. Especially in the Provincial Government Office case the frames shared by the decision makers in the preparation phased and the realisation phase were not aligned. This caused serious difficulties in the selection phase and eventually led to cancellation of the tender. Also in the case of the Faculty Building the connection between the author of the aims and ambitions in the tender brief and the decision process appeared to be tenuous. Therefore I conclude that in line with Factor 5 in Chapter 5.2.3 it is important to align the decision frames throughout the process and to other actor groups. At the same time the observations of all cases show that it is especially difficult to involve the members of the jury panel in the design process of a tender because of their busy schedules.

All cases needed a selection phase to reduce the number of proposals that would be further analysed by the jury panel, although the Faculty Building case did not explicitly distinguish two phases in the procedure. My interpretation of the data suggests that this reduction process can be characterised by 'selection by positive impression' as agreed upon by several of the jury members. In the award phase the decision process can be considered as a process of 'selection by comparison' based discussion among the panel members. These strategies resemble the recognition-based decision theory of Klein (1997) in field of naturalistic decision making, but they also show signs of the use of fast and frugal heuristics in order to limit the time spent on each submission (Gigerenzer & Selten, 2001; Tversky & Kahneman, 1974). From a psychological perspective expert decision making can be explained by the use of implicit patterns that are recognised and applied intuitively during evaluation of a proposal. Intuition is a complex interplay of cognitive and affective processes operating below conscious awareness that improves with experience in a specific domain.

The selection of an architect requires a combination of very specific expertise that is usually not available within the organisation of a public client. This meant for the cases in the research that the majority of members of the jury panel and the project team were hired externally. A comparison of the tender cases and the competition case indicates that the strategies of information perception in the selection process 'first individually, than a discussion' versus 'perception as part of a discussion' may be a result of the level of expertise of the jury members. Experts already have a frame of reference with which they perceive the designs. They are able to use their knowledge and experience in an efficient way (Mieg, 2001) and trust their other panel members not to overlook high quality submissions or make invalid judgements. Therefore for the experts the first strategy of 'individual perception first' seemed appropriate. Less experienced jury members, such as the members of the School board, seemed to have benefited from a 'shared perception during discussion' strategy to build a frame of reference. The data of the tender cases indicated that the expertise of some panel prevented the other members from making biased decisions when using heuristics. In this sense the jury members of the Faculty Building already acknowledged each other's expertise, while the members of the award committee first had to build an atmosphere of trust. The decision task and decision process should therefore be aligned with the level of expertise of the decision makers (Factor 6). By educating decision makers (Factor 7) they can build up their own frame of reference which could enable them to speak the language of the architectural experts involved in the selection process (Bucciarelli, 1994; Jones & Livne-Tarandach, 2008). Aligning decision frames with other decision makers can also be considered as a way to educate each other. Yet, the results also indicate that these factors about alignment and education are hard to implement because of the dynamic character of the decision process and the relatively busy schedules of the experts and responsible officers that are involved in the tender.

A concept design comprises per definition a lot of uncertainties and indistinctness. In all cases decision makers clearly used this indistinctness to their own benefit during decision making. They tried to influence sensemaking (sensegiving) and shape their understanding of the world (sensewrighting) (Balogun, et al., 2008). They sometimes rejected proposals for a possible budget overrun or constructive instability, but also stressed the initial stage of the project to enlarge their decision space in other situations. The data showed that the perception of the proposals or interaction with their designers could lead to emotional responses that influence the course of the process of decision making. In the City Hall case one of the decision makers actually stated that *"it was love at first sight"*. In the School case the personality and competences displayed by the architect seemed to be most important in the final decision. This led to a choice for the *"most convincing party"* and a *"click with the architect"* while in the City Hall case decision makers aimed for the *"most appealing design"*. The expert members justified their decision by putting their findings into a future perspective (the potential of the offer) while the non-architects appeared to phrase their emotional responses in direct relation to their match with the winner. Their domain specific knowledge enabled them to better control their emotions and use intuitive judgements, which is in line with Factor 8.

In all cases the winner of the tender was the design firm which the client believed would fulfil their needs best, and would provide them with the right architectural value for the project. In both the School case and the City Hall building trust between the stakeholders was one of the main motives during a tender procedure. Instead of a systematic and rational process of evaluating alternatives, the course of the process was dominated by the arguments, perspectives and values of the jurors that fitted the intuitive judgement process best. Building trust was part of the sensemaking process that was needed to make a decision about the best architect for the job. The reason why an architect was trusted or not became clear only after a confrontation with the candidates, which is also one of the characteristics of sensemaking (Weick, et al., 2005). The decision process during architect selections therefore conflicts with the assumption in procurement law that compels clients to announce their decision criteria in advance.

In all cases members of a jury panel reached a consensus by several rounds of ranking, discussion and/or voting. The consensus proved not to be the same as an average of opinion but rather the result of a negotiation process, which is in line with the results of Svensson (2008) about jury processes in design competitions. In all cases decision makers needed time to interpret the criteria, the assignment and the brief that was mostly built by others not belonging to the jury panel. The observations confirm that experts were better at seeing significance of information, identifying important cues for risks, estimating consequences and judging autonomously (Hutton & Klein, 1999). Experts also felt the need to discuss and harmonise their preferences with other members of the group, which contributes to legitimization of the decision to the participants and society. Further analysis of the City Hall case showed that the experts addressed more aspects of design quality that the user groups and citizens. In the School case and the City Hall the narratives and additional background information that the architects provided in the award phase proved to support the decision process among the members of the group. Decision makers were able to check their assumptions with the designer of the proposal. This is in line with the statements of Weick (1995) and Vidaillet (2008) about narratives supporting the sensemaking process and Factor 9 in Chapter 5.2.3. The findings further suggest that the ongoing discussions between the decision makers about the criteria and their implications for a certain proposal increased the appreciation of other interests as well as the perception of the submissions. Discussion also seemed to increase satisfaction and acceptance of the process and of the decision itself. Because the process of sensemaking is openly displayed during discussions I conclude that discussions contribute to the transparency of the decision process in tender situations, which is in line with Factor 10.

Based on the results of this research I conclude that the following factors will contribute to successfully writing the decision process in the selection of architects by public clients:

5. Aligning the frame of references of the actors during preparation of the tender process in order to reach a decision at the end of the process that fits the ambitions and aims of the client organisation.

6. Aligning the type of expertise needed for the various decision tasks during the selection process of an architect with the nature and content of the decision task.

7. Educating decision makers in reading architectural designs from a client perspective.

8. Involving (external) experts in the process of decision making about design quality because they have domain specific expertise and are better at controlling product emotions and using intuition than novices.

9. Including a personal explanation by the architect to improve the clients' understanding of the proposal.

10. Creating room and flexibility in the decision making process for discussion and negotiation among the decision makers.

8.2.5 Aggregating different kinds of value judgements

The fourth process of sensemaking as explained in Chapter 5.2.4 relates to the aggregation of different kinds of value judgements in order to reach a final decision about design quality. Simon (1997), Etzioni (1988), and Snellen (1987) address the different rationalities that come to play in decision making. In this process the legal and social rationality of decision making clash when a pseudo-rationality is created by quantification of qualitative judgements. On the one hand, the choice for a winner during a tender is based on the structure that is provided by the pre-announced criteria, but on the other hand, it is part of a process of increasing insight and sensemaking in which value judgements are implicitly aggregated. Structure, system, and expertise were found to be the most important situational characteristics that influence the process of aggregating different kinds of value judgements.

The theoretical framework suggests that people use four basic levels of quality assessment: under-performance, basic performance, added value, and excellence. This scale could only implicitly be retrieved from the data of the cases. The current habit is to use grades or quantitative scales in ranking and measuring preferences. These scales invite calculations while they were not intended for that. The obligation to announce decision criteria and their relative importance implies that there are different aspects of design quality (which I call design qualities) on which the judgement of the proposals is based (see Figure 5.4). In the School case the decision criteria and their relative weight were input for a decision matrix that was used during the decision process of the client. This method can be compared to a multi-criteria analysis (MCA) or situations as described in prospect theory (Kahneman & Tversky, 1979). In the School case, matrix criteria with different measurement scales were treated as equal input for the ranking, while observations showed that aspects were treated differently during the discussions between the decision makers. Observations displayed that the relative importance of the criteria was incorporated in the matrix sheet implicitly while this input had important implications on the selection process. This demonstrates the invalidity of the decision in two ways: the limited power of the decision because of the summation of unequal measurements on an ordinal scale (de Keyzer, 1998), and the untrue image of the relative importance of the criteria.

The matrix did strengthen the structure of the decision process. This structure was used to start the framing process of the decision makers in the beginning of the decision process and to evaluate the decision at the end of the process. It was also used to continue the discussion on a more general level after a potential deadlock and find a balance between the selection of the firms in the selection phase and the personal interests of the decision makers. The School decision makers were not aware of the effect that the aggregation system would have on their decision process. Therefore they implicitly applied a system which helped them to structure the decision process, but did not reflect the true argumentation during the decision process. They mixed up the role of the assessment system and the decision method. I conclude that the validity of an assessment lies in the fit with the decision task, not in the objectivity of the system. The assessment system and aggregation structure should therefore be made explicit in the preparation phase (Factor 11). This means that the assessment system should be part of the tender design, but should not determine the decision process and its outcome. The decision process should be tailored to the characteristics of the actors and the project as part of the tender design.

In analysing the data, a clear difference was exposed between an individual judgement, a judgement of a group, and a decision about the winning level of design quality. Decisions in a tender situation are intentions for action that include an element of choice (Hodgkinson & Starbuck, 2008a): a winner is chosen from all candidates that submitted proposals for a building project. The results of the cases showed that value judgements and decisions include comparisons of alternatives on a holistic as well as on a separated level. This is in line with the relations as shown in Figure 5.4. Aggregation is needed to come from individual judgements to group judgements and from fragmented design qualities to holistic design quality. The decision task during a tender is an ill-defined strategic problem with moral, aesthetic and political judgemental aspects (Dane & Pratt, 2007). In line with Desmet (2002), the results suggest that a judgement provides information about the meaning and the future potential of a design object, the level of pleasure for the individual, and the potential benefits or harm for reaching aims of the object in relation to other objects. A judgement during the selection process of architects usually comprises more subjective than objective assessments about the quality of a design proposal. This assessment information is used as input for decision making. The involvement of experts with their relatively independent frames of reference enables a jury panel to ground a decision on a single holistic and intuitive judgement about the proposals. These judgements are ranked in order to select a winner.

During the analysis of the data, two ways for the group to come from individual judgements to a common judgement were distinguished: discussion and summation. In the discussion approach the differences between the individual judgements are discussed first, and then one judgement for the group is defined. This approach was mainly used in all cases. However, it was used in combination with voting or individual preference statements. These methods belong to a second approach of combining individual statements without interaction. Both these systems are acknowledged in case law and used by judges in case law as the consensus model and the individual assessors model (van Wijngaarden & Chao-Duivis, 2010a). In all tender cases the final decision was based on a discussion, while in the Faculty Building case voting determined the final winners. Both methods can be regarded as inter-subjective. A decision about design quality can therefore be considered as an inter-subjective consensus among the members of a group of decision makers. Inter-subjectivity contributes to the process of sensemaking of the members of a group. I suggest that both methods - discussion and aggregation - are needed to structure the decision process. The main benefits of discussions appeared to be that more information was put on the table and discussions contributed to decision acceptance. At the same time there was more pressure to conform with the other members of the group, more domination of one or two group members, and higher likelihood of group think (Robbins & Judge, 2008). The individual and independent aggregation method eliminates social influences and is therefore often perceived by society as more accurate in current tender practice. Probably this perception contributes greatly to the potential clash of legal and psychological rationalities.

An important contribution to transparency and objectivity in the tender system is provided by publishing of the decision criteria and the assessment system in the calls for participation and proposals. Legally the selection and award criteria are considered to be the complete basis for decisions about the selection of the participant with the best offer. However, in all cases the data showed a shift in priorities within the value system, and thus in the relative importance of the criteria, during the decision process. Criteria that had not been announced explicitly beforehand sometimes became important, or even decisive, in final decision making. This illustrates that from a behavioural perspective actors relate the selection decision to the course of the decision process and involve ethical or political considerations in reaching a decision. In the decision process a compensation strategy is used to resolve internal conflicts between judgements (Hogarth, 1988). A compensatory model for value judgement creates room for compensation on a certain aspect by overachievement on another aspect. This is something that occurred quite often in the cases, even in cases of under performance. In the City Hall case it was for example 'love at first sight' while they were aware of the fact that the proposal would probably exceed the budget. From a legal and rational perspective an under performing proposal is usually excluded from further assessment because compensation is not deemed possible. The results of this research showed that in architect selection processes the assumption that the best offer is a non-compensatory sum of aggregated scores usually does not apply. This means that in line with

Factor 12 allowing compensation in aggregating value judgements will increase the validity of the assessment.

Based on the results for aggregating value judgements I conclude that the following success factors will contribute to the success of a tender in architecture:

11. Making the assessment structure and aggregation system of value judgements explicit in the preparation process.

12. Allowing compensation in aggregating value judgements about design quality.

8.2.6 Justifying against different rationalities

The fifth process of sensemaking deals with the justification of the decision at the end of the process against the different rationalities that are present during selection process for an architect. A client has to justify their final decision to their own organisation, to the public, to society, and to the architects that joined the tender. These multiple responsibilities are described by Bovens (1990): the many hands make it difficult to identify one single person responsible for a tender decision, as also explained in the book of Hodgkinson and Starbuck (2008b). In justifying a decision a decision maker is simultaneously confronted with the legal structure of the decision procedure and the psychological decision process of sensemaking. The situational characteristics of support, trust and control were found to be of influence to the process of justifying against different rationalities.

Although previous results indicate that experts indeed perform more consistently on value judgements in general (Hekkert & van Wieringen, 1996), the question remains who should be considered an expert. On a holistic level most of the decision makers in the cases appeared reasonably capable of making decisions, even if they felt insecure about their decision tasks. They just followed their intuition, which can be considered as an appropriate strategy for ill-defined decisions in situations with a lack of information (Dane & Pratt, 2007). The need for control and understanding of the situation is strongly felt by decision makers in uncertain situations. So exactly in their perceived strength of the decisions - the use of their intuition - a weakness occurred in the need to justify their decisions to their audience.

In the competition tradition juries have shown their ability to express their preferences without raising considerable resistance of the stakeholders. They used a jury report to reflect briefly on all proposals and address the process of decision making. In this situation the credibility of the juror created trust about the decision. Expertise is however often limited to domains. A building project in a public context comprises of several domains, which means a plethora of areas of expertise. The involvement of external advisory experts changes the power balance and culture within an organisation or team (Kieser & Wellstein, 2008). It also increases the difficulty of explaining a decision and therefore the transparency of a tender decision. The cases confirm the trend that jury panels do not only consist of architectural professionals but often include numerous stakeholders with different backgrounds. Stakeholders can be considered as experts on their specific field but might need support on reading and expressing their preferences in a language akin to that of acknowledged experts in the field of architecture. Without strategic aims and suitable means stakeholder involvement could merely increase uncertainty during the decision process and decrease the support of a decision. Therefore decision makers need to be selected based on their competences, or being educated in performing their tasks, which is in line with Factor 13 about carefully addressing the roles and responsibilities of the decision makers and the level of trust among the stakeholders.

Allocating services for large architectural design projects has traditionally been done through ideas competitions, (invited) project competitions, or private arrangements. Main characteristic of publicly competing for design jobs is that the 'award decision' is taken by a panel of experts from the professional field on the basis of an anonymous visual representation of the future building. This panel of experts was allocated a certain amount of trust and power to fulfil the needs of the participants, the profession and of society for a justification of their decision. Their domain specific expertise enables jury members to match the needs of a client to the potential qualities of the proposals on an abstract level within a limited time and information frame (see also section 8.2.4). This was confirmed in the Faculty Building case where a respected panel of experts evaluated 366 ideas in two days time. The data however also suggest that expert judgement does not guarantee that value judgements are not influenced by personal interests, strategic issues, or (ir)relevant emotions. In all cases the more experienced jury members seemed more aware of the voting behaviour and preferences of other members than the less experienced members. These kinds of strategic issues were confirmed in the interviews with the jury members of the Faculty Building case, mainly by the designers who also were partners in architectural firms and often acted as jury members. In this context, they talked about the balance between clients and professionals in a jury and the right moment to express a preference in the group to convince others. The results also show that expertise contributed to the fact that long term interests appeared to be more important than personal differences, which probably would benefit the quality of the decision. In an expert team members can trust the abilities and intentions of their follow team members (Salas, et al., 2004).

The results of the cases indicate that jury panels in Dutch architect selections are not composed nor fully act as expert teams. Yet, especially in the critical feedback and learning curve of jury members and the issue of building trust the concept of expert team has a lot of offer in the context of design tenders. Based on these insights and in line with Factor 14 I state that the performance of jury panels could be improved by carefully addressing the roles and composition of the jury members in the tender design and training the jury panel during the preparation of the tender. This would make it possible for a public commissioning body to control the decision process.

Procurement law offers current clients a new kind of structure that can be used to justify their decisions but not require clients to motivate their decision content wise. Compared to the design tradition the legal structure is rather rationalistic and quantitative, but these kinds of decision motivations are still preferred from a societal perspective (Sinclair & Ashkanasy, 2005). Besides decisions about design quality does not only include an assessment about the proposal, but also the possible implications of a decision, the emotions that are felt and the justification process to the citizens, employees and other stakeholders. A quantitative decision motivation however fails to include the 'story' behind the value judgements and the process of sensemaking that a client went through to reach their decision about design quality. The numbers do not carry any message about the conflicting aims and possibilities. This creates a lack of trust in the client body as a commissioner. Without trust a decision does not find support among the stakeholders. In the cases of the City Hall and the School only a limited motivation was provided to the tenderers about their performances. The clients of the tender cases did not even think of writing a jury report such as provided in the case of the Faculty Building. In this they appeared to have ignored the competition tradition. The School case showed how the structure of a decision support tool determined the motivation for the final decision. This improved the transparency of the decision process to some extent. However, the final decision was based on a discussion in which the output of the system played a role but did not ground the decision. Therefore the matrix sheet did not truly justify the decision to the architect about their offer. This confirms Factor 15.

Based on the results of this research in relation to the process of justifying a decision against different rationalities I conclude that the following success factors will contribute to a successful architect selection process:

13. Addressing the roles and responsibilities of the decision makers cautiously in the design of the tender for increasing the trust between the decision makers and broadening the support for the decision among the stakeholders.

14. Composing, training and guiding the jury panel carefully to benefit optimally from the expertise available in the panel.

15. Using a decision support system to structure the decision process but not using the output of this system to merely justify a decision to the tenderers and the other stakeholders.

8.3 Recommendations

In this section several recommendations are made for a tender design for the process of architect selection under European tendering regulations. The recommendations are based on the insights and success factors of the theoretical framework, the results of the empirical cases, and the findings of the validation workshop. The first section addresses the recommendations that directly relate to the sensemaking processes. In the second part of this section I make a suggestion about a different interpretation of the current regulations that could change the contemporary Dutch practice.

8.3.1 Recommendations for the design of a tender

In line with the empirical results the participants of the validation workshop acknowledged that the preparation and design process of a tender might be even more important than the actual tender process. Figure 8.1 shows three incremental steps that a clients needs to take in the preparation of the tender. These interrelated and iterative activities should results in a tender design. When the tender is made public, the implementation process of the tender has started.

In a different role than currently applied in Dutch tender practice, architects could also help in the process of analysing the aims and ambitions of the client and making estimations about market situations in architecture. In the process of matching supply and demand, architects or specialised consultants could support clients in estimating the effects of certain requirements, such as the kind of reference projects or the heights of the financial requirements in the selection phase. The results of this research showed that the success factors need to be carefully considered in the preparation phase and guarded in the implementation phase of the tender. The importance of interaction with the tenderers and discussion among the jury members could require a shift in the current interpretation of objectivity, equal treatment, and transparency. The aggregation process of the different value judgements that are made during a selection process is mainly grounded on making valid assessment of characteristics of the proposal that is offered by the architect.

Assessing design quality and the affective response that accompany judgements about design quality mean that a decision will never be objective by nature. In justifying the final decision expectations have to be fulfilled from different rationalities. The recommendations that relate to training, alignment of decision frames, and education imply that decision makers would have to be more actively involved in the preparations of the tender and invest in specific training activities. This would require a change of the current practice because of possible conflicts between the relatively busy schedules of the experts and responsible officers that are currently involved.

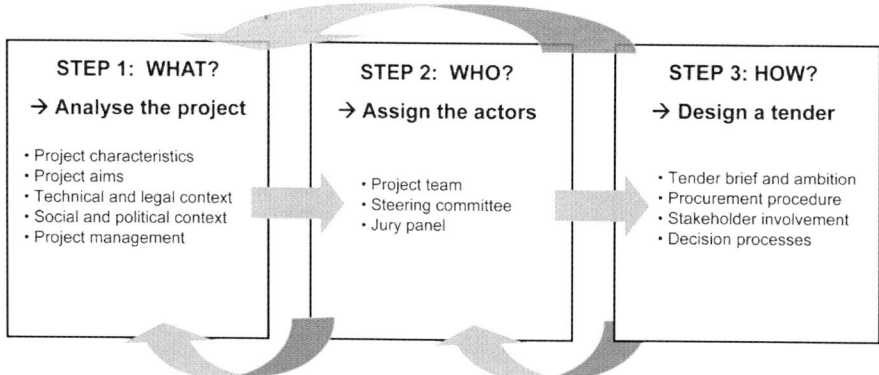

Figure 8.1 Three incremental steps of tender design in the preparation phase

The results of this study lead to the following overview of fifteen recommendations for the selection of architects by public clients under European tendering regulations:

1. Allowing for a holistic judgement in the tender design that incorporates potentially conflicting judgements.

2. Addressing the characteristics of the architect as a person, the proposed design, as well as the firm that they represent in the decision criteria of the tender design.

3. Ensuring a fit between the aims of the selection process and the design of the tender.

4. Ensuring a fit between the position and type of the stakeholders and their role in the decision process.

5. Aligning the frame of references of the actors during preparation of the tender process in order to reach a decision at the end of the process that fits the ambitions and aims of the client organisation.

6. Aligning the type of expertise needed for the various decision tasks during the selection process of an architect with the nature and content of the decision task.

7. Educating decision makers in reading architectural designs from a client perspective.

8. Involving (external) experts in the process of decision making about design quality because they have domain specific expertise and are better at controlling product emotions and using intuition than novices.

9. Including a personal explanation by the architect to improve the clients' understanding of the proposal.

10. Creating room and flexibility in the decision making process for discussion and negotiation among the decision makers.

11. Making the assessment structure and aggregation system of value judgements explicit in the preparation process.

12. Allowing for compensation in aggregating value judgements about design quality.

13. Addressing the roles and responsibilities of the decision makers cautiously in the design of the tender to increase the trust between the decision makers and broaden the support for the decision among the stakeholders.

14. Composing, training and guiding the jury panel carefully to benefit optimally from the expertise available in the panel.

15. Using a decision support system to structure and support the decision process but not using the output of this system to merely justify a decision to the tenderers and the other stakeholders.

8.3.2 Suggestion for change in Dutch tender practice

The results of this research give reason to suggest a change of the current implications of the tender regulations in the Netherlands. In my opinion the composition of the jury panel should be the same in the selection and award phase, the jury should have decisive rights, and the roles and responsibilities of the jury members should differ per phase of the tender. This suggestion is based on the principle that clients should let go of aspects they are not capable of and control the issues that can be prepared and easily supported. The concept is based on the shortlisting principle that the Dutch Chief Government Architect used until a few years ago to assist the Ministries in selecting architects. In general it was found that clients find it hard to select a minimum of five architects that will be capable to fulfil the requirements for the job without ensuring themselves of potential quality by setting unbalanced suitability requirements. Clients appear to lack domain specific knowledge about architecture as well as domain specific knowledge about procurement law, competitions and project management in this area. Yet, the use of an independent expert panel, as is common in the competition tradition, limits the amount of control for a client over the final decision. Therefore clients prefer to not to use jury panels and selection and award committees currently do not consist of a majority of experts. Still domain specific expertise is needed to make judgements about architectural design quality in the context of EU procurement law and to create trust between the architect and the client.

I think that for any tender in which an architect is selected the following procedure should be applied:

A. Assign a diverse jury panel that includes the responsible officer(s) and other representatives of the public commissioning body as well as experts in specific domains that relate to the assignment (e.g. urban planners, architects, sustainability experts, historians etc).

B. Assign decisive rights to the jury panel in both the selection and the award phase.

C. Ensure that jury members trust and support each other before, during, and after the tender. This process could be supported by determining roles and responsibilities among the jury members beforehand for the different phases.

The benefits of assigning a multidisciplinary jury decisive rights in both the selection phase and the award phase are that 1) the potential tenderers can be quality checked by the domain specific experts in the selection phase, and 2) clients can still influence the final decision in the award phase but with the support of domain specific experts. The experts that are involved in the process could monitor the effects of tendering decisions for the professional field. They can agree a certain code of conduct among each other that would increase equality of chances for young and relatively less established firms. Especially during the selection phase the amount of information that needs to be processed is considerable. Domain specific experts will be better at assessing the requests for invitation better than non professional clients or their support staff, even if an open procedure or restricted tender procedure with relatively low technical, financial or organisation suitability requirements would raise the number of potential tender candidates. To align the documentation to the character of the expert judgements, the format of the requests for invitation might need adjustments, for example by adding more visual material, possible on a poster format as is customary in the design competition tradition.

If clients would focus more on the technical and/or professional ability of the candidates in the selection process and less on the financial and organisational competences, unrealised designs could also be used as references. The experts would bear the responsibility to protect the interests of the professional field in the selection phase. By announcing the names of the panel members, potential candidates would know what to expect from their fellow professionals during the decision process. This suggestion meets the principle of transparency, equal treatment, proportionality, and objectivity for both the tender principles and the competition tradition.

In the award phase clients have to select one winner out of a minimum of five comparable tenderers that have been selected by the experts in the selection phase. Clients and representatives of the commissioning body should be included in the jury panel in the sense that 'the one who pays the piper can call the tune' in the final decision phase. I suggest keeping the composition of the jury panel similar to the selection phase but switching the roles and responsibilities between the domain specific experts. The experts can assist the clients in choosing between the alternatives and writing a report to motivate the award decision in the language of the architects. The clients benefit from a learning process within the tender process and will be able to prepare themselves on their decision task in the award phase. The experts could also support other stakeholders that are involved in the decision process in reading and assessing the proposals of the tenderers.

This procedure would benefit from a protected and possible publicly administrated database or other kind of reference system in which clients can check the financial, technical and organisational information (a kind of digital 'passport') if required during a tender. This would substantially decrease the administrative workload of tender candidates and commissioning bodies. By submitting their request for invitation tender candidates should automatically allow clients to obtain a certificate of their suitability that is based on the requirements for that specific tender. Such a database system would however mean that clients and architects have to regain trust in each other.

The suggested change in the implementation of current tender regulations would require a substantially higher involvement of the domain experts and other panel members in the preparation phase of the tender. This could lead to a higher demand for experts need to be educated and trained. Being a jury member could even become a profession or dedicated task for some experts. In my opinion the educational and training activities could be offered by an independent multidisciplinary knowledge centre that focuses on selecting design services, including development competitions and integrated and public-private partnership projects. In this knowledge centre professionals from the domains of architecture, procurement, project management, and process management should collaborate to provide the services and facilities that public clients need to select architects under European tendering regulations.

8.4 Reflection on the research approach

8.4.1 Research tradition

Pragmatism is considered an important characteristic of research in organisational and naturalistic decision making (Gore, et al., 2006). Because of the project-based character of the decision processes studied, this study can be positioned between organisational decision making "mainly looking at social processes which are heavily constrained by organisational goals and norms" (Gore, et al., 2006, p. 929), and naturalistic decision making, which "reflects on cognitive processes to make decision makers more effective" (Lipshitz, et al., 2006, p. 918). The aim of the research is to enable the actors within an organisation (not only the experts) to apply the results by offering a practical understanding of decision making in the context of architect selections. This research responds to the need for interdisciplinary approaches to apply existing knowledge from the more traditional fields of science, such as cognitive and social psychology, to the field of architecture. However, this does not mean that full understanding has been reached. The topic has been approached here from a social psychological rationality, which leaves out the purely legal, technical and economical rationality (Snellen, 1987). The areas of strategic management, project management, organisation science, organisational anthropology, sociology, public administration, and business administration have mainly been left out in the development of the theoretical frame for pragmatic reasons. The final results suggest that during my analysis the influence of power, politics and strategic decision making has not yet been fully explored. Further analysis of the data and additional data collection could provide more insights on these impact factors and the underlying mechanisms. The same applies for issues like the aesthetics and value of design or the concepts of subjectivity, transparency, and integrity, which all could be further elaborated on a theoretical level.

8.4.2 Research methods

The research is entirely based on real life case studies. The method of case study research is part of an ongoing debate in science about the rigour of qualitative research data in terms of reliability, generalisability and validity of the results. Several scholars attempt to create awareness among 'traditional' researchers about the relevance of doing case studies (e.g. Flyvbjerg, 2004; Gerring, 2004; Yin, 2009). Although the importance of laboratory experiments, surveys and simulations was acknowledged during the design of this research, the use of case studies proved to be the most appropriate approach for the explorative and sensitive nature of this research theme. Triangulation of the different research methods strengthened the analysis of complexities underlying the behaviour as shown by the actors. According to Lipshitz, Klein, Orasanu and Salas (2001, p. 343) "field observations are critical to Naturalistic Decision Making (NDM) research because real-world decisions are embedded in and contribute to ongoing tasks". They suggest evaluating studies in naturalistic decision making in terms of credibility and transferability. In this section I discuss the credibility of the research; the transferability is discussed in section 8.5.4 in terms of the generalisation of the research results.

In terms of credibility this study shows that observations open up a possibility to collect scientific insights that would - although commonly known in the field - otherwise be neglected, such as emotions, strategic behaviour and body language. Interviews provided an informative source for interpretation of strategic actions and future related contextual factors influencing the decision processes. Participatory research enhanced both these benefits, but created a different kind of complexity during analysis of the data due to an initial lack of distance from the data. The systematic and narrative approach of evaluating cases offers both scientific and practical benefits for the profession. The observation that public clients appear not to be as public as the term would suggest can be traced back to the uncertainties and lack of clarity about the assessment of design quality, which in the perception of clients conflicts with legal and rational expectations. Public clients already feel pray for the architectural community and they do not want to enlarge their vulnerability. The clients that were willing to participate in this research aimed at fulfilling the principle of transparency to the greatest degree possible. This could imply that other clients might have more to hide. Granting me access as a researcher to decision processes required the same kind of trust and ambition that is expected from participants in a tender. It is also the same kind of trust and ambition needed to act as a professional client during the selection of an architect in the context of European procurement regulations. Hopefully this study will contribute to opening up the mist covering the selection process for architects and create a kind of openness and awareness about the characteristics of assessment processes in design. Public clients who find themselves in a similar situation may feel strengthened by reading about the experiences of others and the scientific explanation of the phenomena that can occur.

8.4.3 Case selection

In answering the research questions I used four different cases: three tender cases and one competition case. The organisation of an ideas competition is not the same as the design of a procurement process to select an architect. However, the combination of several tender cases and an ideas competition appeared to be quite successful for the purpose of this research. The main differences between the Faculty Building case and the restricted tender cases can be related to differences in the legal impact of the decision, the aim of the decision, and the expertise level of the client.

The commissioning body in the case of the Faculty Building can be considered as a very knowledgeable and considerable renowned client. This created several opportunities for collaboration and publicity, such as the exhibition in the Dutch Architecture Institute and the involvement of the Dutch Minister of Education, Culture and Science at the opening of the competition. The future users of the building could be more easily involved in the assessment process and relevance of the competition was generally acknowledged among the employees and students. The downside of this media attention is that the client had to deal with this exemplary function that increased the amount of pressure on the selection process. Yet I assume that this kind of pressure can be compared to the pressure that is felt by public commissioning bodies during a tender process for any public building.

In the Faculty Building case no contract was awarded and only a relatively small amount of money was allocated. Therefore no legal obligations were made and due to the smaller impact of decisions, less pressure from outside might have been felt. This lack of legal obligations can be considered as a weakness of this case but is probably also the main advantage. Because the lower impact of the legal dimension on the behavioural dimension in the competition case, the actors were permitted to follow their professional tradition, to act freely at their own discretion, and to let me in as a researcher. The results suggest that some actors and participants in the competition case might have put less effort into the competition because of the limited impact of the consequences of their actions. The aftermath of the fire caused a certain kind of haste and dynamism that normally would not or in a lesser matter occur. In my opinion the results of this research show that a limited time frame does not automatically lead to poor decision quality. In current practice some clients start a tender procedure while they might have better organised a competition or similar trajectory first. The official aim of the Faculty Building case was not to select an architect for a future building but to collect ideas, to stimulate debate, and to inspire young talent. The final decision about the distribution of the prize money reflected these aims in that a single proposal but multiple winners were chosen. However, reaching a consensus about several winners might be easier than selecting a particular one. This could be a topic for future research.

8.4.4 Generalisation of the results

All of the case studies in this research addressed architectural design but were taken from different contexts. While the situation was very specific, the clash of rationalities and the five sensemaking processes that were identified are likely to apply to other situations in organisational decision making as well. In situations where the probable outcome does not meet the common expectations, the need to justify a decision could cause cognitive dissonance. The findings of this research could therefore be compared to other high-stakes strategic decision making settings that allow for deliberation about options, such as decisions about a portfolio of real estate or projects in new product development. In urban developments the discussion between awarding construction works and design services causes the same multi-dimensional conflicts as found in this research. However, application of the findings in other situations would require additional research.

Public clients represent their users, citizens, visitors, and employees. As decision makers they can be held personally accountable for their decisions. Because a decision about the selection of an architect has significant implications for the community, I found that decision makers not only aim for the best alternative but also for the alternative that would receive strongest support from the stakeholders. This consideration required an intensive analysis of meanings and potential consequences for the decision makers. The results show that the responsibilities of public clients were multifaceted and conflicting by nature, which means that there was no correct answer. Decision makers however used this uncertain character of the decision process to follow their intuition and reach an inter-subjective consensus. Such a decision made sense for the all those involved, but was, unless properly explained, sometimes hard to accept by others. The process of sensemaking only involved the decision makers themselves and not the other stakeholders who did not actively take part in the selection process. Procurement law, on the other hand, argues from the concept of fair competition and equal treatment of all market parties. An award decision therefore has to make sense on a level of the common principles of equal treatment, objectivity, transparency and proportionality. An important practical implication of this research is to explore options for how both legal and psychological rationalities can be reconciled in a decision procedure.

The focus of this research lay on judgement and decision making about architectural design quality as it is embedded in Dutch culture. Dutch culture traditionally tends to show great interest in design quality. Government policy regulations and the office of the Chief Government Architect significantly contribute to this development (see also Chapter 4.2.1). The experts in the validation workshop confirmed the overall analysis of decision processes in the Dutch situation. In terms of transferability there is reason to believe that the findings will hold in other settings with comparable elements in other countries because of the international character of architecture. I assume that the results could be useful for other situations in which a transfer from a buyer to a seller or a selection process takes place, such as the procurement of services, works and deliveries in general and the purchase of other kinds of products. More specifically I would like to address some examples of relatively scarce research that relate to the assessments of research grants (e.g. Lamont, 2009; Langfeldt, 2001), the decision process of judges (e.g. Posner, 2008), the evaluation of student work in design education (e.g. Lans & Volker, 2008), the recruitment process at universities (e.g. van den Brink, 2009), and partner selection in business (e.g. Elsbach & Kramer, 2003). In all these cases there was a considerable time difference between the official request for proposals and the evaluation of the alternatives. Because in daily life the legal context of selection and assessment processes are not as strict as in procurement, the invitation frame and the evaluation frame are usually not as explicitly compared to each other as in tender situations. This makes it possible to adjust to processes of increasing insight and sensemaking and react according to the affective response that accompanies a judgement. Whenever something or somebody is assessed, the assessment system should fit the decision processes and the decision task. If there is no agreement on the value of the outcome of the system, the assessment will always be somewhat arbitrary.

8.4.5 Scientific relevance

The built environment concerns us all. Competitions and tenders contribute a lot to debate and education in the architectural profession and beyond. Therefore it is very important to ensure the element of debate in current procurement practice. Disaffection with the current procurement system poisons the relationship between clients, their (legal) advisors and the design profession. Difference of perceptions appears to trap both parties in a very uncomfortable situation. So far no necessity is felt for the parties to change their attitude or behaviour. A tight financial situation could be an impetus for change but might as easily lead to an exaggerated concern for costs at the expense of quality. This research should be considered within the limitations of the Dutch public sector before the financial crisis. A reduced availability of funds will probably cause a decrease of the number of tenders and an increase of the number of candidates interested in each tender. This could mean more requests for participation per tender and thus more information to be assessed in the selection phase of the tender process. More careful decision making could be beneficial from the perspective of efficiency and effectiveness but it could also damage the unique and emotion driven character of the architectural profession. In any situation more professionalism is required from clients.

Apart from money, knowledge, governmental regulations, and intrinsic motivation of decision makers are important drivers for change. Especially local authorities are short on knowledge and specific experience of tendering a design contract. Universities play an important role in providing scientific knowledge and a critical voice. Motivation can be stimulated by (financial) support from governmental bodies or professional associations. There is an explicit need for the establishment of an independent knowledge centre on architect selections and related issues in the construction industry. Such a knowledge centre should build expertise and awareness on pragmatic issues, such as the level of requirement and selection criteria, format of the tender brief or the selection of jury members. It should also connect the different players in the field, stimulate debate, and set out policy to ensure quality improvements in the field. Because of the influence of consultants on decision making, clients, their representatives and externally hired advisors should be actively involved in developing and exchanging knowledge and experiences. This requires a considerable amount of awareness, transparency, openness, and integrity of the actors. The UK provides an example of awareness processes through the obligation to apply the Design Quality Indicator for all public-private partnership for school buildings and through the establishment of the Office of Government Commerce and the Commission for Architecture and the Built Environment. Internships or workshop programmes could give users, designers, advisors, managers, and other stakeholders the opportunity to try out another person's shoes for a period of time and feel the accompanying roles and responsibilities. The BNA has announced that it prefers a quality system for tender procedures in architecture (Architectuur Lokaal, 2009a). Such a 'quality label' would create some grip and certainty for most parties, but could also create further inflexibility of the system and interdependence with an institute overlooking the selection process. In this stage the need for actors to take on their responsibilities to act as critical professionals appears to be more important than the need for another model, system, guideline or method.

In research there is still the ambitious strive for a perfect world in which decisions are 'evidence based' to reduce uncertainty, power struggles, errors and delays (e.g. Morrell, 2008). The idea persists among certain groups of engineers, psychologists and other disciplines that the world would be better place if we could predict and objectify every activity around us. In this context we should note that we are incapable of truly overseeing all consequences of our actions. Also more popular literature acknowledges the complexity of values in decision making. Pirsig (1999) for example states that the dualism between classical (quantitative) and romantic (qualitative) appraisal systems in public administration is just too complex to judge about quality in the relation between subject and object. Pirsig's search for quality ends at a philosophical level. The question is whether we actually aim for a level of science in the field of management and design that enables us to apply justified knowledge and predict behaviour. Law does not appear to aim for that. From the behavioural perspective we could strive for transparent decision making that can be accepted in other rationalities. This study can be considered as a contribution to this kind of transparency.

8.5 Future research

Every researcher is aware of the fact that a research project often raises more questions than it answers. One could say that in science questions are more important than answers, but it is the often answers that provide good starting points for new ideas. The results of this research inspired me to consider research directions that relate to future research in the area of naturalistic and organisational decision making and a more structured approach for collecting data about selection processes in the build environment.

Until now no systematic data collection has taken place on the local or European level about the number of (design) tenders, their participants, winners, applied criteria, et cetera. Such a database would enable the identification of success and failure factors, which could be of major importance for architects as well as for clients. Just recently similar initiatives have started in the Netherlands and other countries (e.g. Manzoni, et al., 2009; Zheng, 2008), but these initiatives are not yet embedded in an international scientific culture. Dutch architects tend to point to other European countries for successful policies in architect selections while the Dutch situation tends to have a positive reputation in other countries. From a legal and/or professional perspective no comparisons have yet been carried out on the differences and similarities between European countries, let alone between the American, Australian, Asian, and other design cultures. The experts in the validation workshop suggested that the architect selection processes in private developments do not seem as problematic as in the public administration. Future research could explore the differences between selection processes in the private and the public sector, between development competitions and design competitions, and between the different procurement procedures. On a managerial level the validity of budgets, planning and quality estimations in the initial phase of a project and the concept of value for money would also benefit from systematic data collection. The current trend of concern for life cycle costs and sustainability offers a great opportunity to link use and realization to initialization and design of building. Maybe this is the time to answer to the calls of numerous researchers in previous publications for systematic post occupancy and project evaluations (e.g. Bordass, 2003; Green & Moss, 1998).

According to Lipshitz et al. (2006, p. 347) the ultimate theoretical challenge for the field of Naturalistic Decision Making is "to specify the link between the nature of the task, person, and environment on the one hand and the various psychological processes and strategies involved in naturalistic decisions on the other". For the selection of architects there are various characteristics of the task, person and environment that can be further explored in decision making, for example by laboratory experiments and simulations. But also from an organisational decision research perspective more attention could be given to the constraints imposed by the context, distributed information and differences in power and interests, as stated by Gore et al. (2006). Organisational decision making in the context of a tender raises important issues about the influence on the decision, such as different power levels of actors, decision methods (such as Delphi, or MCA's), sources of information (verbal, textual and visual), roles and power levels of external advisors, composition of jury panels, and the influence of training and education on the level of expertise in architecture. Power and participation issues could be linked to organisational anthropological or ethnographical studies, or to more social theories such as discourse. One of the most pressing issues in practice at the moment is to find a way that solves the problem of the right amount of reactions to a call for tenders in the selection phase. The current perception is that selection requirements are set too high, but for clients this is a way to limit the number of reactions. By lowering the requirements the number of (young) participants would likely increase, but the chances to be selected decrease per party. The exchange of experience in this matter could provide input for a set of experiments or modelling of the situation.

In relation to product experience it would be very interesting to know more about the dimensions of complexity and arousal in relation to building characteristics that contribute to the wow-factor and other kinds of emotional responses to different stakeholders in the built environment. Additionally the relevance of functionality compared to aesthetics could be analysed during different phases of the building project, including the tender phase. Findings of such studies can have implications for the design of stakeholder involvement, the format of the procedure, and the development of the tender brief, which cannot be derived in detail from the current results. In a future stage these insights could be incorporated in academic and professional education. Experimental or field studies could also explore the similarities of selecting architects with the selection of job candidates, the allocation of funds in research, art or other design related disciplines and the assessment of student work in educational settings. The findings of this study suggest that clients do not seem to be aware of the possibilities and implications of their decisions. Future research could investigate if education and specific support (psychological, legal, or managerial) in the initial stage of a project would change preferences of clients for certain tender procedures, and if educational support would increase the power of the decisions. Action research could be an appropriate technique for this.

The results suggest that the performance of a jury panel would improve if clients applied the concept of the expert team. For the jury to act as an expert team, they would have to built common understanding, formulate plans on the most effective course of action, execute plans by coordinated team performance and learn by evaluation in order to perform well (Burke, et al., 2006). The chair of the jury panel should solicit ideas and observations of team members, stimulate and enable all kind of feedback and give situational updates (Salas, et al., 2004). Jury members need a similar anchor point, which could be provided by the competition brief and regulations and earlier involvement in the design of the tender. In the current situation every client composes a new team of decision makers that is tailored to the assignment. Using continuing teams and more collective preparations might improve the performance of juries. In Scandinavia jury members are significantly more involved in the preparation process of a competition than in other European countries, and they spend more time on assessing the entries and negotiating about a winner (Kazemian & Rönn, 2009). Especially in terms of learning, critical feedback, room for intuition, and reflection on the decision process would benefit from more time spread across several meetings. Future research could also address the implications of involving stakeholders with different levels and areas of expertise. Because the built environment surrounds us all everybody has some experience in making judgements about buildings. The cases in this research suggest that the procedure to select an architect could be adjusted to the level of expertise of individual decision makers. But decisions are often made in groups, and users, designers, clients, and other stakeholders speak rather different 'languages' (Bucciarelli, 2003). Connected to the different perspectives of individual judgements are the issues of communicating preferences and opinions. So far the question about the required level of expertise for decision making remains unanswered. How should decision process and tender procedures take individual differences into account without making the assessments invalid or unnecessary complicated? How can members of a jury panel support each other in decision making? In my opinion the concept of expert teams in procurement situations would open up an interesting field of research for scholars in organisational decision making.

The potential conflict of human behaviour with the general principles in the EU Directive could be further explored from a legal perspective. It might also be interesting to analyse the process of decision making of judges when they deal with procurement cases. On a more general level it would be interesting to monitor the effect of legislation on preferences for certain procedures or award mechanisms. This might test Winston Churchill's assertion that "we shape our buildings and afterwards our buildings shape us", which could also refer to the shaping of law. Relevance of such research can also be found in the context of innovation and stimulation of young talent in design and the effect of procurement on the national culture. In the same line of reasoning is the discussion about the use of decision support systems in relation to transparency and legitimization of the process. In traditional design competitions it is accepted that the names of the jury members are published beforehand rather than an exact procedure and predefined criteria of decision making. In case of an experienced and well-known jury, the names actually provide a hint for participating architects to decide if they stand a chance in the competition. Traditionally participants trust the jury members in making sound decisions and a consistent jury report. Clients currently hesitate to write a report about their findings. Instead they use the output of a decision support system to justify the decision. Future research will have to investigate the influence of decision support systems on transparency and motivation of a decision. The recent increase of the number of lawsuits indicates a more juristic approach of dealing with disagreements about the decision as made by the contracting authorities in the public sector. More research on the motives of decision makers in relation to the legal perspective could contribute to preventing a further polarization of parties that are involved in creating a built environment that benefits all of us.

References

Allinson, K. (1997). *Getting there by design; An architects' guide to design and project management*. Oxford: Architectural Press.

Allport, G., Vernon, P., & Lindsey, G. (1960). *A study of Values*. New York: Houghton Mifflin.

Amabile, T. M. (1996). *Creativity in Context - update to The social Psychology of Creativity*. Boulder: Westview Press.

Amabile, T. M. (1997). Motivating Creativity in Organizations. *California Management Review*, 42-52.

Architectuur Lokaal (2009a). *EU Aanbestedingendag - publieke opdrachten architectuur & projectontwikkeling (EU tender day - public contracts in architecture and project development)*. Amsterdam: Architectuur Lokaal.

Architectuur Lokaal (2009b). Kompas Light First edition. Retrieved 14 December 2009, from www.architectuuropdrachten.nl.

Arkes, H. R. (1989). Principles in Judgment/Decision Making Research Pertinent to Legal Proceedings. *Behavioral Sciences and the Law, 7*(4), 429-456.

Arkesteijn, M., den Heijer, A., Vande Putte, H., & Volker, L. (2009). Think tank - envisioning the 'faculty of the future'. In TU Delft - faculty of Architecture (Ed.), *Building for Bouwkunde - Open to Ideas* (pp. 64-71). Delft: TU Delft - faculty of Architecture.

Arnstein, S. R. (1969). A ladder of citizen participation. *Journal of the American Institute of Planners, 35*(4), 216-224.

Arrowsmith, S. (2005). *The law of public and utilities procurement* (second ed.). London: Sweet & Maxwell.

Atelier Kempe Thill (2008). *Naar een Nieuwe Aanbestedingscultuur - Europees Aanbesteden van Architectendiensten in Nederland (To a new tender culture in the Netherlands)*. Rotterdam: Atelier Kempe Thill Architects and Planners.

Atelier Rijksbouwmeester (2008). Chief Government Architect Retrieved 22 January 2010, from www.rijksbouwmeester.nl.

Balogun, J , Pye, A., & Hodgkinson, G. P. (2008). Cognitively skilled organizational decision making: making sense of deciding. In G. Hodgkinson & W. H. Starbuck (Eds), *The Oxford Handbook of Organizational Decision Making* (pp. 233-249). New York: Oxford University Press.

Bártolo, H. M. G. (2002). *Value by Design: a qualitative approach*. Paper presented at the Value through Design, Reading.

Bayes, T. (1763). An essay towards solving a problem in the doctrine of chances. *Philosophical Transactions* (53), 370-418.

Beach, L. R. (1990). *Image theory: Decision making in personal and organizational contexts*. Chichester: Wiley.

Beach, L. R., & Connolly, T. (2005). *The psychology of decision making*. Thousand Oaks: Sage Publishers, Inc.

Bechtel, R., & Zeisel, J. (1987). Observation: The world under a glass. In R. Bechtel, R. Marans & W. Michelson (Eds.), *Introduction: Environmental Design Research*. New York: Van Nostrand Reinhold Company.

Behn, R. D., & Vaupel, J. W. (1982). *Quick analysis for busy decision makers.* New York: Basic Books.

Bell, P. A., Greene, T. C., Fisher, J. D., & Baum, A. (1996). *Environmental Psychology* (4th ed.). Fort Worth: Harcourt Brace College Publishers.

Benedikt, M. (2007). Introduction. In W. S. Saunders (Ed.), *Judging architectural value*. Minnesota: University of Minnesota Press.

Bernoulli, D. (1738). Specimen theoriae novae de mensura sortis. *comentarii academia scieniarum imperiales petrolitanae* (5), 175-192.

Best, R., & de Valence, G. (1999). *Building in Value - Pre-Design Values*. London: Arnold.

Betsch, T. (2005). Preference Theory: An Affect-Based Approach to Recurrent Decision Making. In T. H. Betsch, S. (Ed.), *The Routines of Decision Making* (pp. 39-66). Mahwah, NJ: Lawrence Erlbaum Associates.

Bittermann, M. S. (2009). *Intelligent Design Objects (IDO) - A cognitive approach for performance based design.* Delft University of Technology, Delft.

Blau, J. (1984). *Architects and Firms*. Cambridge, MA: The MIT Press.

BNA (2009). Retrieved 26 November 2009 from www.bna.nl.

Bockma, H. (2009). Nieuwe architect is sociaal betrokken en denkt organisch (New architect is socially involved and thinks organically). *De Volkskrant,* p. 15. Retrieved 16 March 2009, from www.volkskrant.nl.

Boland, R. J. j., & Collopy, F. (2004). *Managing as Designing*. Stanford: Stanford University Press.

Bordass, B. (2003). Learning more from our buildings - or just forgetting less? *Building Research & Information,* 31(5), 406-411.

Boudewijn, E. C., & Broekhuizen, R. P. V. (2007). *Bouwen is teamwork - praktijk gids voor succesvol samenwerken in de bouw (Building is teamwork)*. Gouda: Regieraad Bouw & PSIBouw.

Boudewijn, E. C., & Broekhuizen, R. P. V. (2009). *Glashelder bouwen - Dilemma's en transparantie tijdens het bouwproces (Cristal Clear Construction - Dilemmas and transparency during the construction process)*. Gouda: Regieraad Bouw.

Bovens, M. A. P. (1990). *Verantwoordelijkheid en organisatie. Beschouwingen over aansprakelijkheid, institutioneel burgerschap en ambtelijke ongehoorzaamheid. (Responsibility and organisation)*. Zwolle: W.E.J. Tjeenk Willink.

Bowers, K. S., Regehr, G., Balthazard, C., & Parker, K. (1990). Intuition in the Context of Discovery. *Cognitive Psychology,* 22, 72-110.

Boztepe, S. (2007a). Toward a framework of product development for global markets: a user-value-based approach. *Design Studies,* 28(5), 513-533.

Boztepe, S. (2007b). User Value: Competing Theories and Models. *International Journal of Design,* 1(2), 57-65.

BRE Global (2009). Retrieved from http://www.breeam.org.

Brenders, F. (2008). Vitruvius, de Architectural Libri X Retrieved 11 March 2008 from http://www.vitruvius.be.

Brenner, P., Tanner, C. A., & Chesla, C. A. (1996). *Expertise in Nursing Practice: Caring, Clinical judgement, and Ethics*. New York: Springer.

Brown, G., & Gifford, R. (2001). Architects predict lay evaluations of large contemporary buildings: whose conceptual properties? *Journal of Environmental Psychology*, 21(1), 93-99.

Brunsson, N. (1989). *The Organization of Hypocracy: Talk, Decisions and Actions in Organizations*. Chichester UK: Wiley.

Brunsson, N. (2007). *The Consequences of Decision-Making*. Oxford: Oxford University Press.

Brunswik, E. (1947). *Systematic and representative design of psychological experiments, with results in physical and social perception*. Berkeley CA: University of California Press.

Bryner, G. C. (2007). Public Organizations and Public Policies. In B. G. Peters & J. Pierre (Eds.), *Handbook of Public Administration* (pp. 189-198). London: Sage Publications.

Bucciarelli, L. (1994). *Designers Engineering*. Cambridge: MIT Press.

Bucciarelli, L. (2003). *Engineering Philosophy*. Delft: DUP.

Buijs, A. (2009). *Public Natures - Social representation of nature and local practices*. Unpublished PhD, Wageningen University, Wageningen.

Bunderson, J. (2003). Recognizing Utilizing Expertise in Work Groups: A Stats Characteristics Perspective. *Administrative Science Quarterly*, 48, 557-591.

Burke, C. S., Stagl, K., Salas, E., Pierce, L., & Kendall, D. (2006). Understanding team adaptation: a conceptual analysis and model. *Journal of Applied Psychology*, 91(6), 1189-1207.

Burke, L. A., & Miller, M. K. (1999). Taking the Mystery Out of Intuitive Decision-Making. *Academy of Management Executive*, 13(4), 91-99.

Cairns, G., & Beech, N. (1999). User involvement in organisational decision making. *Management Decision*, 37(1), 14-23.

Canter, D. (1977). *The psychology of place*. London: Architectural Press.

Cardellino, P., Leiringer, R., & Clements-Croome, D. (2009). Exploring the Role of Design Quality in the Building Schools for the Future Programme. *Architectural Engineering and Design Management*, 5, 249-262.

Chaiken, S., & Trope, Y. (1999). *Dual-Process Theories in Social Psychology*. New York: Guilford Press.

Chao-Duivis, M. A. B. (2008). *Quickscan contactmomenten in aanbestedingsprocedures (Quick scan interaction moments in tender procedures)*. Den Haag: Instituut voor Bouwrecht.

Chao-Duivis, M. A. B. (2009). Capita selecta uit de geschiedenis van het privaatrechtelijk bouwrecht (Examples from the history of private building law). In R. W. M. Kluitenberg (Ed.), *40 jaar Instituut voor Bouwrecht* (pp. 9-57). Den Haag: Instituut voor Bouwrecht.

Chao-Duivis, M. A. B., Koning, R., Spekkink, D., & Sauerwein, L. (2007). *BNA Werkmap Europees Aanbesteden voor Opdrachtgevers (Guidelines European Tendering for Clients)*. Amsterdam: BNA.

Chase, W. G., & Simon, H. A. (1973). Perception in chess. *Cognitive Psychology,* 4, 55-81.

Cheung, F. K. T., Kuen, J. L. F., & Skitmore, M. (2002). Multi-criteria evaluation model for the selection of architectural consultants. *Construction Management and Economics,* 20, 569-580.

Christianson, M. K., Farkas, M. T., Sutcliffe, K. M., & Weick, K. E. (2009). Learning Through Rare Events: Significant Interruptions at the Baltimore & Ohio Railroad Museum. *Organization Science,* 20(5), 846–860.

City of Amsterdam (2009). Retrieved 9 May 2009 from www.iamsterdam.com.

Clark, T. (1995). *Managing Consultants - Consultancy as the Management of Impressions*. Buckingham: Open University Press.

Clements-Croome, D. (2005). Designing the interior environment for people. *Architectural Engineering and Design Management,* 1(1), 45-55.

Cohen, M. D., March, J. G., & Olson, J. P. (1972). The Garbage Can Model of Organizational Choice. *Administrative Science Quarterly* (17), 1-25.

Cohen, M. S., Freeman, J. T., & Wolf, S. (1996). Meta-Recognition in Time Stresses Decision Making. *Human Factors,* 38(2), 206-219.

Cold, B. (1993). Quality in Architecture. In B. Farmer & H. Louw (Eds.), *Companion to contemporary architectural thought*. London: Routledge.

Collins, P. (1971). *Architectural Judgement*. Montreal: McGill-Queen's University Press.

Collyer, G. S. (2004). *Competing Globally in Architecture Competitions*. Chichester, England: Wiley Academic.

Construction Industry Council (2009). Retrieved 22 November 2009, from http://www.dqi.org.uk.

Coxe, W., Hartung, N. F., Hochberg, H., Lewis, B. J., Maister, D. H., Mattox, R. F., et al. (1987). *Success strategies for design professionals - superpositioning for Architecture & Engineering Firms*. New York: McGraw-Hill Book Company.

Creswell, J. W. (1994). *Research design; qualitative and quantitative approaches*. Thousand Oaks: Sage.

Crosby, N., Kelley, J. M., & Schaefer, P. (1986). Citizens Panels: A New Approach to Citizen Participation. *Public Administration Review,* 46(2), 170-178.

CROW, & Balance and Results (2009). Decision Support System bij het Leidraad Aanbesteden, retrieved from www.leidraadaanbesteden.nl.

Cyert, R. M., & March, J. G. (1963). *A behavioral theory of the firm*. Englewood Cliffs NJ: Prentice-Hall.

Daake, D., Dawley, D. D., & Anthony, W. P. (2004). Formal Data Use in Strategic Planning: An organizational field experiment. *Journal of Managerial Issue,* 16(2), 232-247.

Damasio, A. R. (1994). *Descartes' Errors: Emotion, Reason and Human Brain.* New York: Harper Collins.

Damasio, A. R. (1999). *The Feelings of What Happens: Body, Emotion and the Making of Consciousness.* London: Vintage.

Dane, E., & Pratt, M. (2007). Exploring intuition and its role in managerial decision making. *Academy of Management Review,* 32(1), 33-54.

Davies, M., Stankov, L., & Roberts, R. D. (1998). Emotional Intelligence: In Search of an Elusive Construct. *Journal of Personality and Social Psychology,* 75(4), 998-1015.

Day, E., & Barksdale, H. C. (1992). How Firms Select Professional Services. *Industrial Marketing Management,* 21(2), 85-91.

de Groot, A. D. (1946). *Thought and Choice in Chess.* New York: Mouton.

de Haan, H., & Haagsma, I. (1988). *Architecten als Rivalen (Architects as Rivals).* Naarden: Meulenhoff/Landshoff.

de Jong, T. M., & van der Voordt, D. J. M. (Eds.). (2002). *Ways to study and research architectural, urban and technical design.* Delft: Delft University Press.

de Jonge, J. M. (2009). *Landscape Architecture between Politics and Science - An integrative perspective on landscape planning and design in the network society.* Wageningen: Uitgeverij Blauwdruk.

de Keyzer, W. (1998). *Meten gewikt & gewogen - een humoristische en kritische kijk op meten en het verwerken van meetresultaten (Measurement weighed up).* Brussel: Ministerie van de Vlaamse Gemeenschap.

Dean, J. W. J., & Sharfman, M. P. (1993). The Relationship between Procedural Rationality and Political Behavior in Strategic Decision Making. *Decision Sciences,* 24(6), 1069-1083.

Desmet, P. (2002). *Designing emotions.* Unpublished PhD, Delft University of Technology, Delft.

Desmet, P., & Hekkert, P. (2007). Framework of Product Experience. *International Journal of Design,* 1(1), 57-66.

Desmet, P., Porcelijn, R., & van Dijk, M. B. (2007). Emotional design; application of a research-based design approach. *Knowledge Technology and Policy,* 20(3), 141-155.

Devine-Wright, H., Thomson, D. S., & Austin, S. A. (2003). *Matching values and value in construction and design.* Paper presented at the Crossing Boundaries - The Value of Interdisciplinary Research, Proceedings of the Third Conference of the EPUK (Environmental Psychology in the UK) Network.

Devlin, K., & Nasar, J. L. (1989). The beauty and the beast: some preliminary comparisons of 'high' versus 'popular' residential architecture and public versus architect judgments of same. *Journal of Environmental Psychology,* 9(4), 333-344.

Dewulf, G., & van Meel, J. (2004). Sense and nonsense of measuring design quality. *Building Research & Information,* 32(3), 247-250.

Dietz, T., & Stern, P. C. (Eds.). (2008). *Panel on Public Participation in Environmental Assessment and Decision Making*. Washington, D.C.: The National Academies Press.

Dijksterhuis, A. (2007). *Het slimme onbewuste - Denken met gevoel (The smart unconsciousness, thinking with a feeling)*. Amsterdam: Uitgeverij Bert Bakker.

Dijksterhuis, A., Bos, M. W., Nordgreen, M. F., & Van Baaren, R. B. (2006). On Making the Right Choice: The Deliberation Without Attention Effect. *Science*, 311, 1005-1007.

Dijkstra, T. (2001). *Architectonische kwaliteit (Architectural Quality)*. Rotterdam: 010.

Dijkstra, T., Rijksgebouwendienst, & Ministerie van VROM (1985). *Architectonische kwaliteit; een notitie over architectuurbeleid (Architectural Quality - a memo about the policy)*. 's-Gravenhage: Rijksgebouwendienst.

Doree, A. (1996). *Gemeentelijk aanbesteden - een onderzoek naar de samenwerking tussen diensten, gemeentewerken en aannemers in de grond-, weg- en waterbouw (Tendering at municipalities)*. Unpublished PhD, University of Twente, Enschede.

Dreschler, M. (2008). *Analysis of price correction award mechanisms applied in the Dutch construction industry*. Paper presented at the 3rd International Public Procurement Conference, Amsterdam.

Dreschler, M. (2009). *Fair competition - How to apply the 'Economically Most Advantageous Tender' (EMAT) award mechanism in the Dutch construction industry*. Unpublished PhD, Delft University of Technology, Delft.

Dreschler, M., Beheshti, R. M., & de Ridder, H. A. J. (2005). *An analysis of Value Determination in the Building and Construction Industry*. Paper presented at the CIB Helsinki Joint Symposium, Helsinki.

Dreyfus, H. L., & Dreyfus, S. E. (1986). *Mind over Machine*. New York: Free Press.

Edelenbos, J., & Klijn, E.-H. (2005). Managing Stakeholder Involvement in Decision Making: A comprative Analysis of Six Interactive Processes in the Netherlands. *Journal of Public Administration Research and Theory*, 16, 417-446.

Edmondson, A. C., & McManus, S. E. (2007). Methodological fit in organizational field research *Academy of Management Review*, 32(4), 1155–1179.

Edwards, W. (1961). Behavioral Decision Theory. *Annual Review of Psychology* (12), 473-498.

Eisenhardt, K. M. (1989). Building Theories from Case Study Research. *Academy of Management Review*, 14(4), 532-550.

Eisenhardt, K. M., & Graebner, M. E. (2007). Theory Building from Cases: Opportunities and Challenges. *Academy of Management Journal*, 50(1), 25-32.

Eley, J. (2004). Design quality in buildings *Building Research and Information*, 32(3), 255-260.

Ellsworth, P. C., & Scherer, K. R. (2003). Appraisal processes in emotion. In R. J. Davidson, K. R. Scherer & H. H. Goldsmith (Eds.), *Handbook of affective sciences* (pp. 572-595). New York: Oxford University Press.

Elsbach, K. D., & Kramer, R. M. (2003). Assessing Creativity in Hollywood Pitch Meetings: Evidence for a Dual-Process Model of Creativity Judgements. *The Academy of Management Journal, 46*(3), 283-301.

Emmitt, S., Sander, D., & Christoffersen, K. (2005, July 2005). *The Value Universe: Defining a value based approach to lean construction.* Paper presented at the IGLC-13, Sydney, Australia.

Epstein, S. (1994). Integration of the cognitive and the psychodynamic unconscious. *American Psychologist, 49*, 709-724.

Epstein, S., Lipson, A., Holstein, C., & Huh, E. (1992). Irrational reactions to negative outcomes: Evidence for two conceptual systems. *Journal of Personality and Social Psychology, 62*, 328-339.

Epstein, S., Pacini, R., Denes-Raj, V., & Heir, H. (1996). Individual Differences in Intuitive-Experiential and Analytical-Rational Thinking Styles. *Journal of Personality and Social Psychology, 71*, 390-405.

Ernst, B., & Kieser, A. (2002). In search of explanations for the consulting explosion. In L. Engwall & Sahlin-Andersson (Eds.), *The Expansion of Management Knowledge: Carriers, Flows, and Sources.* Stanford CA: Stanford University Press.

Essers, M. J. J. M. (2009). *Aanbestedingsrecht voor overheden (Procurement Law for Government).* 's Gravenhage: Reed Business Information bv - Elsevier Overheid.

Etzioni, A. (1988). *The moral dimension - towards a new economics.* New York: The free press.

Directive 2004/18/EC (Council of the European Union 2004).

Evers, A. A. M. (1995). Architectuurwedstrijden (Design competitions). *Bouwrecht, 32*(10), 824-836.

Evers, A. A. M. (2008). Uitschrijven van prijsvragen voor bouwkundige diensten (Organising design competitions for services in architecture) *Succesvol aanbesteden*: Euroforum.

Evers, A. A. M. (2010). Europese selectie van architecten, noodzaak tot eenduidigheid? (European selection of architects, need for univocal policy). *Tijdschrift voor Bouwrecht, 3*(22), 146-152.

Ewenstein, B., & Whyte, J. (2007). Beyond Words: Aesthetic Knowledge and Knowing in Organizations. *Organization Studies, 28*, 689-708.

Faculty of Architecture - TU Delft (2009). *Building for Bouwkunde - Open to Ideas.* Delft: TU Delft.

Faculty of Architecture (2008). Open International Ideas Competition Building for Bouwkunde, from www.buildingforbouwkunde.nl.

Fawcett, W., Ellingham, I., & Platt, S. (2008). Reconciling the Architectural Preferences of Architects and the Public - The Ordered Preference Model. *Environment and Behavior, 40*(5), 599-618.

Feldman, D. (2003). The limits of law: can laws regulate public administration? In B. G. Peters & J. Pierre (Eds.), *Handbook of Public Administration* (pp. 279-292). London: Sage Publications.

Feldman, M. S., & March, J. G. (1981). Information in Organizations as Signal and Symbol. *Administrative Science Quarterly* (26), 171-186.

Fisher, P., Robson, S., & Todd, S. (2007). The disposal of public sector sites by "development competition". *Property Management,* 25(4), 381-399.

Flyvbjerg, B. (2004). Five Misunderstandings About Case-Study Research. In C. Seale, G. Gobo, J. F. Gubrium & D. Silverman (Eds.), *Qualitative Research Practice* (pp. 420-434). London and Thousand Oaks, CA: Sage.

Foote, N., Matson, E., Weiss, L., & Wenger, E. (2002). Leveraging Group Knowledge for High-Performance Decision-Making. *Organizational Dynamics,* 31(2), 280-295.

Frijda, N. H. (1986). *The Emotions.* Cambridge: Cambridge University Press.

Gann, D. M., Salter, A. J., & Whyte, J. K. (2003). Design Quality Indicator as a tool for thinking. *Building Research & Information,* 31(5), 318-333.

Gann, D. M., & Whyte, J. K. (2003). Editorial: Design quality, its measurement and management in the built environment. *Building Research & Information,* 31(5), 314-317.

Geertse, M. (2010). Aanbestedingen architectuuropdrachten in 2009 (Tenders for architectural design services in 2009). Retrieved 11 January 2010, from www.ontwerpwedstrijden.nl.

Geertse, M., Talman, B., & Jansen, C. (2009). Aanbesteding van architectuuropdrachten sinds het BAO (Tenders for architectural services since the BAO). Steunpunt Architectuuropdrachten & Ontwerpwedstrijden.

Gehner, E. (2008). *Knowlingly taking risk - Investment decision making in real estate development.* Delft: Eburon Academic Publishers.

George, E., & Chattopadhyay, P. (2008). Group composition and decision making. In G. Hodgkinson & W. H. Starbuck (Eds.), *The Oxford Handbook of Organizational Decision Making* (pp. 361-379). New York: Oxford University Press.

Gerring, J. (2004). What is a case study and what is it good for? *American Political Science Review,* 98(2), 341-354.

Gerritse, K. (2008). *Controlling costs and quality in the early phases of the accommodation process.* Delft: VSSD.

Gifford, R. (2002). *Environmental psychology: principles and practice.* Coleville, WA: Optimal books.

Gifford, R., Hine, D. W., Muller-Clemm, W., Reynolds, D. A. J., & Shaw, K. T. (2000). Decoding Modern Architecture - A Lens Model Approach for Understanding the Aesthetic Differences of Architects and Laypersons. *Environment and behavior,* 32(2), 163-187.

Gifford, R., Hine, D. W., Muller-Clemm, W., & Shaw, K. T. (2002). Why architect and laypersons judge buildings differently: cognitive properties and physical bases. *Journal of Architectural and Planning Research,* 19(2), 131-148.

Gigerenzer, G. (1991). How to make cognitive illusions disappear: beyond heuristics and biases. In W. Stroebe & M. Hewstone (Eds.), *European Review of Social Psychology - Volume 2* (pp. 83-115). Chichester: Wiley.

Gigerenzer, G. (2007). *Gut feelings. The intelligence of the unconscious.* New York: Viking.

Gigerenzer, G., & Selten, R. (2001). *Bounded rationality - the adaptive toolkit.* Cambridge, Massachusetts: MIT Press.

Gigerenzer, G., Todd, P. M., & ABC research group (1999). *Simple heuristics that make us smart.* New York: Oxford University Press.

Giuliani, M. V., & Scopelliti, M. (2009). Empirical research in environmental psychology: Past, present, and future. *Journal of Environmental Psychology,* 29(3), 375-386.

Gladwell, M. (2005). *Blink - the Power of Thinking without Thinking.* London: Penguin Books Ltd.

Glunk, U., & Olie, R. (2008). *De sleutel tot succes - Cultuur, samenwerking en innovatie in de bouw (The key to success).* Gouda: Regieraad Bouw & PSIBouw.

Glusberg, J. (Ed.). (1992). *A decade of RIBA Student Competitions.* London: Academy Group Ltd.

Gore, J., Banks, A., Millward, L., & Kyriakidou, O. (2006). Naturalistic Decision Making and Organizations: Reviewing Pragmatic Science. *Organization Studies,* 27(7), 925-942.

Green, S. D., & Moss, G. W. (1998). Value management and post-occupancy evaluation: closing the loop. *Facilities,* 16(1/2), 34-39.

Groat, L. (1982). Meaning in post-modern architecture: An examination using the multiple sorting task. *Journal of Environmental Psychology,* 2(1), 3-22.

Groat, L., & Wang, D. (2002). *Architectural research methods.* New York: John Wiley & Sons.

Gutman, R. (1988). *Architectural Practice - a critical view.* New York: Princeton Architectural Press.

Hackman, J. R. (2003). Learning more by crossing levels: evidence from airplanes, hospitals, and orchestras. *Journal of Organizational Behavior,* 24, 905-922.

Hahn, M., Lawson, R., & Lee, Y. G. (1992). The effects of time pressure and information load on decision quality. *Psychology and Marketing,* 9(5), 365-378.

Hamel, R. (1990). *Over het denken van de architect; een cognitief psychologische beschrijving van het ontwerpproces bij architecten (On designing by architects; a cognitive psychological description of the architectural design process).* Amsterdam: AHA Books/ University of Amsterdam.

Harrison, E. F. (1999). *The Managerial Decision-Making Process.* Boston: Houghton Mifflin.

Heijbrock, F. (2008, 5 August). Aanbestedingsregels verarmen architectuur. *Cobouw,* retrieved from www.cobouw.nl.

Hekkert, P. (2006). Design aesthetics: Principles of pleasure in product design. *Psychology Science,* 48(2), 157-172.

Hekkert, P., & van Wieringen, P. C. W. (1993). *Oordeel over kunst - Kwaliteitsbeoordelingen in de beeldende kunst* (Research report). Amsterdam: University of Amsterdam.

Hekkert, P., & van Wieringen, P. C. W. (1996). Beauty in the eye of expert and nonexpert beholders: A study in the appraisal of art. *American Journal of psychology,* 109(3), 389-407.

Hubbard, P. (1996). Conflicting interpretations of architecture: an empirical investigation. *Journal of Environmental Psychology,* 16, 75-92.

Herzberg, F., Mausner, B., & Snyderman, B. (1959). *The Motivation to Work.* New York: Wiley.

Heynen, H. (Ed.). (2001). *Overheidsopdrachten architectuur. Strategieën voor kwaliteit (Procurement in architecture. Strategies for quality).* Brussel: Politeia.

Hijdra, T. A. (2007). *Transparantie & doelgerichte competitie - Het efficient selecteren van een marktpartij op basis van een meervoudige ontwikkel/ontwerpopdracht waarbij de selectie zoals op prijs als kwaliteit vindt (Transparency & purposeful competition).* Unpublished MSc, Delft University of Technology, Delft.

Hodgkinson, G., & Starbuck, W. H. (2008a). Organizational decision making: mapping terrains on different planets. In G. Hodgkinson & W. H. Starbuck (Eds.), *The Oxford Handbook of Organizational Decision Making* (pp. 1-29). Oxford: Oxford University Press.

Hodgkinson, G., & Starbuck, W. H. (Eds.). (2008b). *The Oxford Handbook of Organizational Decision Making.* Oxford: Oxford University Press.

Hodgkinson, G., & Wright, G. (2006). Neither Completing the Practice Turn, nor Enriching the Process Tradition: Secondary Misinterpretations of a Case Analysis Reconsidered. *Organization Studies, 27,* 1895-1901.

Hogarth, R. M. (1988). *Judgement and Choice - The psychology of decision* (2 ed.). Chichester: John Wiley & Sons.

Hogarth, R. M. (2002). Deciding analytically or trusting your intuition? The advantages and disadvantages of analytic and intuitive thought. Unpublished Chapter. ICREA and Pompeu Fabra University.

Hogarth, R. M. (2005). Deciding analytically or trusting your intuition? The advantages and disadvantages of analytic and intuitive thought. In T. H. Betsch, S. (Ed.), *The Routines of Decision Making* (pp. 67-82). Mahwah, NJ: Lawrence Erlbaum Associates.

Hogarth, R. M., & Schoemaker, P. (2005). Beyond Blink: A Challenge to Behavioral Decision Making. *Journal of Behavioral Decision Making,* 18, 305-309.

Houben, F. (2007, 23 July). Opdrachtgever vindt zijn architect (Client finds its architect). *Financieel Dagblad,* p. 5, retrieved from www.fd.nl.

Howes, Y., & Gifford, R. (2009). Stable or Dynamic Value Importance?: The Interaction Between Decision-Making in Environmental Issues Value Endorsement Level and Situational Differences on. *Environment and Behavior,* 41(4), 549-582.

Hubbard, P. (1996). Conflicting interpretations of architecture: an empirical investigation. *Journal of Environmental Psychology,* 16(2), 75-92.

Hutton, R. J. B., & Klein, G. (1999). Expert Decision making. *Systems Engineering,* 2(1), 32-45.

ICOP (2006). ArchiSelect*: een nieuwe methode voor architectenselecties (ArchiSelect: a new method for architect selections). Rotterdam, retrieved from www.icop.nl.

Ivory, C. (2005). The cult of customer responsiveness: is design innovation the price of a client-focused construction industry? *Construction Management and Economics, 23*(8), 861-870.

Jackson, P. (1983). Principles and problems of participant observation. *Geografiska Annaler, 65*(B 1), 39-46.

Jackson, T. (1997). Survey - Management Consultancy: Growth and Revenues Seem to be Unstoppable. *Financial Times.*

Jacobsen, T., & Hofel, L. (2002). Aesthetic judgments of novel graphic patterns: analyses of individual judgments. *Perceptual and Motor Skills, 95*(3), 755-766.

Janis, I. L. (1982). *Groupthink: Psychological studies of policy decisions and fiascos.* Boston: Houghton Mufflin.

Jansen, A., Kolkman, S., Kuijpers, P., Pries, F., van Reeuwijk, T., Witteveen, B., et al. (2007). *Gunnen op waarde (Value based contract allocation).* Gouda: PSIBouw.

Jansen, C. E. C. (2009). *Leidraad Aanbesteden (Procurement Guide).* Gouda: Regieraad Bouw.

Jansen, R., Gössling, T., Merks, I., & Geurts, J. (2005). *The Quality of Organizational Decision-making: Collecting Decision Makers' Perception.* Paper presented at the EGOS Colloquium, Berlin, Germany.

Jencks, C. (Ed.). (1987). *The Architecture of Democracy - the Phoenix municipal government center design competition.* London: Academy Groups Ltd.

Jones, C., & Livne-Tarandach, R. (2008). Designing a frame: rhetorical strategies of architects. *Journal of Organizational Behavior, 29,* 1075–1099.

Kahneman, D. (2003). A Perspective on Judgment and Choice - Mapping Bounded Rationality. *American Psychologist, 58*(9), 697-720.

Kahneman, D., Slovic, P., & Tversky, A. (1982). *Judgement under Uncertainty; Heuristics and Biases.* Cambridge: Cambridge University Press.

Kahneman, D., & Tversky, A. (1979). Prospect Theory: an analysis of decision under risk. *Econometrica, 47*(2), 263-291.

Kahneman, D., & Tversky, A. (1984). Choices, values and frames. *American Psychologist* (39), 341-350.

Kahtri, N., & Ng, H. A. (2000). The Role of Intuition in Strategic Decision Making. *Human Relations, 53*(1), 57-86.

Kano, N. (1984). Attractive quality and must-be quality. *The Journal of the Japanese Society for Quality Control* (April), 39-48.

Kaplan, S. (1987). Aesthetics, Affect, and Cognition: Environmental Preference from an Evolutionary Perspective. *Environment and Behavior, 19*(1), 3-32.

Karmanov, D. (2009). *Feeling the Landscape: Six Psychological Studies into Landscape Experience.* Unpublished PhD, Wageningen University, Wageningen.

Kazemian, R., & Rönn, M. (2009). Finnish architectural *competitions: structure, criteria and judgement process. Building Research & Information, 37*(2), 176-186.

Kelly, J. (2007). Making client values explicit in value management workshops. *Construction Management and Economics,* 25(4), 435-442.

Kelly, J., Male, S., & Graham, D. (2004). *Value Management of Construction Projects.* Oxford: Blackwell Publishing.

Kennisportal Europese Aanbesteding (2009). Retrieved 14 May 2009 from http://www.europeseaanbestedingen.eu.

Kersten, R. A. E. M., Wolting, A., ter Bekke, M. G. A., & Bregman, A. G. (2009). *Reiswijzer Gebiedsontwikkeling 2009 - een praktische routebeschrijving voor marktpartijen en overheden (Guidebook Urban Area Development 2009).* Den Haag: Ministerie van Volkshuisvesting, Ruimtelijke Ordening en Milieubeheer.

Khalid, H. M., & Helander, M. G. (2006). Customer Emotional Needs in Product Design. *Concurrent Engineering,* 14(3), 197-206.

Kieser, A., & Wellstein, K. (2008). Do Activities of Consultants and Management Scientists Affect Decision Making by Managers? In G. Hodgkinson & W. H. Starbuck (Eds.), *The Oxford Handbook of Organizational Decision Making* (pp. 495-516). New York: Oxford University Press.

King, S. (1983). *Co-Design: A Process of Design Participation.* New York: Van Nostrand Reinhold.

Klein, G. (1993). A recognition-primed decision (RPD) model of rapid decision making. In G. Klein, J. Orasanu, R. Calderwood & C. E. Zsambok (Eds.), *Decision Making in Action: Models and Methods* (pp. 138-147). Norwood: Ablex.

Klein, G. (1997). The current status of the naturalistic decision making framework. In R. S. Flin, Eduardo; Strub, Michael; Martin, Lynne (Ed.), *Decision making under stress - emerging themes and applications.* Aldershot: Ashgate.

Klein, G. (1998). *Sources of power - how people make decisions.* Cambridge, Massachusetts: The MIT Press.

Klein, G. (2004). *The Power of Intuition - How to use your gut feelings to make better decisions at work.* New York: Currency Doubleday.

Koenen, I. (2009, 25 September). Aanbestedingsklimaat in b&u stuk gunstiger dan in gww (Tender situation in utilities less negative than in infrastructure). *Cobouw,* retrieved from www.cobouw.nl.

Koolwijk, J., Geraedts, R., & Chao-Duivis, M. A. B. (2005). *Prequalification of contractors based on past performance.* Delft: Kenniscentrum Bouwprocesinnovatie.

Kreiner, K. (2006). *Architectural Competitions - A case-study.* Copenhagen: Center for Management Studies of the Building Process.

Kreiner, K. (2007a). *Constructing the client in architectural competition - an ethnographic study of revealed strategies.* Paper presented at the EGOS 2007.

Kreiner, K. (2007b). *Strategic Choices in Unknowable Worlds.* Copenhagen: Center for Management Studies of the Building Process.

Kreiner, K. (2008). *Architectural Competitions - Empirical Observations and Strategic Implications for Architectural Firms.* Paper presented at the Architectural Competitions Nordic Symposium.

Kroese, R. J., Meijer, F., & Visscher, H. (2008). *De toepassing van Europese aanbestedings-regels bij architectenselecties (Implementation of procurement rules at architect selections)*. Delft: Onderzoeksinstituut OTB.

Kubr, M. (2002). *Management Consulting: A guide to the profession*. Geneva: International Labour Office.

Lamont, M. (2009). *How professors think - inside the curious world of academic judgment*. Cambridge, Massachusetts: Harvard University Press.

Lang, J. (1987). *Creating architectural theory. The role of behavioral sciences in environmental design*. New York: Van Nostrand Reinhold Company.

Langfeldt, L. (2001). The Decision-Making Constraints and Processes of Grant Peer Review, and Their Effects on the Review Outcome. *Social Studies of Science, 31*(6), 820-841.

Langley, A., Mintzberg, H., Pitcher, P., Posada, E., & Saint-Macary, J. (1995). Opening Up Decision-Making: The view from the Black Stool. *Organization Science, 6*(3), 260-279.

Lans, W., & Volker, L. (2008). *Exploring the assessment of a jury panel in architectural design education and practice* Paper presented at the ICERI 2008. International Conference of Education, Research and Innovation, Madrid.

Larson, M. S. (1994). Architectural Competitions and Discursive Events. *Theory and Society, 23*(4), 469-504.

Lazarus, R. S. (1990). *Emotion and Adaptation*. Oxford: Oxford University Press.

Le Dantec, C. A., & Y-Luen Do, E. (2009). The mechanisms of value transfer in design meetings. *Design Studies, 30*(2), 119-137.

Lerner, S., & Keltner, D. (2000). Fear, anger and risk. *Journal of Personality and Social Psychology, 81*(1), 146-159.

Lerner, S., & Tiedens, Z. (2006). Portrait of the angry decision maker. How appraisal tendencies shape anger's influence on cognition. *Journal of Behavioral Decision Making, 19*(2), 115-137.

Lipshitz, R., Klein, G., & Carroll, J. S. (2006). Naturalistic Decision Making and Organizational Decision Making: Exploring the Intersections. *Organization Studies, 27*(7), 917-923.

Lipshitz, R., Klein, G., Orasanu, J., & Salas, E. (2001). Taking stock of Naturalistic Decision Making (Focus article). *Journal of Behavioral Decision Making, 14*(5), 331-352.

Lipstadt, H. (2005, 17 and 18 November). *The Competition in the Region's Past, the Region in the Competitions Future*. Paper presented at the The Politics of Design: Competitions for Public Projects, New York.

Loe, E. (2000). *The Value of Architecture: Context and Current Thinking*. London: RIBA Future Studies.

Love, P. E. D., Davis, P. R., Edwards, D. J., & Baccarini, D. (2008). Uncertainty avoidance: public sector clients and procurement selection. *International Journal of Public Sector Management, 21*(7), 753-776.

Lybaart, J. (2008, 23 August). Foutje: architect mag toch niet Utrechtse bibliotheek bouwen (Slip-up: architect is not allowed to design library after all). *de Volkskrant,* p. 1, retrieved from www.volkskrant.nl.

Maandag, B. (2007). Eten met een accountantsverklaring (Having dinner with an financial statement). *De Architect* (June), 30-35.

Maandag, B. (2008). *Bouwkunde - portrait of the Faculty of Architecture of Delft University of Technology 1970-2008.* Delft: TU Delft, faculty of Architecture.

Maarleveld, M., Volker, L., & van der Voordt, T. J. M. (2009). Measuring employee satisfaction in new offices – the WODI toolkit. *Journal of Facilities Management, 7*(3), 181 - 197.

Macmillan, S. (2005). *Better designed building: improving the valuation of intangibles* (Summary). Cambridge: Eclipse Research Consultants.

Macmillan, S. (2006). Added value of good design. *Building Research & Information, 34*(3), 257-271.

Macmillan, S. (Ed.). (2004). *Designing Better Buildings.* London: Spon Press.

Manzoni, B. (2010). *A Content Analysis of 35 Years of Cross-Disciplinary Research on Architectural Competitions.* Paper presented at the Construction Matters Conference.

Manzoni, B., Morris, P., & Smyth, H. (2009). *Equipping project teams to win tenders: an insight into Italian architecture projects.* Paper presented at the 25th Annual ARCOM Conference, Nottingham, UK.

Marans, R. W., & Spreckelmeyer, K. F. (1982). Measuring Overall Architectural Quality: A Component of Building Evaluation. *Environment and behavior, 14*(6), 652-670.

March, J. G. (1997). Understanding how decisions happen in organizations. In Z. Shapira (Ed.), *Organizational Decision Making.* Cambridge UK: Cambridge University Press.

March, J. G., & Simon, H. A. (1958). *Organizations.* New York: John Wiley.

Markus, T. A. (2003). Lessons from the Design Quality Indicator. *Building Research & Information, 31*(5), 399-405.

Mieg, H. A. (2001). *The social psychology of experience - Case studies in research, professional domains, and expert roles.* Mahwah: Erlbaum.

Mieg, H. A. (2006). System experts and decision making in transdisciplinary projects. *International Journal of Sustainability in Higher Education, 7*(3), 341-351.

Mills, G. R., Austin, S. A., Thomson, D. S., & Devine-Wright, H. (2009). Applying a Universal Content and Structure of Values in Construction Management. *Journal of Business Ethics, 90*(2), 473-501.

Mintzberg, H., Ahlstrand, B., & Lampel, J. (1998). *Strategy Safari: A guided tour through the Wilds of Strategic Management.* New York: Free Press.

Moore, G. E. (1903). *Principa Ethica.* Cambridge: Cambridge University Press.

Morrell, K. (2008). The Narrative of 'Evidence Based' Management: A polemic. *Journal of Management Studies, 45*(3), 613-635.

Mosier, K. L., & Fischer, U. M. (2009). *Does Affect Matter in Naturalistic Decision Making?* Paper presented at the NDM9, the Ninth International Conference on Naturalistic Decision Making, London.

Muramatsu, R., & Hanoch, Y. (2005). Emotions as a Mechanism for Boundedly Rational Agents: The Fast and Frugal Way. *Journal of Economic Psychology,* 26(2), 201-221.

Nabi, R. L. (2003). Exploring the framing effects of emotion: Do discrete emotions differentially influence information accessibility, information seeking and policy preference. *Communication Research,* 30(2), 224-247.

Nasar, J. L. (1994). Urban Design Aesthetics: the evaluative qualities of building exteriors. *Environment and Behavior,* 26(3), 377-401.

Nasar, J. L. (1999). *Design by Competition: Making Design Competitions Work* Cambridge University Press.

Nasar, J. L., & Kang, J. (1989). A Post-Jury Evaluation: The Ohio State University Design Competition for a Center for the Visual Arts. *Environment and behavior,* 21(4), 464-484.

Nasar, J. L., & Purcell, A. T. (1990). *Beauty and the beast extended: knowledge, structure and evaluation of houses by australian architects and non-architects.* Paper presented at the IAPS, Ankara, Turkey.

Newcombe, R. (2003). From client to project stakeholders: a stakeholder mapping approach. *Construction Management and Economics,* 21(8), 841-848.

Nieuwenhuis, J. H. (1976). Legitimatie en heuristiek van het rechterlijk oordeel (Legitimation and heuristic of the legal judgement). *Rechtsgeleerde Magazijn Themis,* 494-515.

Norman, D. A. (2002). *The design of everyday things.* New York: Basic Books.

Nowee, V. (2008). *Architectenselectie op basis van proceskwaliteit - de ontwikkeling van een selectie-instrument voor Europese aanbestedingen (Architect selection based on process qualities).* Unpublished MSc, Delft University of Technology, Delft.

Ontario Association of Architects (2008). How to Find, Select and Engage an Architect Using Quality Based Selection (QBS) Retrieved 1 December 2009, from www.oaa.on.ca.

Orasanu, J., & Strauch, B. (1994). *Temporal factors in aviation decision making* Paper presented at the Human Factors and Ergonomics Society Meeting.

Osgood, C. E., Suci, G. J., & Tannenbaum, P. H. (1957). *The measurement of meaning.* Urbana: University of Illinois Press.

Overheid.nl (2009). Retrieved 9 May 2009, from www.overheid.nl.

Parsons, M. J. (1987). *How we understand art: a cognitive developmental account of aesthetic experience.* Cambridge: Cambridge University Press.

Patijn, W. (2000). *Tips van de rijksbouwmeester bij de selectie van architecten in het kader van de Europese aanbesteding (Tips of the Chief Government Architect).* Rotterdam: Uitgeverij 010.

Paulus, P. B., & Dzindolet, M. T. (1993). Social influence processes in group brainstorming. *Journal of Personality and Social Psychology,* 64(4), 575-586.

Peters, E., Vastfjall, D., Garling, T., & Slovic, P. (2006). The roles of affect in decision making. *Journal of Behavioral Decision Making*, 19(2), 79-85.

Philip, D. (1996). Essay: The practical failure of architectural psychology. *Journal of Environmental Psychology*, 16(3), 277-284.

Phillips, S., Martin, J., Dainty, A. R. J., & Price, A. D. F. (2007). Uncertainty in best value decision making. *Journal of Financial Management in Property and Construction*, 12(2), 63-72.

Phillips, S., Martin, J., Dainty, A. R. J., & Price, A. D. F. (2008). Analysis of the quality attributes used in establishing best value tenders in the UK social housing sector. *Engineering, Construction and Architectural Management*, 15(4), 307-320.

Phillips, T., Mannix, E., Neale, M., & Gruenfeld, D. (2004). Diverse Groups and Information Sharing: the Effects of Congruent Ties. *Journal of Experimental Social Psychology*, 40, 497-510.

PIANOo-vakgroep Aanbestedingsrecht (2009). *Afwijzingsberichten en motiveringsplicht (Rejection notifications and motivation duty)*. Den Haag: PIANOo Expertisecentrum Aanbesteden.

Pijnacker Hordijk, E. H., van der Bend, G. W., & van Nouhuys, J. F. (2009). *Aanbestedingsrecht; handboek van het Europese en het Nederlandse aanbestedingsrecht (Procurement Law - Handbook of European and Dutch procurement law)* (4 ed.). The Hague: SDU Uitgeverij.

Pirsig, R. M. (1999). *Zen and the Art of Motorcycle Maintenance - An Inquiry into Values* (25th Anniversary ed.). New York: Quill William Morrow.

Plous, S. (1993). *The psychology of judgment and decision making*. New York: McGraw-Hill Inc.

Pongpeng, J., & Liston, J. (2003). TenSeM: a multicriteria and multidecision-makers' model in tender evaluation. *Construction Management and Economics*, 21(1), 21-30.

Posner, R. A. (2008). *How judges think*. Cambridge, Massachusetts: Harvard University Press.

Postel, D. J. (2001, 12 October). Een goede architect heeft geen punten nodig (A good architect does not need points). *Volkskrant*. Retrieved from www.volkskrant.nl.

Prasad, S. (2004a). Clarifying intentions: the Design Quality Indicator. *Building Research & Information*, 32(6), 548-551.

Prasad, S. (2004b). Inclusive maps. In S. Macmillan (Ed.), *Designing Better Buildings* (pp. 175-184). London UK: Spon Press.

Preiser, W. F. E., Rabinowitz, H. Z., & White, E. (1988). *Post-occupancy evaluation*. New York: Van Nostrand Reinhold.

Prins, M. (2009). Architectural value. In S. Emmitt, M. Prins & A. Den Otter (Eds.), *Architectural management - International research and practice*. Chichester: Wiley-Blackwell.

Province of Utrecht (2009). Retrieved 9 May 2009, from www.provincie-utrecht.nl.

Pultar, M. (1996). *A conceptual framework for value in the built environment*. Paper presented at the IAPS, Stockholm.

Purcell, A. T. (1986). Environmental Perception and Affect: A Schema Discrepancy Model. *Environment and behavior,* 18(3), 3-30.

Regieraad-Bouw (2005). *Opdrachtgevers aan het woord (Clients speak).* Gouda: Regieraad Bouw.

Renier, B., & Volker, L. (2008). *The Architect as a System Integrator?* Paper presented at the 24th Annual ARCOM Conference, Cardiff, UK.

Robbins, S. P., & Judge, T. A. (2008). *Essentials of Organizational Behavior* (ninth ed.). Upper Saddle River: Pearson Prentice Hall Inc.

Robinson, I. M. (Ed.). (1972). *Decision-Making in Urban Planning - an introduction to new methodologies.* Beverly Hills/London: Sage Publications.

Rokeach, M. (1973). *The Nature of Human Values.* New York: Free Press.

Rönn, M. (2008). *Architectural Policies and the Dilemmas of Architectural Competitions.* Paper presented at the Architectural Competitions Nordic Symposium.

Rönn, M. (2010). *Expertise and judgment in architectural competitions - A theory for assessing architecture quality.* Paper presented at the Construction Matters Conference.

Roozenburg, N. F. M., & Eekels, J. (1995). *Product Design: Fundamentals and Methods.* Chichester: Wiley.

Rosen, M., Salas, E., Lyons, R., & Fiore, S. M. (2008). Expertise and naturalistic decision making: mechanisms of effective decision making. In G. Hodgkinson & W. H. Starbuck (Eds.), *The Oxford Handbook of Organizational Decision Making* (pp. 211-230). New York: Oxford University Press.

Ruiter, H., Boer, N.-I., Buijs, S., Versendaal, J., de Bruin, A., Kroese, E., et al. (2009). *Feitenonderzoek voorwaarden in de aanbestedingspraktijk (Inquiry requirement in procurement practice).* Utrecht: Berenschot.

Runeson, G., & Skitmore, M. (1999). Tendering theory revisited. *Construction Management and Economics,* 17(3), 285-296.

Russell, J. A., Ward, L. M., & Pratt, G. (1981). Affective quality attributed to environments: A factor analytic study. *Environment and behavior,* 13(3), 259-288.

Russo, J. E., & Schoemaker, P. H. (2002). *Decisions - Getting it Right the first Time.* New York: Currency Doubleday/Random House Inc.

Sadler-Smith, E., & Sparrow, P. R. (2008). Intuition in Organizational Decision Making. In G. Hodgkinson & W. H. Starbuck (Eds.), *The Oxford Handbook of Organizational Decision Making* (pp. 305-324). New York: Oxford University Press.

Sagalyn, L. B. (2005). *The Political Fabric of Design Competitions.* Paper presented at the The Politics of Design: Competitions for Public Projects, New York.

Salas, E., Burke, C. S., & Stagl, K. C. (2004). Developing teams and team leaders: strategies and principles. In D. Day, S. J. Zaccaro & S. M. Halpin (Eds.), *Leader development for transforming organizations: growing leaders for tomorrow* (pp. 325-355). Mahwah, New York: Lawrence Erlbaum Associates Inc.

Salas, E., Rosen, M. A., Burke, C. S., Goodwin, G. F., & Fiore, S. (2006). The Making of a Dream Team: When expert teams do best. In K. A. Ericsson, N. Charness, P. J. Feltovich & R. R. Hoffman (Eds.), *Cambridge Handbook of Expertise and Expert Performance*. New York: Cambridge University Press.

Sánchez, M., Prats, F., Agell, N., & Ormazabal, G. (2005). Multi-Criteria Evaluation for Value Management in Civil Engineering. *Journal of Management in Engineering*, 21(3), 131-137.

Sanger, J. (1996). *The compleat observer? A field research guide to observation*. London: The Falmer Press.

Sanoff, H. (2006). Multiple views of participatory design. *METU Journal of the Faculty of Architecture*, 23(2), 131-143.

Saunders, W. S. (2007). From Taste to Judgment: Multiple Criteria in the Evaluation of Architecture. In W. S. Saunders (Ed.), *Judging architectural value* (pp. 129-149). Minnesota: University of Minnesota Press.

Schifferstein, H. N. J., & Hekkert, P. (Eds.). (2008). *Product experience*. Amsterdam: Elsevier.

Schön, D. A. (1991). *The Reflective Practitioner: How Professionals Think in Action*. Aldershot UK: Avebury.

Schwartz, S. H., & Boehnke, K. (2004). Evaluating the Structure of Human Values with Confirmatory Factor Analysis. *Journal of Research in Personality*, 38, 230-255.

Senter Novem (2009). *Innovation Intelligence: Verkenning Creatieve Industrie (Innovation Intelligence: enquiries of the creative industry)*. Den Haag: Senter Novem.

Shapiro, S., & Spence, M. T. (1997). Managerial Intuition: A conceptual and Operational Framework. *Business Horizons*, 40(1), 63-68.

Silverberger, J. (2010). *In or Out - Following a controversial architectural project trhough three days of jury sessions*. Paper presented at the Construction Matters.

Silverman, D. (2007). *A very short, fairly interesting and reasonably cheap book about qualitative research*. Los Angeles: Sage.

Simon, H. A. (1987). Making management decisions: the role of intuition and emotion. *Academy of Management Executive*, 1(1), 57-64.

Simon, H. A. (1997). *Administrative Behavior: A Study of Decision-Making Processes in Strategic Decision Making Processes in Administrative Organizations*. (4th ed.). New York: MacMillan.

Sinclair, M., & Ashkanasy, N. M. (2005). Intuition: Myth or a Decision-making tool? *Management Learning*, 36(3), 353-370.

Slaughter, E. S. (2004). DQI: the dynamics of design values and assessment. *Building Research & Information*, 32(3), 245-246.

Snellen, I. T. M. (1987). *Boeiend en geboeid - ambivalenties en ambities in de bestuurskunde (Fascinating and Chained - ambivalence and ambitions in public administration)*. Alphen aan den Rijn: Samson H.D. Tjeenk Willink.

Sniezek, J. A., Paese, P. W., & Switzer III, F. S. (1990). The effect of choosing on confidence in choice. *Organizational Behavior and Human Decision Processes, 46*(2), 264-282.

Soane, E., & Nicholson, N. (2008). Individual Differences and Decision Making. In G. Hodgkinson & W. H. Starbuck (Eds.), *The Oxford Handbook of Organizational Decision Making* (pp. 342-360). New York: Oxford University Press.

Speier, C., Valacich, J. S., & Vessey, I. (1999). The influence of task *interruption on individual decision making: an information overload perspective. Decision Sciences, 30*, 337-359.

Sporrong, J., & Bröchner, J. (2009). Public Procurement Incentives for Sustainable Design Services: Swedish Experiences. *Architectural Engineering and Design Management*, 5, 24-35.

Spreiregen, P. D. (1979). *Design Competitions*. New York: McGraw-Hill Book Company.

Spreiregen, P. D. (2008). *The Vietnam Veterans Memorial Design Competition*. Paper presented at the Architectural Competitions Nordic Symposium, Stockholm.

Stake, R. (1995). *The art of case research*. Thousand Oaks, CA: Sage Publications.

Stanovich, K. E., & West, R. F. (2000). Individual differences in Reasoning: Implications for the Rationality Debate? *Behavioral and Brain Sciences, 25*(5), 645-665.

Stasser, G., & Titus, W. (1985). Pooling of Unshared Information in Group Decision Making: Biased Information Sampling During Discussions. *Journal of Personality and Social Psychology*, 48, 1467-1478.

Stichting Aanbestedingsinstituut Bouwend Nederland (2009). Resultaten aanbestedingsregistratie eerste helft 2008 (Results tender registration first half of the year 2008), Retreived 25 February 2009, from http://www.aanbestedingsinstituut. nl/overzicht-statistieken.

Stichting Bouwresearch (1980). *Architectuurwedstrijden nader bekeken (A closer look at Architectural Design Competitions)*. Deventer: Kluwer Technische Boeken BV.

Stichting Bureau Architectenregister (2007). *Jaarverslag 2007 (Annual Report 2007)*. Den Haag: Stichting Bureau Architectenregister.

Stichting Bureau Architectenregister (2009). Wat doet het SBA? (What does the SBA do?) Retrieved 27 November 2009, from http://www.architectenregister.nl/.

Stroebe, W., & Diehl, M. (1994). Why Groups are less Effective than their Members: On Productivity Losses in Idea-generating Groups. *European Review of Social Psychology*, 5(1479-277X), 271 - 303.

Strong, J. (1976). *Participating in architectural competitions: A guide for competitors, promotors and assessors*. London: The Architectural Press Ltd.

Strong, J. (1996). *Winning by Design - Architectural Competitions*. Oxford: Butterworth Architecture.

Sudjic, D. (2005). *Competitions: the Pitfalls and the Potentials*. Paper presented at the The Politics of Design: Competitions for Public Projects, New York.

Sutcliffe, K. M., & Weick, K. E. (2008). Information overload revisited. In G. Hodgkinson & W. H. Starbuck (Eds.), *The Oxford Handbook of Organizational Decision Making* (pp. 56-75). New York: Oxford University Press.

Svensson, C. (2008). *Speaking of Architecture - a study of the jury's assessment in the invited competition of an educational centre in Hagfors, Sweden.* Paper presented at the Architectural Competitions Nordic Symposium, Stockholm.

Svensson, C. (2010). *On Quality Assessment in an Architectural Competition.* Paper presented at the Construction Matters Conference, Copenhagen.

Tetlock, P. (1983). Accountability and the Complexity of Thought. *Journal of Personality and Social Psychology,* 45, 74-83.

Tetlock, P. (1992). The Impact of Accountability on Judgement and Choice: Toward a Social Contingency Model. *Advances in Experimental Social Psychology,* 25, 331-376.

Teunissen, L. (2009). Een andere kijk op de praktijk (Another look at practice). *De Architect,* 40(4), 44-49.

Thompson, D. F. (1980). Moral responsibility of public officials: The problem of many hands. *The American Political Science Review,* 74, 905-916.

Thomson, D. S., Austin, S. A., Devine-Wright, H., & Mills, G. R. (2003). Managing value and quality in design. *Building Research & Information,* 31(5), 334-345.

Thyssen, M. H., Emmitt, S., Bonke, S., & Kirk-Christoffersen, A. (2010). Facilitating Client Value Creation in the Conceptual Design Phase of Construction Projects: A Workshop Approach. *Architectural Engineering and Design Management,* 6, 18-30.

Tombesi, P. (2006). Good thinking and poor value: on the socialization of knowledge in construction. *Building Research & Information,* 34(3), 272-286.

Tversky, A., & Kahneman, D. (1974). Judgement under uncertainty: heuristics and biases *Science* (185), 1124-1131.

Tversky, A., & Kahneman, D. (1981). The framing of decisions and the psychology of choice. *Science* (221), 1124-1131.

Valadez, J. J. (1984). Diverging meanings of development among architects and three other professional groups. *Journal of Environmental Psychology,* 4(3), 223-228.

van Campen, J., & Hendrikse, J. (1997). *Kompas - handleiding en voorbeeldmodellen bij het uitschrijven van prijsvragen en meervoudige opdrachten op het gebied van architectuur, stedebouw en landschapsarchitectuur ('Kompas' - Guidelines and models for design competitions).* Rotterdam: Uitgeverij 010.

van den Brink, M. (2009). *Behind the Scenes of Science. Gender practices in the recruitment and selection of professors in the Netherlands.* Unpublished PhD, Radboud University Nijmegen, Nijmegen.

Van den Dobbelsteen, A. A. J. F. (2004). *The Sustainable Office. An exploration of the potential for factor 20 environmental improvement of office accommodation* Unpublished PhD, Delft University of Technology, Delft.

van den Hurk, R. (2008). Resultaten Bedrijfsvergelijkend onderzoek (BVO) 2008 (Results Firm Comparison 2008). *BNABLAD* (9), 10-11.

van den Hurk, R. (2009). Rendementen architectenbranche onder druk (Returns architecture sector under pressure). *BNABLAD* (1), 10-11.

van der Pol, L., Brouwer, J., Jansen, C., Mensink, J., & Geertse, M. (2009). *Europa en de architecten - Stand van zaken in de discussie over Europese aanbestedingen van architectendiensten (Europe and the architects - the state of affairs)*. Den Haag: Ministerie van VROM/Atelier Rijksbouwmeester & Architectuur Lokaal.

van der Voordt, D. J. M., & van Wegen, H. B. R. (2005). *Architecture in Use: an introduction to the programming, design and evaluation of buildings*. Oxford: Architectural Press.

van Eldonk, J. (2008). Projectontwikkeling en de architect (Project development and the architect) *Handboek projectontwikkeling* (pp. 320-329): Reed Business.

van Geels, A. H. P., & Kriens, J. (2009). Voordracht ontwerp stadskantoor (Recommendation design City hall), *Letter to the City Council*. Rotterdam: Executive Board City of Rotterdam.

van Hardevelt, I. D., & Schönau, W. F. (2009, 30 October). Architectenselectie als polderoplossing (Architect selection as polder solution). *Cobouw*, retrieved from http://www.cobouw.nl.

van Loon, P. P., Heurkens, E., & Bronkhorst, S. (2008). *The Urban Decision Room - An Urban Management Instrument*. Amsterdam: Uitgeverij IOS Press.

van Romburgh, H. D. (2005). *Op weg naar een nieuw aanbestedingsrechtelijk kader in Nederland. Een proeve van een wet voor het verstrekken van overheidsopdrachten. (Towards a new procurement frame in the Netherlands)*. Deventer: Kluwer.

van Rossum, H., & de Wildt, R. (1996). *Rijkshuisvesting in ontwikkeling (Governmental Real Estate in development)*. Rotterdam: NAI Publishers.

van Wezemael, J. E. (2008). *The Complexity of Competitions*. Paper presented at the Architectural Competitions Nordic Symposium, Stockholm.

van Wijngaarden, M. A., & Chao-Duivis, M. A. B. (2010a). *Deel 17 Bouw-en Aanbestedingsrecht (Building and Procurement Law Part 17)* (6 ed.). Zutphen: Uitgeverij Paris.

van Wijngaarden, M. A., & Chao-Duivis, M. A. B. (2010b). *Deel 18 Bouw- en Aanbestedingsrecht (Building and Procurement Law Part 18)* (6 ed.). Zutphen: Uitgeverij Paris.

Vergu, V. (2008). Retrieved 20 June 2008 from http://www.buildingforbouwkunde.nl/creativefestival/.

Vidaillet, B. (2008). When 'decision outcomes' are not the outcomes of decisions. In G. Hodgkinson & W. H. Starbuck (Eds.), *The Oxford Handbook of Organizational Decision Making* (pp. 418-436). New York: Oxford University Press.

Vischer, J., & Preiser, W. F. E. (2004). *Assessing Building Performances*. London: Butterworth-Heinemann.

Visser, D. (2009). De rechterlijke beslissing (in IE-zaken) (The decision of the judge). *Boek9.nl,* B98189 (16 September 2009).

Vitruvius, P., & Morgan, M. H. (1960). *Vitruvius: the ten books on architecture* (M. H. Morgan, Trans.). New York,: Dover Publications.

Vogels, R., Mooibroek, M., & de Vries, N. (2008). *Brancheonderzoek BNA 2007 (Branch Research BNA 2007)*. Amsterdam: BNA - Stratus.

Volker, L., & Chao-Duivis, M. A. B. (2010). *Potential conflicts with procurement law during architect selection*. Paper presented at the CIB World Congress 2010, Salford Quays - UK.

Volker, L., & de Jonge, H. (2007). *Methodological reflections on case-research for partner selection in architecture*. Paper presented at the Third Scottish Conference for Postgraduate Researchers of the Built and Natural Environment (PRoBE), Glasgow.

Volker, L., & Heintz, J. (2007). *Negotiating design collaboration in architecture: a case study of a European design tender*. Paper presented at the 2007 Construction Research Congress, Grand Bahama Island, Bahamas.

Volker, L., & Lauche, K. (2008). *Designing a decision making framework for judging design quality*. Paper presented at the 24th EGOS (European Group of Organisation Studies) colloquium, Amsterdam.

Volker, L., Lauche, K., Heintz, J. L., & de Jonge, H. (2008). Deciding about design quality: design perception during a European tendering procedure. *Design Studies, 29*(4), 387-409.

Vollaard, P. (2009a). Kompas Light: aanbestedingsleidraad nu online (Kompas Light: tender manual now online) Retrieved 15 December 2009, from www.archined.nl.

Vollaard, P. (2009b). 'We zitten goed op de Julianalaan, vooral geen iconen en leve de jaren zeventig' ('Our current accommodation on the Julianalaan functions well, particularly no icons and long live the seventies'). Retrieved 15 March 2009, from www.archined.nl.

von Neumann, J., & Morgenstern, O. (1947). *Theory of games and economic behavior*. Princeton: Princeton University Press.

von Winterfeldt, D., & Edwards, W. (1986). *Decision analysis and behavioral research*. Cambridge UK: Cambridge University Press.

Vroom, V. H., & Jago, A. G. (1988). *The new leadership, managing participation in organizations*. Englewood Cliffs, New Jersey: Prentice Hall.

Walden, D., Berger, C., Blauth, R., Boger, D., Bolster, C., Burchill, G., et al. (1993). Kano's Methods for Understanding Customer-defined Quality. *Center for Quality Management Journal, 4*(2), 1-37.

Wandahl, S. (2004). *Value carriers in a construction project - how different are they?* Paper presented at the 12th Annual Conference on Lean Construction Helsingør, Danmark.

Wandahl, S. (2005). *Value in Building*. Unpublished PhD, Alborg University, Denmark, Aalborg.

Watt, D. J., Kayis, B., & Willey, K. (2010). The relative importance of tender evaluation and contractor selection criteria. *International Journal of Project Management, 28*(1), 51-60.

Webler, T., Tuler, S., & Krueger, R. (2001). What Is a Good Public Participation Process? Five Perspectives from the Public. *Environmental Management, 27*(3), 435-450.

Weick, K. E. (1969). *The social psychology of organizing.* New York: McGraw Hill.

Weick, K. E. (1995). *Sensemaking in Organizations.* Thousand Oaks CA: Sage Publications.

Weick, K. E., & Sutcliffe, K. M. (2001). *Managing the Unexpected.* San Francisco, CA: Jossey-Bass.

Weick, K. E., Sutcliffe, K. M., & Obstfeld, D. (1999). Organizing for high reliability: Processes of collective mindfulness *Research in Organizational Behavior,* 21, 81-123.

Weick, K. E., Sutcliffe, K. M., & Obstfeld, D. (2005). Organizing and the Process of Sensemaking. *Organization Science,* 16(4), 409-421.

Weijnen, T., & Berdowski, Z. (2009). *Het totale inkoopvolume van Nederlandse overheden (Total Purchasing Volume of the Dutch Government).* Zoetermeer: Instituut voor Onderzoek van Overheidsuitgaven (IOO bv).

Weiss, H. M., & Cropanzano, R. (1996). Affective Events Theory: A theoretical discussion of the structure, cause and consequences of affective experiences at work. In B. M. Staw & L. L. Cummmings (Eds.), *Research in Organizational Behavior* (Vol. 18, pp. 17-19). Greenwich: JAI Press.

Wendte, R. (2004). *Schoonheidsbeoordelingen van gebouwen, systematische verschillen tussen leken en architecten (Systematic differences between architects and novices in judging the aesthetics of buildings).* Unpublished MSc, Universiteit van Amsterdam, Amsterdam.

Whyte, J. K., & Gann, D. M. (2003). Design Quality Indicators: work in progress. *Building Research & Information,* 31(5), 387-398.

Whyte, J. K., Gann, D. M., & Salter, A. J. (2004). Building indicators of design quality. In S. MacMillan (Ed.), *Designing Better Buildings* (pp. 195-205). London: Spon Press.

Wierzbicki, A. P. (1997). On the Role of Intuition in Decision Making and Some Ways of Multicriteria Aid of Intuition. *Journal of Multi-Criteria Decision Analysis,* 6(2), 65-76.

Wilson, M. A. (1996). The Socialization of Architectural Preference. *Journal of Environmental Psychology,* 16(1), 33-44.

Wilson, T. D., & Schooler, J. W. (1991). Thinking too much: Introspection can reduce the quality of preferences and decisions. *Journal of Personality and Social Psychology,* 60, 181-192.

Wong, C. H., Holt, G. D., & Cooper, P. A. (2000). Lowest price of value? Investigation of UK construction clients' tender selection process. *Construction Management and Economics,* 18(7), 767-774.

World Trade Organization (2009). Retrieved 15 October 2009, from www.wto.org.

Wright, G., & Goodwin, P. (2008). Structuring the decision process: an evaluation of methods. In G. Hodgkinson & W. H. Starbuck (Eds.), *The Oxford Handbook of Organizational Decision Making* (pp. 534-551). New York: Oxford University Press.

Wright, P. C., & Geroy, G. D. (1991). Experience, Judgement and Intuition: Qualitative Data-gathering Methods as Aids to Strategic Planning. *Leadership & Organization Development Journal,* 12(3), 2-32.

Wulz, F. (1986). The concept of participation. *Design Studies,* 7(3), 153-162.

Yaneva, A. (2008). How Buildings 'Surprise': The renovations of the Alte Aula in Vienna. *Science Studies,* 21(1), 8-28.

Yin, R. K. (2009). *Case Study Research: Design and Methods* (4 ed. Vol. 5). Beverly Hill, Cal.: Sage Publications.

Zheng, L. (2008). *An overview of trajectories of change of design competitions across Europe.* Paper presented at the Architectural Competitions Nordic Symposium, Stockholm.

Zsambok, C. E. (1997). Naturalistic Decision Making: Where are we now? In C. E. Zsambok & G. A. Klein (Eds.), *Naturalistic Decision Making.* Mahwah NJ: Lawrence Erlbaum.

Publications related to this research

Volker, L. (2010b). *Designing a design competition: the client perspective.* Paper presented at the Design Research Society 2010 - Design & Complexity, Montreal - Canada.

Volker, L. (2010a). *Architect selection: rational or intuitive decision-making.* Paper presented at the ARCOM research workshop on decision-making across levels, time and space: exploring theories, methods and practices, Manchester - UK.

Volker, L., & Chao-Duivis, M. A. B. (2010). *Potential conflicts with procurement law during architect selection.* Paper presented at the CIB World Congress 2010, Salford Quays - UK.

Volker, L., & van Meel, J. (2010). *Dutch design competitions: lost in EU Directives? Architect selections for public commissions in the Netherlands.* Paper presented at the Constructions Matter: Managing Complexities, Decisions and Actions in the Building Process, Copenhagen - Denmark.

Chao-Duivis, M. A. B., Koolwijk, J., & Volker, L. (2009). De aanbesteding (Tendering). In J. W. F. Wamelink (Ed.), *Inleiding Bouwmanagement (Introduction Construction Management)* (pp. 73-131). Delft: VSSD.

Faculty of Architecture - TU Delft (2009). *Building for Bouwkunde - Open to Ideas.* Delft: TU Delft.

Volker, L. (2009). Architectenselecties: You never get a second chance to make a first impression. *de Architect* (April), 36.

Volker, L., & Heintz, J. L. (2009). Kwaliteit (Quality). In J. W. F. Wamelink (Ed.), *Inleiding Bouwmanagement (Introduction Construction Management)* (pp. 229-261). Delft: VSSD.

Volker, L., & Lauche, K. (2009). Decision making during a tendering procedure: case studies of restricted European tenders in architecture. *Construction Information Quarterly, 11*(2), 60-65.

Volker, L., Zijlstra, H., de Jong, P., & van Dorst, M. (2009). Analysis of ideas competition entries. In TU Delft - faculty of Architecture (Ed.), *Building for Bouwkunde - Open to Ideas* (pp. 53-58). Delft: TU Delft - faculty of Architecture.

Faber, E., & Volker, L. (2008). Opdrachtgever: een eurotender light graag! (Client: a Euro tender light please!). *Real Estate Magazine, 61*(6).

Lans, W., & Volker, L. (2008). *Exploring the assessment of a jury panel in architectural design education and practice* Paper presented at the ICERI 2008. International Conference of Education, Research and Innovation, Madrid - Spain.

Volker, L. (2008). Early design management in architecture: partner selection and value judgement in design. In S. D. Pryke & H. J. Smyth (Eds.), *Collaborative Relationships in Construction* (pp. 179-196). Oxford: Blackwell Publishing.

Volker, L., & Lauche, K. (2008a). *Decision making during a tendering procedure: case studies of restricted European tenders in architecture.* Paper presented at the 24th Annual ARCOM Conference, Cardiff - UK.

Volker, L., & Lauche, K. (2008b). *Designing a decision making framework for judging design quality.* Paper presented at the 24th EGOS (European Group of Organisation Studies) colloquium, Amsterdam.

Volker, L., Lauche, K., Heintz, J. L., & de Jonge, H. (2008). Deciding about design quality: design perception during a European tendering procedure. *Design Studies,* 29(4), 387-409.

Volker, L., & de Jonge, H. (2007). *Methodological reflections on case-research for partner selection in architecture.* Paper presented at the Third Scottish Conference for Postgraduate Researchers of the Built and Natural Environment (PRoBE), Glasgow - UK.

Volker, L., & Heintz, J. (2007). *Negotiating design collaboration in architecture: a case study of a European design tender.* Paper presented at the 2007 Construction Research Congress, Grand Bahama Island, Bahamas.

Volker, L., & Prins, M. (2006a). *Critical reflections in measuring the effect of steering activities in architectural design.* Paper presented at the 22nd Annual ARCOM Conference 2006, Birmingham - UK.

Volker, L., & Prins, M. (2006b). *Linking design management to value perception in architectural building design.* Paper presented at the International Built & Human Environment Research Week - 6th International Postgraduate Research Conference -Delft.

Volker, L., & Prins, M. (2006c). *Measuring the effect of steering techniques on value creation in architectural design.* Paper presented at the Annual research conference of the royal institution of chartered surveyors (COBRA), London - UK.

Volker, L., & Prins, M. (2005). *Exploring the possibilities of correlating management with value in architectural design.* Paper presented at the Designing Value: New Directions in Architectural Management CIB W096 Architectural Design Management, Lyngby - Denmark.

Summary

Deciding about Design Quality

Value judgements and decision making in the selection of architects by public clients under European tendering regulations

PhD Thesis of Leentje Volker

The search for an architect can be characterized as an interactive selection process in which a client tries to find an architect who can visualize and implement the clients' needs and ambitions best. This process is not without problems. "Architect soap opera", "Slip-up, architect cannot build library after all", and "Public does not want A but B to build town office" are just some newspaper headlines that have dominated the image of tender procedures in which Dutch public clients selected an architect the past few years. It is thus a challenging decision process of surprises and unforeseen circumstances in which legal and social obligations have to be considered. The current practice of architect selection by public clients has its roots in three distinct systems: 1) tendering for services and works, 2) the selective search to identify a suitable architect or design team, and 3) the architectural competition. It is these diverse roots of the selection process that appear to cause conflicts between the legal rationality and the psychological rationality of decision making.

In this research I explore the origin of these problems as currently experienced by public commissioning clients in architect selection in order to propose implications for future practice. In Chapter 1 the research topic, research focus, and research approach are introduced. I also describe the knowledge gaps, scientific challenges, and contribution to the field. This research focuses on the complete process of decision making from the perspective of public clients willing to select an architect in the context of European Union procurement law. The aim of the research is to describe, understand, and explain the design and implementation of procedures by means of which the quality of design proposals is judged in order to award a contract to the architect who will deliver design services for a particular project. The research questions in the study are:

1. How do public commissioning bodies decide on the selection of an architect in the context of EU procurement law?

2. Which situational characteristics influence the process of decision making of public commissioning bodies in this context?

3. What are the implications for the design of procedures for the selection of architects?

In Chapter 2 I address the concept of design quality across the fields of architectural design, environmental psychology, product experience, and value management. Based on these perspectives I define design quality as an overall value judgement of an individual stakeholder that is based on the interaction between the person and an (representation of an) object in the built environment. As a result of the interaction between the individual and the product, a value judgement is always accompanied by an affective response and an assessment about the level of quality or value of a product.

Chapter 3 provides an overview of the psychological aspects that seem most relevant for judgement and decision making in the context of selecting an architect. It elaborates on the definition of design quality as a value judgement from Chapter 2. The chapter starts with an overview of three generations of decision theory: rational decision models, behavioural decision models, and naturalistic decision making. The current practice of architect selections appears to be based on two conflicting models about decision making: the legal and the psychological model. The legal model assumes a rational and sequential decision process in which alternatives are compared based on pre-announced criteria. The naturalistic decision model attributes an important role to the use of intuition and affect. The origin of the current problems in practice could consequently be found in these different rationalities. I adopt the concept of sensemaking as the main concept for decision making to explain how clients deal with these different rationalities.

An architect selection is not an isolated event. Public clients operate in a context of governance and have to consider this organisational structure in their decisions. Chapter 4 addresses four contextual elements that in my view are essential to understand the environment in which architect selections take place: the political, cultural, legal and economical context. I found that the choices made during the preparation phase determine to a considerable extent the results and appropriateness of the tender, as well as the style of the architectural design. Existing models and guidelines can be divided into procurement models, competitions models, decision support systems, and project management tools. Yet, existing knowledge remains scattered and is not used adequately by the contracting authorities. Only the procurement models have an obligatory nature and could actually be enforced. It appears that it is the perception of these legal obligations rather than the actual procurement law that prevent a selection process based on open dialogue between the client and the architects about design quality. There are no open discussions about the difficulties experienced by clients as well as architects. Professions tend to search for solutions within their own domains while an architect selection is in fact a multidisciplinary phenomenon by nature. A gap exists between the existing structures provided to support decision making for architect selection processes and actual decision making of public clients.

In Chapter 5 I formulate fifteen possible success factors based on the insights from theories about assessing design quality, value judgements, and decision making. The theoretical framework shows a structure of five sensemaking processes: 1) reading the decision task, 2) searching for a match between aims, ambitions, needs and opportunities, 3) writing the decision process, 4) aggregating different kinds of value judgements, and 5) justifying against different rationalities. To account for the fact that the research field on architect selections is nascent and neither empirical studies nor theories exist that address processes of decision making in this context, the case study method was chosen to gather empirical data and validate these possible success factors. I conducted three instrumental cases in the context of a restricted tendering procedure: a School, a City Hall, and a Provincial Government Office. The cases differed in the scope of the brief, the type of tender, and the characteristics of the selection process. Additionally I performed one case about an ideas competition for a new Faculty Building. A variety of different forms of data was collected for each case to allow for triangulation between self-report, observed behaviour, and official justifications. The results of the research were successfully tested in a workshop with experts, in which also implications were discussed for the design of the selection procedures.

The empirical cases as described in Chapter 6 and Chapter 7 show that the decision process of selecting an architect is indeed a result of the interacting of the decision makers with the alternatives once they are confronted with them and start to make sense of the proposed designs. It is, therefore, almost impossible for clients to design a selection procedure and announce the criteria and weighting factors up front, as required by procurement law. In this respect the rationality of the legal requirements clashes with the psychological rationality of decision making. On the other hand, both rationalities strengthen each other by providing a public client with the structure and room needed for successful decision making. In Chapter 8 the five sensemaking processes and their underlying situational characteristics are explained from the theoretical insights and empirical findings. Based on the results recommendations are done for the design of a tender procedure.

The first sensemaking process of *reading the decision task* is based on the concepts of 'sensereading' and 'framing' and deals with the translation of the aims of the client into the tender procedure. Because a public commissioning client acts as a client rather than a customer, distinctive dimensions of architectural and legal language has to be analysed by the decision makers during the process of decision making in order to know what to expect. The development of the brief and the analysis of the project environment are important parts of this sensemaking process. The most important dilemma that clients faced during this process was a distinction between the search for the right solution for their design problem, as suggested by the tradition of design competitions, and the search for the right partner in designing a solution for their accommodation needs, as suggested by the tender principles. The results of the study suggest that complexity, uncertainty and time are the main situational characteristics that influence the process of reading a decision task.

Tendering is a way of granting contracts for projects based on the principle of an open market. The second sensemaking process of *searching for a match between aims, ambitions, needs and opportunities* relates to the fact that during the selection process the values of a client about architecture are connected to the opportunities that are offered by the architects. Although European procurement law aims at opening the market across the EU member states, experiences show that Dutch clients prefer to work with Dutch architects. The results indicate that the decision makers apply existing knowledge about the architects in order to create a sense of control over the situation and the quality of the architects that participate in the tender. The high degree of uncertainty is increased by the stakeholders that have to be involved in decision making. The results of this study suggest that control, affect, and time are the main situational characteristics that underlie the matching process of aims, ambitions, needs and opportunities.

The third sensemaking process that I identified is the process of *writing the decision process*. This process is based on the concepts of 'sensewrighting', 'sensegiving' and 'framing' and entails the writing of the selection process of an architect by the client during a project. Observations showed that decision makers have to deal with a lot of uncertainty during a tender process due to incomplete understanding, lack of information and conflicting alternatives. In these kinds of situations expert judgement and intuitive decision making rather than a rational evaluation of alternatives are needed to reach a decision. In all cases the procedure of the selection process (restricted or open) determined the amount of phases in decision making but not the interpretation of these phases. Both procedures showed similar iterative decision processes of goal setting, perception, individual value judgement, group decision making and evaluation. In general a distinction could be made between the preparation of the tender in which the brief, procedure, stakeholder involvement and decision process was designed, and the execution of the tender in which the design was applied. The main situational characteristics that I distinguished as influencing the process of writing a decision process are time, intuition, and expertise.

The fourth process of sensemaking relates to the *aggregation of different kinds of value judgements* that is needed to reach a final decision about design quality. In this process the legal and social rationality of decision making clash when a pseudo-rationality is created by quantification of qualitative judgements. The choice for a winner during a tender is on the one hand based on the structure that is provided by the pre-announced criteria, but on the other hand part of a process of increasing insight and sensemaking in which value judgements are implicitly aggregated. Structure, system, and expertise were found to be the most important situational characteristics that influence the process of aggregating different kinds of value judgements.

The fifth process of sensemaking deals with *the justification of the decision against the different rationalities* at the end of the selection process. A client had to justify the final decision to the own organisation, to the public, to society, and to the architects that joined the tender. In justifying a decision a decision maker is simultaneously confronted with the legal structure of the decision procedure and the psychological decision process of sensemaking. The situations characteristics of support, trust and control were found to be of influence to the process of justifying against different rationalities.

Based on the results of the research, fifteen recommendations were derived for the selection of architects by public clients under European tendering regulations. These are based on the success factors identified in the theoretical framework in Chapter 5. A few examples are:

- Allowing for a holistic judgement in the tender design that incorporates potentially conflicting judgements within itself.

- Ensuring a fit between the position and type of the stakeholders and their role in the decision process.

- Aligning the type of expertise needed for the various decision tasks during the selection process of an architect with the nature and content of the decision task.

- Allowing compensation in aggregating value judgements about design quality.

- Addressing the roles and responsibilities of the decision makers cautiously in the design of the tender to increase the trust between the decision makers and broaden the support for the decision among the stakeholders.

The results of this research give reason to suggest a change of the current implications of the tender regulations in the Netherlands. In my opinion the composition of the jury panel should be the same in the selection and award phase, the jury should have decisive rights, and the roles and responsibilities of the jury members should differ per phase of the tender. This means that for any tender in which an architect is selected the following procedure should be applied:

A. Assign a diverse jury panel that includes the responsible officer(s) and other representatives of the public commissioning as well as experts in specific domains that relate to the assignment (e.g. urban planners, architects, sustainability experts, historians etc).

B. Assign decisive rights to the jury panel in both the selection and the award phase.

C. Ensure that jury members trust and support each other before, during and after the tender. This process could be supported by determining roles and responsibilities among the jury members beforehand for the different phases.

The benefits of assigning decisive rights to a multidisciplinary jury in both the selection phase and the award phase are that 1) the potential tenderers can be quality checked by the domain specific experts in the selection phase, and 2) clients can still influence the final decision in the award phase but with the support of domain specific experts. The experts that are involved in the process could monitor the effects of tendering decisions for the professional field. The suggested procedure would require a substantially higher involvement of the domain experts and other panel members in the preparation phase of the tender, the establishment of a specialized multi disciplinary knowledge centre and the development database with suitability information of the tender candidates.

In Chapter 8 I also reflect on the research approach and the scientific relevance of the study and make suggestions for further research. I think that this research answers to the need for interdisciplinary approaches to apply existing knowledge from the more traditional fields of science, such as cognitive and social psychology, to the field of architecture. Triangulation of the different research methods strengthened the analysis of complexities underlying the behaviour as shown by the actors. In terms of credibility this study shows for instance that observations open up a possibility to collect scientific insights which would - although commonly known in the field - otherwise be neglected, such as emotions, and strategic behaviour. The main differences between the ideas competition and the tender cases related to the differences in the legal impact of the decision, the aim of the decision and the expertise level of the client. Generalisations of the research can be found in comparable selection processes which allow for deliberations about options, such as decisions about a real estate portfolio, new product development, research funds allocations, student assessments, or awarding contracts in other sectors.

The study highlights several directions for further research. Both the scientific and the professional field would benefit greatly of a more structural data collection on tenders in architectural design. The results indicate that the role of expert teams, the strategies for winning, the underlying situational characteristics of the sensemaking processes, and the role of decision support systems deserve to be investigated further. Future research could also compare different sectors, different client characteristics, and different tendering procedures and include more theories from the fields of strategic management, public administration, and process management.

This research contributes to knowledge in the areas of architectural design, the psychology of making judgements, and organisational decision making. It is therefore of interest to public commissioning clients, management consultants, architects, policy makers and legal advisors in practice, but also to scholars in the field of design management, product experience, environmental psychology, or decision making. The main audience of this thesis is public commissioning bodies that have to organise a tender, their advisors, and governmental authorities that develop and implement regulations and policies, and scholars in this (multidisciplinary) area. Because the research shows insights into the client perspective that have never been studied before and are usually not open to the public, I believe that the results of this research also offer an interesting story for those interested in decision making for the built environment in general.

Leentje Volker
August 2010

Samenvatting

Beslissen over ontwerpkwaliteit

Waardeoordelen en besluitvorming tijdens architectenselecties van publieke opdrachtgevers onder Europese aanbestedingsregelgeving

Dissertatie van Leentje Volker

Het zoeken naar een architect kan worden beschreven als een interactief selectie-proces waarin een opdrachtgever een architect probeert te vinden die de behoeften en de ambities van de opdrachtgever het beste kan visualiseren en realiseren. Dit proces is niet zonder problemen. "Architectensoap in Delft", "Foutje, architect mag toch niet Utrechtse bibliotheek bouwen", en "Publiek Rotterdam wil Search, maar OMA bouwt Stadskantoor" zijn enkele krantenkoppen die overheersen in het beeld van aanbestedingen waarin Nederlandse publieke opdrachtgevers in de afgelopen jaren een architect selecteerden. Het gaat dan ook om een uitdagend besluitvormingsproces vol verrassingen en onvoorziene omstandigheden waarin aan verscheidene juridische en maatschappelijke voorwaarden moet worden vol-daan. De huidige praktijk van de architectenselectie door publieke opdrachtgevers heeft wortels in drie verschillende systemen: 1) het aanbesteden van diensten en werken, 2) het zoeken naar een geschikte architect of ontwerpteam, en 3) de ar-chitectonische ontwerpwedstrijd. Het zijn deze verschillen in de herkomst van de selectieprocedure, die conflicten tussen de juridische rationaliteit en de psycholo-gische rationaliteit van het besluitvormingsproces lijken te veroorzaken.

In dit onderzoek verken ik de oorsprong van deze problemen zoals die door publieke opdrachtgevers momenteel ervaren worden in het selectieproces. In Hoofdstuk 1 worden het onderwerp, de focus van het onderzoek en de onder-zoeksaanpak geïntroduceerd. Ik bespreek daar ook de kennishiaten, de weten-schappelijke uitdagingen en de bijdrage van het onderzoek aan het werkveld. Het onderzoek richt zich op het volledige besluitvormingsproces vanuit het oogpunt van een publieke opdrachtgever die een architect wil selecteren in het kader van de Europese aanbestedingswet- en regelgeving. Doel van het onderzoek is het be-schrijven, begrijpen en verklaren van het ontwerp en de implementatie van pro-cedures waarmee de kwaliteit van ontwerpvoorstellen worden beoordeeld om een architect voor een specifiek project te contracteren. Op basis van de resultaten heb ik een aantal aanbevelingen voor toekomstige architectenselectieprocedures opgesteld. De onderzoeksvragen in deze studie zijn:

1. Hoe beslissen aanbestedende diensten over de selectie van een architect in de context van de EU aanbestedingswetgeving?

2. Welke situationele kenmerken zijn van invloed op het proces van besluitvor-ming van de publieke opdrachtgevers in dit verband?

3. Wat zijn de implicaties voor het ontwerpen van procedures voor de selectie van architecten?

In Hoofdstuk 2 richt ik mij op het concept van de ontwerpkwaliteit vanuit de domeinen van de architectonische vormgeving, omgevingspsychologie, productemotie en value management. Op basis van deze inzichten definieer ik ontwerpkwaliteit als een alomvattend waardeoordeel van een individuele belanghebbende, dat is gebaseerd op de interactie tussen een persoon en (de representatie van) een object in de gebouwde omgeving. Als gevolg van deze interactie tussen het individu en het product wordt een waardeoordeel altijd vergezeld van een affectieve reactie en een evaluatie over het niveau van kwaliteit of de waarde van een product.

Hoofdstuk 3 geeft een overzicht van de psychologische aspecten die het meest relevant zijn voor de beoordeling en besluitvorming in het kader van de architectenselectie. Het bouwt voort op de definitie uit Hoofdstuk 2 van ontwerpkwaliteit als een waardeoordeel. Het hoofdstuk begint met een overzicht van drie generaties van besluitvormingstheorieën: rationele beslissingsmodellen, modellen over besluitvormingsgedrag en naturalistische besluitvormingsconcepten. De huidige praktijk van de architectenselectie lijkt te zijn gebaseerd op twee tegenstrijdige modellen over besluitvorming. Het juridische model gaat uit van een rationeel en lineair besluitvormingsproces waarin alternatieven worden vergeleken op basis van vooraf bekendgemaakte criteria. Het naturalistische model dicht een belangrijke rol toe aan het gebruik van intuïtie en gevoel. De oorsprong van de huidige problemen in de aanbestedingspraktijk kan gevonden worden in deze verschillende rationaliteiten. Ik gebruik vooral het concept van 'sensemaking' (betekenisgeving) om uit te leggen hoe publieke opdrachtgevers tijdens de besluitvorming omgaan met deze verschillende rationaliteiten.

Een architectenselectie is geen op zich zelf staande gebeurtenis. Publieke opdrachtgevers opereren in een bestuurlijk kader en nemen dit kader onvermijdelijk mee in hun beslissingen. Hoofdstuk 4 behandelt vier contextuele elementen die in mijn ogen essentieel zijn voor de omgeving waarin architectenselecties plaatsvinden: de politieke, culturele, juridische en economische context. De resultaten van deze contextuele verkenning laten zien dat de keuzes die gemaakt worden tijdens de voorbereidingsfase in belangrijke mate de resultaten en het succes van de aanbesteding bepalen, evenals de stijl van het architectonisch ontwerp dat voorgesteld wordt door de architecten. Bestaande modellen en richtlijnen ter ondersteuning van het selectieproces kunnen worden onderverdeeld in aanbestedingsmodellen, wedstrijdmodellen, beslissingsondersteunende systemen en projectmanagement tools. Het losstaande karakter van deze modellen leidt ertoe dat bestaande kennis versnipperd blijft en niet adequaat gebruikt wordt door de aanbestedende diensten. Alleen de aanbestedingsmodellen hebben een verplichtend karakter en daarvan kan het gebruik daadwerkelijk worden afgedwongen. Momenteel vinden er weinig open discussies plaats over de moeilijkheden die ervaren worden door opdrachtgevers en architecten. Het lijkt erop dat vooral de interpretatie van de wettelijke verplichtingen is in plaats van het aanbestedingsreglement zelf dat ervoor zorgt dat het huidige selectieproces niet gebaseerd wordt op een open dialoog tussen de opdrachtgever en de architecten over de kwaliteit van een ontwerpvoorstel. Professionals hebben de neiging om naar oplossingen te zoeken binnen hun eigen domein, terwijl de architectenselectie in feite een multidisciplinair fenomeen is. Tijdens architectenselecties bestaat er daardoor een kloof tussen de bestaande structuren die besluitvormingsprocessen ondersteunen en de daadwerkelijke besluitvormingsprocessen van de publieke opdrachtgevers.

In Hoofdstuk 5 formuleer ik vijftien mogelijke succesfactoren op basis van de inzichten uit theorieën over het meten van ontwerpkwaliteit, het maken van waardebeoordelingen en de kenmerken van besluitvormingsprocessen. Het theoretische kader is gebaseerd op vijf sensemaking processen: 1) het lezen van de besluitvormingstaak, 2) het zoeken naar een match tussen doelstellingen, ambities, behoeften en mogelijkheden, 3) het schrijven van het besluitvormingsproces, 4) het optellen van verschillende soorten van waardeoordelen, en 5) rechtvaardigen van het besluit naar verschillende rationaliteiten. Omdat onderzoek op het gebied van architectenselecties schaars is, zowel in empirische studies als binnen besluitvormingstheorieën, heb ik ervoor gekozen om de case studie methode te gebruiken voor het verzamelen van empirische gegevens en om de mogelijke succesfactoren te valideren. Ik heb drie case studies uitgevoerd naar niet-openbare aanbestedingen: een school, een gemeentehuis en een provinciehuis. De cases verschilden in grootte, aard en kenmerken van de selectieprocedure. Daarnaast heb ik een case studie uitgevoerd naar een internationale ideeënprijsvraag voor een nieuw faculteitsgebouw. In deze cases heb ik verschillende soorten gegevens verzameld (met behulp van interviews, observaties, documenten) om triangulatie mogelijk te maken tussen zelfrapportage, waargenomen gedrag, en de officiële documentatie. De resultaten van het onderzoek zijn met succes getest in een workshop met deskundigen. In deze workshop zijn ook de mogelijke implicaties besproken voor het toekomstige ontwerp van selectieprocedures.

Uit de empirische data als beschreven in Hoofdstuk 6 en Hoofdstuk 7 blijkt dat het besluitvormingsproces van het selecteren van een architect een gevolg is van de directe confrontatie van de besluitvormers met de (ontwerp)alternatieven. Dientengevolge worden de besluitvormers geconfronteerd met de opties en ontdekken zij de betekenis en mogelijke implicaties van de voorgestelde ontwerpen. Het is hierdoor bijna onmogelijk voor publieke opdrachtgevers om de procedure, criteria en wegingsfactoren vooraf aan te kondigen, zoals vereist is in de aanbestedingswetgeving. In dit opzicht botst de rationaliteit van de wettelijke voorschriften met de psychologische rationaliteit van de besluitvorming. Aan de andere kant versterken de rationaliteiten elkaar door de opdrachtgever de structuur en de ruimte aan te reiken die nodig is voor een succesvol besluitvormingstraject. In Hoofdstuk 8 worden de vijf sensemaking processen en hun onderliggende situationele kenmerken verklaard uit de theoretische inzichten en empirische bevindingen. Gebaseerd op de resultaten worden aanbevelingen gedaan voor het ontwerp van een aanbestedingsprocedure.

Het eerste sensemaking proces van het *lezen van de besluitvormingstaak* is gebaseerd op de begrippen 'sensereading' en 'framing' en richt zich op de vertaling van de doelstellingen van de opdrachtgever in de aanbestedingsprocedure. Omdat een publieke opdrachtgever als een opdrachtgever fungeert in plaats van een klant, moeten verschillende aspecten van bouwkundige en juridische taal tijdens het proces van de besluitvorming door de besluitvormers geanalyseerd worden om te weten wat ze kunnen verwachten. De ontwikkeling van het programma van eisen en de analyse van de context van het project zijn daarom belangrijke onderdelen van het sensemaking proces. Het belangrijkste dilemma waarmee cliënten geconfronteerd worden tijdens dit proces is het onderscheiden van 1) de juiste oplossing voor hun ontwerpprobleem binnen de wedstrijdcultuur in de architectuur en 2) het zoeken naar de juiste partner in het ontwerpen van een oplossing voor hun huisvestingsbehoefte als onderdeel van de aanbestedingsbeginselen. De resultaten van het onderzoek suggereren dat complexiteit, onzekerheid en tijd de belangrijkste situationele kenmerken zijn die het proces van het lezen van de besluitvormingstaak beïnvloeden.

Aanbesteden is een manier om contracten te verlenen voor projecten van publieke opdrachtgevers. Het is gebaseerd op het principe van de open markt. Het tweede sensemaking proces van *het zoeken naar een match tussen doelen, ambities, behoeften en mogelijkheden* heeft betrekking op het feit dat tijdens het selectieproces de waarden van een cliënt over architectuur gekoppeld worden aan de mogelijkheden die door de architecten worden aangeboden. Hoewel het Europese aanbestedingrecht gericht is op openstelling van de markt in de EU-lidstaten, lijkt het erop dat Nederlandse opdrachtgevers in de huidige praktijk een voorkeur hebben om met Nederlandse architecten te werken. De resultaten van dit onderzoek geven aan dat besluitvormers bestaande kennis over de architecten toepassen tijdens het selectieproces om een gevoel van controle te creëren over de situatie en de potentiële kwaliteit van de deelnemende architecten. De grote mate van onzekerheid wordt vergroot door de verschillende belanghebbenden die moeten worden betrokken bij de besluitvorming. De resultaten van deze studie suggereren dat controle, invloed en tijd de belangrijkste situationele kenmerken zijn die ten grondslag liggen aan de koppeling van de doelstellingen, ambities, behoeften en kansen bij publieke opdrachtgevers tijdens een architectenselectie.

Het derde sensemaking proces dat ik onderscheid is het proces van *het schrijven van het besluitvormingsproces*. Dit proces is gebaseerd op de concepten van 'sensewrighting', 'sensegiving' en 'framing' en houdt in dat de selectie van een architect vormgegeven (geschreven) wordt door de opdrachtgever tijdens de aanbesteding. De observaties wekken de indruk dat de besluitvormers door onvolledige kennis, gebrek aan informatie en tegenstrijdige alternatieven veel onzekerheid ervaren tijdens een aanbestedingsprocedure. In dit soort situaties zijn expertoordelen en intuïtieve besluitvorming in plaats van een rationele evaluatie van alternatieven nodig om tot een besluit te komen. In alle cases bepaalde de aanbestedingsprocedure (niet-openbaar of openbaar) het aantal fasen in de besluitvorming, maar niet de interpretatie van deze fasen. Beide procedures vertoonden vergelijkbare iteratieve besluitvormingsprocessen van het stellen van doelen, perceptie, individuele beoordeling, groepsbesluitvorming en evaluatie van het besluit. Over het algemeen kan er een onderscheid worden gemaakt tussen 1) de voorbereiding van de aanbesteding waarin het programma van eisen, de aanbestedingsprocedure, de betrokkenheid van de belanghebbenden en het besluitvormingsproces ontworpen worden, en 2) de uitvoering van de aanbesteding waarin het ontwerp wordt toegepast. De belangrijkste situationele kenmerken die het schrijfproces van het besluit beïnvloeden lijken tijd, intuïtie en deskundigheid te zijn.

Het vierde proces van sensemaking heeft te maken met *het optellen van verschillende soorten van waardeoordelen* dat nodig is om tot een definitieve beslissing over de kwaliteit van de architect te komen. In dit proces botsen de juridische en sociale rationaliteit van de besluitvorming omdat een pseudo-rationaliteit wordt gecreëerd door het kwantificeren van de kwalitatieve uitspraken. De keuze voor een winnaar tijdens een aanbesteding is enerzijds gebaseerd op de structuur die wordt geleverd door de vooraf bekendgemaakte criteria, maar is aan de andere kant ook onderdeel van een proces van voortschrijdend inzicht en betekenisgeving waarin verschillende waardeoordelen impliciet worden opgeteld. Structuur, systeem, en expertise lijken de belangrijkste situationele kenmerken te zijn die van invloed zijn op het optellingsproces.

Het vijfde proces van sensemaking heeft betrekking op *de motivering van het besluit aan het einde van het selectieproces naar de verschillende rationaliteiten.* Een opdrachtgever moet het definitieve besluit rechtvaardigen naar de eigen organisatie, het publiek, de samenleving en de deelnemende architecten. In de rechtvaardiging van een besluit wordt een besluitvormer gelijktijdig geconfronteerd met de juridische structuur van de besluitvormingsprocedure en het psychologische proces van betekenisgeving. De situationele kenmerken draagvlak, vertrouwen en controle lijken van invloed te zijn op het proces van het rechtvaardigen naar de verschillende rationaliteiten.

Op basis van de resultaten van het onderzoek worden vijftien punten voorgesteld die eraan kunnen bijdragen dat de architectenselectie van een publieke opdrachtgever onder Europese aanbestedingsregelgeving een succes wordt. Deze zijn gebaseerd op de succesfactoren zoals geïntroduceerd in het theoretische kader van Hoofdstuk 5. Enkele voorbeelden zijn:

- Het toelaten van een holistisch oordeel dat potentieel tegenstrijdige uitspraken verenigt in het ontwerp van de aanbestedingprocedure.

- Zorgen dat de positie en het type belanghebbenden afgestemd zijn op hun rol in het besluitvormingsproces.

- Het aanpassen van de aard van de expertise aan de aard en inhoud van de besluitvormingstaak die nodig is voor de selectie van een architect.

- Het toelaten van compensatie in het optellen van de waardeoordelen over de kwaliteit(en) van de offertes.

- Het vergroten van het vertrouwen tussen de besluitvormers en het draagvlak van het besluit onder de belanghebbenden door in het ontwerp van de aanbesteding de rollen en verantwoordelijkheden van de besluitvormers zorgvuldig te adresseren.

De resultaten van dit onderzoek geven aanleiding tot een wijzigingssuggestie voor de huidige aanbestedingspraktijk in Nederland. Naar mijn mening zou de samenstelling van de selectie - en gunningcommissie gelijk moeten zijn voor de selectie- en de gunningsfase, zou deze commissie (annex jury) beslissingsrecht moeten hebben, en zouden er per fase verschillen kunnen zijn in de rollen en verantwoordelijkheden van de juryleden. Dit betekent dat voor elke aanbesteding waarbij een architect wordt gekozen de volgende procedure van toepassing zou kunnen zijn:

A. Wijs een gevarieerde jury aan waarin zowel de verantwoordelijke bestuurder(s) en andere vertegenwoordigers van de aanbestedende dienst als een aantal domeinspecifieke deskundigen die gerelateerd zijn aan het karakter van de opdracht (bijv. stedenbouwkundigen, architecten, duurzaamheid experts, historici, enzovoort) zitting nemen.

B. Geef de jury zowel in de selectiefase als in de gunningsfase beslissingsrecht.

C. Zorg dat de juryleden elkaar ondersteunen en vertrouwen, zowel voor, tijdens als na de aanbestedingsprocedure. Dit kan bevorderd worden door van te voren de rollen, verantwoordelijkheden en taken voor de verschillende fasen van het selectieproces binnen de jury vast te leggen.

De voordelen van het betrekken van domeinspecifieke deskundigen voor opdrachtgevers is dat 1) in de selectiefase een kwaliteitscontrole plaatsvindt van de potentiële inschrijvers door deskundigen uit het betrokken werkveld, en 2) dat de opdrachtgevers in de gunningsfase met de ondersteuning van domeinspecifieke deskundigen invloed kunnen uitoefenen op de uiteindelijke beslissing. De deskundigen die betrokken zijn bij het proces kunnen toezien op de potentiële gevolgen van de aanbestedingsbeslissingen op het werkveld. Het voorgestelde concept zou aanzienlijk meer betrokkenheid van de deskundigen en andere panelleden vereisen in de voorbereidende fase van de aanbesteding. De oprichting van een gespecialiseerd multidisciplinair kenniscentrum en de ontwikkeling van een database met informatie over de potentiële geschiktheid van de inschrijvers kan deze verandering ondersteunen.

In Hoofdstuk 8 reflecteer ik ook op de aanpak en wetenschappelijke relevantie van het onderzoek en worden suggesties gedaan voor toekomstig onderzoek. Ik denk dat dit onderzoek een belangrijke bijdrage levert aan de vraag naar een interdisciplinaire benadering van onderzoek in dit werkveld door het toepassen van bestaande kennis uit de meer fundamentele gebieden van de wetenschap zoals de cognitieve en sociale psychologie. Triangulatie van de verschillende methoden van onderzoek versterkt de analyse van de complexiteit die ten grondslag ligt aan het gedrag van de actoren. In de zin van de betrouwbaarheid van dit onderzoek blijkt bijvoorbeeld dat het observeren een mogelijkheid biedt om wetenschappelijke inzichten te verzamelen over gedrag die anders niet opgemerkt zouden worden, zoals de rol van emoties en strategisch gedrag. De belangrijkste verschillen tussen de ideeënwedstrijd en de aanbestedingscases in dit onderzoek houden verband met verschillen in de juridische gevolgen van het besluit, het doel van het besluit en het expertiseniveau van de opdrachtgever. De resultaten van het onderzoek kunnen worden gegeneraliseerd naar vergelijkbare selectieprocessen waarin alternatieven worden afgewogen, zoals besluiten over een vastgoedportefeuille, de ontwikkeling van nieuwe producten, toewijzingen van onderzoeksgelden, selecteren